Louis Starr

Diseases of the Digestive Organs in Infancy and Childhood

Louis Starr

Diseases of the Digestive Organs in Infancy and Childhood

ISBN/EAN: 9783337372880

Printed in Europe, USA, Canada, Australia, Japan

Cover: Foto ©berggeist007 / pixelio.de

More available books at **www.hansebooks.com**

DISEASES

OF THE

DIGESTIVE ORGANS

IN

INFANCY AND CHILDHOOD,

WITH

CHAPTERS ON THE INVESTIGATION OF DISEASE; THE DIET
AND GENERAL MANAGEMENT OF CHILDREN,
AND MASSAGE IN PÆDIATRICS.

BY

LOUIS STARR, M.D.,

LATE CLINICAL PROFESSOR OF DISEASES OF CHILDREN IN THE HOSPITAL OF THE UNI-
VERSITY OF PENNSYLVANIA; PHYSICIAN TO THE CHILDREN'S HOSPITAL,
PHILADELPHIA; CONSULTING PÆDIATRIST TO THE
MATERNITY HOSPITAL, PHILADELPHIA,
ETC., ETC.

SECOND EDITION—ILLUSTRATED.

PHILADELPHIA:
P. BLAKISTON, SON & CO.,
No. 1012 WALNUT STREET.
1891.

TO

PROFESSOR JOHN ASHHURST, Jr., M.D.,

THIS VOLUME IS DEDICATED,

AS A

TRIBUTE TO HIS GENIUS AS A SURGEON AND AUTHOR,

AND IN

GRATEFUL REMEMBRANCE OF

MANY ACTS OF KINDNESS.

PREFACE TO THE SECOND EDITION.

In preparing this issue of "The Diseases of the Digestive Organs in Infancy and Childhood," the author, while endeavoring to bring the general subject-matter thoroughly abreast with the times, has deemed it advisable to make some re-arrangement of the original text and to add a quantity of new material. The chief additions consist of a section on alterations in the odor of the breath in disease ; a section on urine alterations ; a chapter on massage in pædiatrics, and a detailed account of second dentition and its influence on the health in late childhood—a subject heretofore greatly neglected.

The author wishes to thank the critics of the first edition of his book for many valuable suggestions, from which he has profited greatly. His thanks are also due to Dr. Wm. M. Powell for his untiring assistance in preparing the copy and in making the index ; to Dr. Robert J. Hess for his aid in proof-reading, and to Prof. Charles B. Nancrede for the diagram illustrating the extended connections of the dental nerves.

LOUIS STARR.

1818 South Rittenhouse Square, Philadelphia.
January 1st, 1891.

PREFACE TO THE FIRST EDITION.

It is the author's object, in this book, to give prominence to a class of disorders constituting a large proportion of the ailments of childhood, but often too briefly considered in works on pædiatrics. For the successful treatment of the diseases of the digestive organs in infancy and childhood, attention to the general regimen is quite as important as the administration of drugs, and it is upon the former that the student and young practitioner are usually the least thoroughly instructed.

So much may be done by the selection of suitable food, by artificial digestion, by regulating the clothing, bathing and other elements of hygiene, that the author, without neglecting therapeutics, has given greater prominence to these points.

The chapter on the investigation of disease does not necessarily belong to a work on disorders of the digestive organs, but as so much difficulty is experienced by students in the study of disease in children, it has been incorporated as an aid to such. In the article on the general management of children, the effort has been made to present to the inexperienced results that can only be obtained by much study and practical work.

The author is indebted to Dr. Henry D. Harvey for his aid in preparing the index, and to the pencil of Dr. John Madison Taylor for the illustrations.

LOUIS STARR.

Philadelphia, April, 1886.

CONTENTS.

PART I.
INTRODUCTION— PAGE
The Investigation of Disease, ... 17
1. Questioning the Attendants, ... 18
2. Inspecting the Child, ... 20
3. Physical Examination, ... 39

PART II.
THE GENERAL MANAGEMENT OF CHILDREN—
1. Feeding, ... 60
2. Bathing, ... 102
3. Clothing, ... 105
4. Sleep, ... 106
5. Exercise, ... 108

PART III.
MASSAGE IN PÆDIATRICS, ... 116

PART IV.
DISEASES OF THE DIGESTIVE ORGANS.

CHAPTER I.
Affections of the Mouth and Throat, ... 124
1. Catarrhal Stomatitis, ... 124
2. Aphthous Stomatitis, ... 126
3. Ulcerative Stomatitis, ... 131
4. Gangrenous Stomatitis—Noma, ... 136

	PAGE
5. Parasitic Stomatitis—Thrush,	141
6. Dentition,	148
7. Simple Pharyngitis,	183
8. Superficial Catarrh of the Tonsils,	186
9. Follicular Tonsillitis,	187
10. Suppurative Tonsillitis,	190
11. Hypertrophy of the Tonsils,	194
12. Retropharyngeal Abscess,	197

CHAPTER II.

Affections of the Stomach and Intestines,	199
1. Acute Gastric Catarrh,	199
2. Chronic Gastric Catarrh,	202
3. Ulcer of the Stomach,	211
4. Softening of the Stomach (Gastro-Malacia),	212
5. Chronic Gastro-Intestinal Catarrh,	213
6. Acute Intestinal Catarrh,	227
7. Chronic Intestinal Catarrh—Chronic Entero-Colitis,	235
8. Entero-Colitis,	248
9. Cholera Infantum,	258
10. Inflammation of the Colon and Rectum—Dysentery,	264
11. Tubercular Ulceration of the Intestines,	268
12. Colic,	270
13. Habitual Constipation,	273
14. Simple Atrophy,	279
15. Typhlitis and Perityphlitis,	287
16. Intussusception,	296
17. Intestinal Worms,	311

CHAPTER III.

Caseous Degeneration and Tuberculosis of the Mesenteric Glands— Tabes Mesenterica,	329

CHAPTER IV.

Affections of the Liver,	337
1. Jaundice,	337
2. Congestion of the Liver,	343
3. Fatty Liver,	346
4. Amyloid Liver,	347

	PAGE
5. Syphilitic Inflammation of the Liver,	351
6. Cirrhosis of the Liver,	352
7. Suppurative Hepatitis,	357

CHAPTER V.

Affections of the Peritoneum,	364
1. Peritonitis,	364
2. Tubercular Peritonitis,	371
3. Ascites,	377
INDEX,	383

DISEASES
OF
THE DIGESTIVE ORGANS
IN
INFANCY AND CHILDHOOD.

PART I.—INTRODUCTION.

THE INVESTIGATION OF DISEASE.

The clinical investigation of disease in children, usually considered so difficult, is in some respects easier than the same study in adults.

It is easier because in the child disease is commonly uncomplicated, rarely has its course and symptoms modified by tissue lesions the result of previous affections, and never by vicious habits, such as the abuse of stimulants and narcotics, or by mental over-work and nerve-strain. The confusing element of mis-stated subjective symptoms is also absent, while correct diagnosis is greatly aided by the facility with which physical examination of the whole body may be practiced.

That there are difficulties to be encountered, and very grave ones too, is equally certain. The absence of speech in the infant deprives us of the important assistance afforded by correctly described subjective symptoms, and renders it necessary to look to the mother or nurse for the history of an illness. In older children the case is not much better, since with them words are not prompted by sufficient knowledge or judgment to be of great service. Further, the wilfulness, dislikes, fear and agitation of the child are impediments which must be overcome before a satisfactory examination can be made, and which will often tax the skill and patience of the physician to the utmost in the overcoming. Another source of difficulty lies in the activity of growth and development in infants, which renders them liable to be affected by slight causes, and makes disease sudden in its attack, short in its course and intense in its symptoms. The rapid development of the nervous system especially leads to confusion. The nerves bind every portion of the frame in a sympathy so close that an affection of a single part may cause marked general disturbance, and local symptoms are often reflected, directing attention to organs very distant from those really diseased. Finally, the extreme excitability of the nervous system of healthy children often causes a trifling illness to assume an aspect of the greatest gravity, while, on the contrary, the depression of nervous sensibility that attends chronic wasting diseases so obscures the symptoms that a dangerous intercurrent affection may appear trifling or remain altogether latent.

The plan of conducting the clinical investigation in children differs materially from the method adopted in adults. It is best to proceed in three regular stages, as follows: 1st. Questioning the attendants; 2d. Inspecting the child; 3d. Physical examination.

1. Questioning the Attendants.

When the patient is under eight or ten years of age, the only way of obtaining a knowledge of the previous history and of what may occur between visits, is carefully to question the mother or nurse. The account must be patiently elicited and listened to, and credited with due reference to the narrator's intelligence. It is well never entirely to discredit a statement without good reason, for many women, though weak and foolish in other respects, are excellent observers when their powers are guided by affection. Besides, being thoroughly acquainted with their children's habits and dispositions, they will often detect deviations from health that the physician might overlook entirely. This part of the examination, particularly when the acquaintance and good will of the child has not previously been obtained, should, if possible, be made before entering the sick-room. By taking this precaution the agitation produced by the prolonged presence of a stranger, and its consequent trouble and delay, will be avoided to a great degree.

As there are certain points about which it is always necessary to be informed, the adoption of a definite order of questioning is advisable.

The family history as far back as the parents should first be ascertained. Inquiry being chiefly directed to the detection of chronic maladies and transmissible diseases, as tuberculosis and syphilis. If any deaths have occurred, their causation should be investigated, and an inquiry into the occurrence, or the reverse, of previous still-births is often important.

Next, an outline of the child's life from birth up to the date of the illness in question must be obtained. This should include the following items: The manner of feeding during infancy; whether at the breast, or from a bottle, and if the latter, whether cow's milk, condensed milk or the farinacea have formed the basis of the diet. The date of commencement and the regularity of dentition. The general state of health in regard to strength or weakness and liability to illness. The time of occurrence and the nature of any prominent attack of illness, especially of the eruptive fevers. Whether vaccination has been performed or no. The hygienic surroundings; for instance the healthfulness of the locality of residence, the sort of house and room occupied—if large, well ventilated, light and dry or the reverse, and the character of the clothing and food. In older children, if at school, the time devoted to study, and if at labor, the nature and the hours of work.

After this it is necessary to fix the time the attack in hand began. The occurrence of some striking symptom, as convulsions or violent vomiting, often establishes this point beyond a doubt, but when there is any uncertainty the best plan is to question back, day by day, until a time is reached at which the child was perfectly well, and to date the onset from this period. The most common of the general indications of commencing illness are disturbed sleep and irritability of temper. A perfectly well child sleeps quietly and continuously at night, and is never cross.

Having determined, as nearly as possible, the exact time of onset, the next step is to learn the mode of attack and the symptoms and course of the disease prior to the first visit. The questions now must be general, never leading. They must be sufficiently exhaustive to touch upon all the functions of the body, and when a trail is started it must be patiently followed to the end. Alterations in sleep, bodily strength, surface temperature, appetite, digestion, urine elimination, respiration and so on, must be sought for, and the account of such deviations from the normal state as vomiting, diarrhœa or cough, will suggest further questions as well as point out the path to be followed in the future examination.

This portion of the investigation is closed by an inquiry into the treatment that may have been already adopted.

2. Inspecting the Child.

When the eye and ear of the physician are trained to their work, valuable information can be obtained by simply looking at an ill child and listening to its cry or spoken words. Even while the child is lying asleep or sitting quietly in the nurse's lap many facts may be learned, but this portion of the examination is never complete without an inspection of the naked body. The points thus ascertained consist in alterations in the expression of the face, in decubitus, in the appearances of the body and so on, and may be designated the *features* of disease. The relative position of the observer and patient during inspection is of importance. If possible the former should stand with his back to, and the latter be so placed that his face is toward, a window or lamp. The light must never be strong enough to dazzle when the countenance is the object of inspection, as this causes distortion of the features.

For convenience, the *features* of disease will be studied under different headings, and since to appreciate them it is necessary to have a knowledge of the healthy aspect, both the normal and abnormal appearances will be described.

FACE.—The face of a healthy, sleeping child wears an expression of perfect repose. The eyelids are completely closed, the lips slightly parted, and while a faint sound of regular breathing may be heard, there is no perceptible movement of the nostrils. Incomplete closure of the lids with more or less exposure of the whites of the eyes is noted when sleep is rendered unsound by moderate pain and during the course of all acute and chronic diseases, particularly when they assume a grave type. Twitching of the lids heralds the approach of a convulsion, and at such times, too, there is often oscillation of the eyeballs, or squinting. A marked smile, due to contraction of the muscles about the mouth, signifies abdominal pain or colic, and pursing out of the lips and chewing motions of the jaw, gastro-intestinal irritation. Dilatation of the alæ nasi, with or without noisy breathing, points to embarrassed respiration, the result of extensive bronchial catarrh, pneumonia or pleurisy with effusion.

When awake and passive the healthy infant's face has a look of wondering observation of whatever is going on about it. As age advances the expression of intelligence increases, and every one is familiar with the bright, round, happy face of perfect childhood, so indicative of careless contentment, and so mobile in response to emotions.

The picture is altered by the onset of any illness, the change being in proportion to the severity of the attack. An expression of anxiety or of suffering appears, or the features become pinched and lines are seen about the eyes and mouth. Pain most of all sets its mark upon the countenance, and by noting the feature affected it is often possible to fix the seat of serious disease. Thus, contraction of the brows denotes pain in the head; sharpness of the nostrils, pain in the chest; and a drawing of the upper lip, pain in the abdomen. As a rule, the upper third of the face is modified in expression in affections of the brain, the middle third in diseases of the chest, and the lower third in lesions of the abdominal viscera.

M. Jadelot has drawn attention to certain furrows that appear on the face in serious cases, and to the indications that these afford as to the part of the body to be further examined. There are three sets of furrows. First, the *oculo-zygomatic*, beginning at the inner canthus of the eye and passing outward beneath the lower lid, to be lost a little below the most prominent portion of the cheek. This points to primary or secondary disorder of the cerebro-nervous system. Second, the *nasal*, starts above the ala of the nose, and, passing downward, forms a semicircle around the angle of the mouth. This may be associated with another line, the *genal*, which extends from its middle almost to the malar bone. These indicate disease of the gastro-intestinal tract, or other abdominal viscera. Third, the *labial*, commencing at the angle of the mouth and running outward, to be lost in the lower part of the face. This furrow is more shallow than the others. It directs attention to the lungs. These furrows are often present, and when met with are worthy of consideration, but their constancy and value have been over-estimated by their discoverer.

Puffiness of the eyelids and a fulness of the bridge of the nose, indicate dropsy and should direct attention to the kidneys as the seat of disease. Each of the two prominent diatheses is distinguished by a peculiar physiognomy. When there is a tuberculous tendency the face is oval and the features delicate; the hair is fine and silky; the skin smooth and transparent; the temporal veins are visible; the eyelashes are long and curving, the irides large and deep-colored and the sclerotics pearly white or bluish; finally, a growth of fine hair is often noticeable on the temples and in front of the ears. The general expression is most intelligent. In the strumous diathesis, on the contrary, the face is round and heavy ; the complexion doughy ; the upper lip swollen ; the nostrils wide and the alæ of the nose thick ; the eyelids are thickened and reddened at their edges ; the hair coarse, and the lymphatic glands of the neck enlarged.

A marked disfigurement of the face may indicate one of several diseases, according to its character. For example, broadness of the bridge of the nose, or complete flatness at this point, is significant of constitutional syphilis. A large, square head and projecting forehead with a face of natural size or smaller, shows that the child has suffered from rickets. An immense globular head, overhanging forehead, and diminutive face with eyeballs projected downward and irides almost concealed by the lower lids, are pathognomonic signs of chronic hydrocephalus.

DECUBITUS.—The complete repose depicted on the countenance of a sleeping child when free from illness is shown also by the posture of the body. The head lies easy on the pillow, the trunk rests on the side slightly inclined backward, the limbs assume various but always most graceful attitudes, and no movement is observable but the gentle rise and fall of the abdomen in respiration. In the waking state the child, after early infancy, is rarely still. The movements of the arms, at first awkward, soon become full of purpose as he reaches to handle and examine various objects about him. The legs are idle longer, though these, too, soon begin to be moved about with method, feeling the ground, in preparation, as it were, for creeping and walking.

With the onset of disease the scene changes. In acute attacks attended with pain, sleep is no longer restful. The infant is content only when rocked, fondled or "walked" in the nurse's arms. The older child tosses about uneasily in bed, or demands a constant change from the bed to the lap. During the waking hours the movements are purposeless, quick and impatient, the position is constantly shifted and frequent whining complaints are made. As a contrast to this condition of jactitation, at the beginning of the specific fevers, children often lie for hours quiet and drowsy upon the bed or lap.

In chronic affections attended with debility, the movements become slow and languid, and in stupor and coma there is perfect stillness and immobility.

There are certain positions and gestures which have especial significance.

Sleeping with the head thrown back, and the mouth open, is a frequent accompaniment of chronic enlargement of the tonsils. A tendency to "sleep high," that is with the head and shoulders elevated by the pillow, indicates impaired pulmonary or cardiac function. So, too, does an upright position in the nurse's arms, with the chest against her breast and the head hanging over her shoulder—a posture assumed by young children. "Sleeping cool," namely, resting only after all the bed-clothing has been kicked off, is an early symptom of rickets.

The position termed "*en chien de fusil*," is a symptom of the advanced stages of cerebral disease, especially tubercular meningitis. The child lies upon one side, with the head stretched far back, the arms pressed close to the sides and folded across the chest, the thighs drawn up toward the abdomen, the legs flexed on the thighs and the feet crossed. Restless movements of the head or boring of the head into the pillow also point to cerebral disease.

When there is an evident desire to retain one position, as on the back or one side, together with short, quick breathing, some inflammatory change in the respiratory or abdominal organs may be suspected. Persistent lying on the face is an evidence of photophobia.

Of the gestures, the frequent carrying of the hand to the head, ear or mouth indicates headache, earache or the pain of dentition respectively, and constant rubbing of the nose is a feature of gastro-intestinal irritation.

If the thumbs be drawn into the palms of the hands, and the fingers tightly clasped over them, or if the toes be strongly flexed or extended, a convulsion may be expected. The presence of clonic contractions of the muscles, with unconsciousness, indicates, of course, a convulsion; while irregular, badly coördinated, jerky movements—consciousness being retained—attend chorea.

In infants the existence of colic is shown by repeated extension and retraction of the legs, clenching of the hands into fists, flexion and extension of the forearms, and a writhing movement of the trunk.

The fact of one limb remaining passive while the others are actively moved about, naturally suggests motor paralysis.

THE SKIN.—In the new-born infant the color of the skin varies from a deep to a light shade of red. After the lapse of a week this redness fades away, leaving the surface yellowish-white. Sometimes the yellow hue is so deep that it might readily be mistaken for jaundice were it not for the whiteness of the conjunctivæ, and the absence of disordered digestion and other symptoms of ill-health. Usually in a fortnight all discoloration disappears, and the skin assumes its typical appearance. Allowing for the natural variations in complexion, the skin of a healthy child is beautifully white, transparent and velvety. The cheeks, palms of the hands and soles of the feet have a delicate pink color, and the general surface is rosy in a warm atmosphere, marbled with faint blue spots or lines, in a cool one. As age advances, the coloring becomes more pronounced, and until the completion of childhood the complexion is much fresher than in adult life.

In the inspection alterations of the skin of the face are chiefly noticeable. Lividity of the eyelids and lips is a sign of imperfect aeration of the blood, and points to pulmonary or cardiac disease. Marked blueness of the whole face is a symptom of *morbus cæruleus*, and indicates a congenital malformation of the heart. On the other hand, a faint purple tint of the eyelids and around the mouth shows weak circulation merely, or, more frequently, deranged digestion.

A decided yellow hue of the skin and conjunctivæ is seen in jaundice; an earthy tinge of the face in chronic intestinal diseases; a waxy pallor in renal diseases, and paleness in any acute or chronic affection attended with exhaustion.

Brownish-yellow discoloration of the forehead is significant of inherited syphilis; a bright, circumscribed flush on one or both cheeks, of inflammation of the lungs or pleura, or of gastro-intestinal catarrh, according to its occurrence with or without an elevated surface temperature.

The cutaneous lesions of certain of the eruptive fevers appear first upon the face; each of these has its special characteristics. An eruption of herpetic vesicles on the lips may be mentioned as present both in pneumonia and in malarial fevers.

Some information may be obtained from the hands. Slight want of proper aeration of the blood is shown by blueness of the finger nails, a greater degree, by cyanosis of the whole hand. Deformity of the nails is a symptom of syphilis: clubbing of the finger tips of chronic lung disease; and redness, swelling and suppuration about the nails of struma. The dropsy of scarlatinal nephritis causes a puffiness and cushiony appearance of the dorsum of the hands. Often, too, in this condition, the finger ends are glossy, as if smeared with oil, and there is an exfoliation of the epidermis about the nails. The last two symptoms frequently serve to confirm a retrospective diagnosis of scarlet fever.

MODE OF DRINKING.—By watching a child taking the breast or bottle, some knowledge can be obtained of the condition both of the mouth and throat, and of the respiratory organs.

If there be any soreness of the mouth the nipple is held only for a moment, and then dropped with a cry of pain. When the throat is affected deglutition is performed in a gulping manner, an expression of pain passes over the face, and no more efforts are made than required to satisfy the first pangs of hunger. Under similar circumstances older children drink little and refuse solid food entirely.

An infant suffering from the oppression of pneumonia or severe bronchitis, seizes the nipple with avidity, swallows quickly several times, and then pauses for breath. In older patients the act of drinking, which should be continuous, is interrupted in the same way.

If the finger be put into the mouth of a healthy baby it will be vigorously sucked for some little time. The diminution of the act of suction during a severe illness is a sign of danger; its reëstablishment a good omen. In conditions of stupor and coma it is noticeably absent.

THE CRY.—The vocal sound, termed crying, is the chief if not the only means that the young infant possesses of indicating his displeasure, discomfort or suffering. Even long after the powers of speech have been developed, the cry continues to be the main channel of complaint. It may be accepted, as a rule, that a healthy child rarely cries. Of course, some acute pain, as from a fall or accident or blow, will cause crying in the most healthy child, but the storm is quickly over. Nothing like frequent, peevish crying or fretfulness is compatible with health, consequently, when this disposition exists, the cause must be looked for in some disease.

Incessant, unappeasable crying is due to one of two causes, namely, earache or hunger, and the distinction may readily be made by putting the child to the breast or offering a properly prepared bottle. The *hydrencephalic* cry, denoting pain in the head, is a sudden, sharp and very loud shriek, occurring at intervals and audible at a considerable distance. Crying during an attack of coughing, or for a brief time afterwards, and attended with distortion of the features, indicates pneumonia. In acute pleuritis, the cry also accompanies the cough, but it is produced too by movements of the body and by pressure on the affected side. It is louder, indicative of greater suffering, and sometimes most difficult to check. Intestinal pain causes crying just before or after an evacuation of the bowels, and is associated with wriggling movements of the body and pelvis, and with the eructation or passage of flatus. Conditions of general distress or malaise predispose to fits of fretful crying, the paroxysms being excited by any disturbing influence, or even by merely looking at the little sufferer.

When the cry has a nasal tone, it indicates swelling of the mucous membrane of the nares, or other obstructing condition. Thickening and indistinctness occurs with pharyngeal affections. A loud, brazen cry is a precursor of spasmodic croup. Hoarseness points to a lesion of the laryngeal mucous membrane, either catarrhal or syphilitic in nature. In membranous croup, and in some cases of extreme exhaustion, the cry is faint and inaudible. Finally, in severe croupous pneumonia, in extensive pleural effusion and in rickets, ordinary disturbing causes are inoperative for the production of fits of crying, and there is a seeming unwillingness to cry, on account of the action interfering with the respiratory function.

The conditions of altered tone apply equally to the articulate voice in children who are old enough to speak.

The cough, too, must not be disregarded. Many of its characters correspond with the voice and cry. It is brazen in spasmodic croup, suppressed in true croup, hoarse in laryngeal catarrh, and so on. But it has certain features of its own. In bronchitis it is more or less paroxysmal, evidently dry in the early stages, loose and rattling as the catarrh "breaks up." In the painful pulmonary affections, pneumonia and pleurisy, it is choked back, and whenever it occurs, an expression of pain passes like a cloud over the face. In pertussis, the peculiar spasmodic cough is the pathognomonic symptom, and when once heard, immediately stamps the case. Cough is always unproductive, that is, unattended by expectoration, in children under seven years of age

The formation of tears rarely begins before the third or fourth month of life. Subsequently, an alteration in this secretion may be of aid in forecasting the result of disease. The prognosis is bad when the tears become suppressed; good when the secretion continues during an illness, or when it reappears after being suppressed.

There are three other sources of information which can and should be investigated before proceeding to the physical examination, although, strictly speaking, they do not come under the head of inspection of the child. These are the characters of the fæcal evacuations, of the urine, and of the material ejected by vomiting.

THE BREATH.—The breath of a healthy infant or child should be odorless, or as the nurse will say, "sweet," except perhaps immediately after taking nourishment, when it may, for a short time, have the smell of milk or any special food eaten. The persistent presence of an odor, therefore, is abnormal and indicates disease.

Any morbid condition of the system that prevents the elimination of metamorphosed nitrogenous tissue through the mucous membrane of the intestines, or retards the passage of decomposing detritus along the bowels, will cause an offensive breath. Under this head are conditions characterized by high temperature, catarrhal inflammations of the gastro-intestinal tract, chronic debilitating diseases, etc. The same result, also, frequently attends structural lesions of the kidneys. The reason for this is, that the system, in order to get rid of poisonous matter—for accumulated waste is poison—and to maintain the balance between the constant construction and destruction of tissue, must throw off elsewhere what the intestinal glands and the kidneys fail to excrete; so the lungs take on vicarious activity and the expired air becomes tainted with the products of waste. Very often, by the way, the skin takes a part in the abnormal excretory process, and a similar odor is noticed in the perspiration.

Purely local causes of *halitosis* also exist. These are decayed teeth, caries of the nasal and maxillary bones, ulceration of the mucous membrane of the mouth, nose, larynx, trachea and bronchial tubes, and gangrene of the cheeks.

Chronic poisoning by lead, arsenic and mercury, though not very common in childhood, is another cause of ill-smelling breath.

To speak in general terms, the breath may become sour, catarrhal, fetid, gangrenous, ammoniacal and stercoraceous. This classification is a rude one, and many subdivisions can be made of some of the odors. Thus, there are many varieties of catarrhal and fetid breath, which, while more or less distinctive of different conditions, cannot be differentiated in words and must be experienced by the observer's sense of smell to be recognized; once this is done they become valuable symptoms.

Sour breath is present, in infants more especially, when there is gastric fermentation. The variety of food, whether milk or farinaceous substances, makes little difference in the odor, and it is most perceptible in cases attended by eructation and vomiting. In chronic vomiting, chronic entero-colitis and thrush the intensely acid odor exhaled from the mouth and, in fact, from the whole body, is a most prominent feature.

What I have classed as catarrhal breath has, as already stated, numerous shades of difference.

In chronic catarrh of the pharynx there is always a "heavy" breath, not noticeable far from the patient's face. The odor is always more marked during sleep and is greatest after the long sleep of night, as then the mucus, to which the odor is due, not being removed by acts of swallowing, collects in larger masses.

Should the catarrh invade the follicles deeply, and, especially, should there be associated follicular tonsillitis, the breath, while still having the quality of heaviness, becomes extremely offensive, with a scent somewhat like that of decaying cheese, and is very penetrating. This odor, too, is worse after sleeping.

At the onset of acute catarrh of the stomach the breath becomes decidedly tainted. Sometimes it has a vinous odor, at others it is sweetish, and I have attended a number of cases in which it had the same quality as after an inhalation of æther. Later in the attack it becomes sour or has the odor of sulphuretted hydrogen. The former is apt to be the case with infants, the latter with older children, who have a more solid albuminoid diet.

What is known as a "feverish breath" has a heavy sweetish smell. It is met with in diseases of high temperature and depends partly upon catarrh of the gastro-intestinal mucous membrane, the common attendant of fevers, and partly upon the elimination of fever waste. It is very marked and rapid in appearance in scarlatina.

In chronic intestinal catarrh with obstinate constipation the breath often has a slightly fæcal odor.

Simple catarrh of the nasal mucosa when of any standing, gives rise to moderate heaviness, and the same is true of catarrh of the mucous membrane of the mouth—stomatitis—though in the latter affection, mastication and swallowing being difficult, small quantities of food collect in the mouth, and there undergoing decomposition add an element of fetor to the breath.

Fetor of the breath is observed in its mildest form in such affections as aphthæ and ulcerative stomatitis. It is better developed in ozænæ and necrosis of the maxillary bones, when the well-known stench of dying bony tissue is added. Decaying teeth give much the same odor, though it is less strong and penetrating. In all these conditions, however, the fetor differs not only in degree, but in kind.

Noma gives rise to a gangrenous odor, and a patient affected with this disease will fill the ward of a hospital, the room in which he lies, or even a whole dwelling, with the most sickening stench. Cases of empyema, with ulceration of the lung and discharge of pus through the bronchial tubes, have an almost equally offensive breath, but here there is often a flavor of garlic combined with that ordinarily due to tissue necrosis.

Ammoniacal breath is observed only in patients suffering with uræmic poisoning.

A purely stercoraceous breath is rare, and when met with is an accompaniment of fæcal tumor or of intussusception.

The metallic poisons while giving rise to fetor of the breath have no individual characteristics, and it is necessary to look to the history and symptoms of the individual case to determine the special poison.

THE FÆCAL EVACUATIONS.—The daily number of evacuations natural for a child varies greatly with its age. For the first six weeks there should be three or four stools every twenty-four hours.

After this time up to the end of the second year, two movements a day is the normal average. Subsequently, the frequency of defecation is the same as in adults—once per diem—though two or three movements in the same interval often occur, especially after over-feeding or after eating food difficult of digestion, and must be looked upon as conservative rather than as the evidence of ill-health.

During the first period the stools have the consistence of thick soup, are yellowish-white, or orange-yellow in color, with sometimes a tinge of green, have a faint fæcal, slightly sour odor, and are acid in reaction. In the second, they are mushy or imperfectly *formed*, of uniform consistence throughout, brownish-yellow in color, and have a more fæcal odor. The last two characters become more marked as additions are made to the diet. After the completion of the first dentition the motions have the same appearance as in adult life, they are *formed*, and brownish in color, with a decided fæcal odor.

Many alterations occur in disease. The frequency of the movements may be increased, constituting diarrhœa, or lessened, constituting constipation. In the former condition the consistency is diminished, in the latter increased. Instead of being uniform throughout, the stool may be mixed, partly liquid partly solid, indicating imperfect digestion, and curds of milk and pieces of undigested solid food may be mingled with the mass. Flaky, yellowish or yellowish-green evacuations, containing whitish, cheesy lumps, are also met with in cases of indigestion. Scanty, scybalous stools, dark brown or black in color, and mixed with mucus, are characteristic of intestinal catarrh. Doughy, grayish, or clay-colored motions show a deficiency of bile. An intermixture of blood, altered blood clots, and shreds of mucous membrane, indicate some breach of continuity in the intestinal lining, such as occurs in follicular enteritis, typhoid fever, dysentery and tubercular disease. Watery, almost odorless stools occur in the latter stages of entero-colitis; most offensive, carrion-like motions, in both catarrhal and tuberculous ulceration of the intestines, and sour-smelling evacuations in the diarrhœa of sucklings. The discovery of worms or their ova in the stools is the certain evidence of the existence of intestinal parasites.

This mere outline of the changes that may take place will serve to show how much may be learned from the stools, and the importance of making a personal examination of them.

THE URINE.—It is impossible to make a definite statement as to the number of times the urine is voided by a healthy infant, in each twenty-four hours. In any given case the frequency will differ very much from day to day, depending upon the temperature of the surrounding air, the amount of moisture that it contains, and so on. Sometimes it will be necessary to change the diaper every hour during the day and three or four times at night. Again it may remain dry for six, eight, or even ten hours. Neither condition indicates disease, and between the two extremes there is a wide range of variation. Should the urine not be passed for twelve hours or more, a careful examination should be made to discover and remedy retention.

As the child grows older the frequency diminishes, and at the age of three years the number of voidings will be reduced to six or eight during the waking hours, and perhaps one at night. When the desire does arise during sleep, the child, if in a normal state, wakes up and demands the chamber, and never passes urine unconsciously. Wetting the bed, therefore, or the involuntary passage of the urine during sleep, is indicative of an abnormal condition and requires investigation.

Painful micturition points to inflammation of the urethra, a narrow preputial orifice, a highly acid condition of the excretion, or stone in the bladder.

The urine of a healthy infant, while it wets, should not *stain* the diaper, the fluid being clear and almost colorless. It has a low specific gravity—1.003 to 1.006—and an acid reaction produced by the considerable amount of uric acid it contains.* As

---

* The specific gravity falls markedly during the first few days after birth, on account of the ingestion of food. Alantoin is present in abundance during the

childhood advances, the adult characters are more and more nearly approached.

The normal daily amount excreted cannot be stated absolutely, since it is difficult to collect the urine in infants and very young children, and since, in children of all ages, the flow depends so much upon circumstances quite compatible with average health. Thus it is influenced by the state of the weather, the condition of the various emunctories, the amount of blood pressure in the renal vessels, the state of the nervous system, and the quantity of solids and liquids consumed. However, from a few observations, I am led to believe that the quantity of urine voided by healthy children from the fourth to the seventh years is not nearly so large as supposed; eighteen to twenty ounces being the average in several cases in which I have lately made measurements.

Increased secretion of urine is a prominent symptom in diabetes insipidus and mellitus, and as a transient event is encountered after an attack of abdominal pain, an epileptic fit, a paroxysm of ague and a convulsive seizure. Diminution may result from diarrhœa and vomiting, from extreme prostration due to deficient nutrition or other causes, and from renal congestion, whether occurring in Bright's disease or in diseases of the heart and liver. In febrile conditions the flow is diminished, while the proportion of solids excreted is normal or increased; the specific gravity, in consequence, is high. Complete suppression may occur when general prostration and renal congestion become intense.

The quantity of solids excreted in health is also subject to great variation. The amount of urea passed by a child is relatively greater (1.7) than in the adult. Between the ages of three and six years, according to Uhle, one gramme of urea for every kilogramme of weight is voided every twenty-four hours. Eustace Smith, from a rough calculation based on the specific gravity of the urine, estimates that the solids excreted daily between the

---

first weeks of life. Pyrocatechin (Ebstein and Müller) is also present, but indican (Senator) is not found in the urine of the newborn.

ages of four and ten years, amount to five grains to each pound of weight.

The normal acidity of the urine is increased by trifling agencies. A urine so affected deposits urates on cooling, or may indeed be turbid when passed and while still warm. Often the urates are so abundant as to render the fluid thick and milky-looking. In addition to increased acidity, this excess may depend upon an augmented secretion of salts. Over-feeding is the cause of the latter, and this relation of cause and effect must be borne in mind in the treatment of convalescence from acute diseases, during which a turbid urine is often seen. Free uric acid in the form of fine red sand is sometimes observed. Specific gravity and color vary with quantity.

Two abnormal ingredients—albumen and blood—are frequently present in the urine of children.

Albumen, though a frequent attendant of organic kidney disease, by no means always indicates the existence of such a condition. It attends many febrile and inflammatory affections; is always to be detected where blood or pus are present, and appears where there is passive congestion of the kidney from chronic disease of the heart, liver or lungs.

Again, transient albuminuria may arise from very slight causes. School children often have it during an examination or throughout the time given to preparing for it. Dr. Kinnicutt ascribes this to passing oxaluria or lithuria. It is also seen in children living in ague districts. Sometimes over-fatigue or the mere ingestion of a hearty meal will produce it, and some patients have it habitually after eating.

Intermittent albuminuria—albumen being present one day and absent the next—is generally due to an admixture of secretions, and often indicates the habit of masturbation.

The source of blood in the urine may be the ureters, the bladder or the urethra, as well as the kidneys. In the first three cases the blood and urine are passed separately, while in renal hemorrhage the two liquids are intimately blended. When large quantities of blood are voided the cause of bleeding is, as a rule, either purpura hemorrhagica or a renal calculus. Scarlatina frequently, and the other exanthemata occasionally, produce bloody urine, through intense renal congestion, and the same result is sometimes brought about by the severe diarrhœa of entero-colitis or cholera infantum, the appearance of blood following suppression of the urine.

Many special diseases are attended by alterations in the urine. Affections of the kidneys stand, of course, at the head of the list.

Acute Bright's disease and scarlatinal nephritis have a urine diminished in quantity; of high specific gravity; a smoky, blackish hue, as if there had been an admixture of soot; with albumen, blood and epithelial, granular and hyaline tube-casts as abnormal constituents. In chronic Bright's disease the quantity varies, being either about normal or excessive; the specific gravity is low and the amount of solids diminished; the reaction is acid; uric acid crystals may appear; hyaline and granular casts are quite constant, but albumen is often absent. Sudden exacerbations produce the characteristics of the acute form of the disease.

Passive congestion of the kidneys causes albuminuria with epithelial and blood casts; renal calculus and lithæmia—great acidity, with the deposit of uric acid sand, and sometimes blood and albumen, and sarcoma of the kidney—albumen and blood at times.

Simple catarrh of the bladder is attended by an albuminous urine, which is sometimes very offensive and may contain pus, vesical epithelium and phosphates; the reaction is usually alkaline. Tubercular cystitis gives rise to a cloudy, thick urine, containing a trace of albumen, blood or pus.

In incontinence the secretion is, in some cases, highly acid and, on standing, deposits crystals of uric acid.

In hydronephrosis the specific gravity is low, the fluid may be either clear or turbid, and is faintly alkaline. The urine salts are reduced in quantity, and crystals of oxalate of lime are often detected by the microscope.

In malarial fever there is usually a profuse discharge of limpid fluid at the conclusion of the hot stage. During it, according to Gee, the urea and chloride of sodium are increased, the phosphates diminished; after the temperature falls, however, the phosphates are increased and the urea and chloride of sodium are diminished. Patients living in highly malarious districts often show albumen and blood or its coloring matter in their urine.

Throughout the first stage of typhoid fever the urine is scanty, has a high specific gravity and contains an excess of urea and uric acid, but few chlorides. Later it is freer, with diminished density, and may, if the temperature be high, contain albumen.

Fevers as a class produce scanty, dense, high-colored urine, cloudy with lithates, and albuminous when an elevated temperature is maintained.

Croupous pneumonia gives well-marked changes, the quantity is diminished, the specific gravity high, the urea and uric acid above the average, the chlorides diminished or entirely absent at the extremity of the disease, and albumen often present. After the crisis the chlorides reappear.

In diphtheria the urine is usually clear but may be smoky, urea is increased, and albumen and hyaline and granular casts may be discovered; in membranous croup, on the other hand, it is generally normal.

A scanty, high-colored urine, and one which deposits a whitish or pinkish sediment (lithates) on standing, is symptomatic of acute digestive disorder. Uric acid sand is sometimes seen in acute gastric catarrh, an excess of indican in inflammatory diarrhœa when the small intestines are chiefly involved, and albumen in severe cases of thrush. Suppression attends grave enterocolitis and cholera infantum, while acute peritonitis and, occasionally, dysentery and the irritation of seat worms induce retention.

In icterus neonatorum the urine is yellow in color, but contains no biliary coloring matter. Panot and Robin detected yellow, amorphous masses having different chemical reactions from bile pigment, with uric acid; urates; oxalate of lime; hyaline, epithelial and fatty casts and white blood corpuscles. In catarrhal jaundice it is dark yellow, or even brownish, and contains bile coloring matter. Cirrhosis of the liver, in the hypertrophic stage, shows a yellow urine; in the atrophic, a fluid highly acid and filled with lithates and uric acid crystals. Amyloid degeneration is often associated with a similar disease of the kidneys, and the urine is altered in consequence, being copious, pale, lemon-yellow in color, of low specific gravity and containing albumen and hyaline tube casts; the latter do not present the ordinary color reactions of amyloid material with iodine. Ascites has a scanty, high-colored and sometimes albuminous urine.

Acute articular rheumatism presents a febrile urine, and in rheumatism of the abdominal muscles the urine is reduced in quantity, high colored and very acid.

VOMITING.—Both vomiting and regurgitation are of ready production and frequent occurrence in infancy, on account of the vertical position and cylindrical outline of the stomach at this period of life.

Babies suckled at an abundant breast, and who are in the best possible state of health, often vomit habitually. In these cases, the supply of food being large, the infant as it lies at the breast is apt to draw more than it needs and more than it can digest. The stomach rids itself of this over-supply by an act which more nearly resembles regurgitation than vomiting, and which must be regarded as an evidence of health rather than the reverse. There is no violent effort or retching, the material ejected is the breast milk alone, either entirely unaltered or slightly curdled, and there are no symptoms of nausea, such as paleness, languor and faintness.

In older children, vomiting may also occur after the stomach has been overladen. If the act be followed by relief from a feeling of general distress, headache and epigastric pain, it must not be regarded as a symptom of disease.

Vomiting attended with the train of symptoms embraced under the term *nausea*, is not a pathognomonic symptom. It may indicate disease of the stomach, of the intestines, of the lungs and pleura, and of the brain, or it may be a prodrome of one of the eruptive fevers. Which condition is present can only be determined by watching the case, and by a careful study of the rational symptoms and physical signs.

The character of the ejecta is more definite. For instance, the expulsion of mucus is a symptom of gastric catarrh. The regurgitation of mouthfuls of curdled milk, partially digested food and liquid so sour that it causes a grimace to pass over the face, is an indication of dyspepsia with fermentation and the formation of acid. The appearance of lumbricoid worms in the vomit, a not infrequent occurrence, of course shows conclusively the existence of these parasites in the alimentary canal.

3. Physical Examination.

The methods of physical exploration in children are identical with those employed in adults, and the results do not differ in kind. Since, however, the object of exploration is to elicit the greatest amount of information with the least possible disturbance of the child, and as this very disturbance alters the character of some of the information obtained, it is well to adopt a somewhat different order of examination, and one which at first sight may seem irregular. Thus it is best first, to ascertain the character of the respiration and the pulse, then to strip the body to determine the degree of muscular development and the condition of the skin, next, to investigate the physical condition of the lungs, heart and abdominal organs, and last of all to examine the mouth and throat. In this order, then, the normal, as well as the more prominent abnormal features connected with the different organs will be considered.

THE RESPIRATION.—In children the respiration is chiefly *abdominal* in type, irrespective of sex, and it is not until just before the age of puberty that the movements in the female change, becoming *superior costal*. Consequently, in estimating the number of movements per minute, it is best to place the fingers lightly on the epigastrium. The count should always be made by the watch, and the most convenient time for the observation is while the child sleeps.

Soon after birth the number of movements per minute is 44, between the ages of two months and two years, 35, and between two and twelve years, 23. During sleep the frequency is reduced about twenty per cent.

Children under two years, while awake, breathe unevenly and irregularly; there are frequent pauses followed by hurry and precipitation, and some of the movements are shallow, others deep. In sleep there is greater regularity. After the second year the movements become steady and even, like those of adults. All children, however, but particularly the very young, are subject to a great increase in the rapidity of respiration under the excitement of muscular movement and mental emotion.

Accelerated breathing occurs during the course of diseases attended by severe febrile symptoms, such as the acute exanthemata, the inflammatory and other affections of the thoracic viscera, and in rickets. Acute pulmonary lesions are especially characterized by this alteration, and the more the breathing area is lessened the greater is the increase. Thus, in pneumonia, 60, 80 or 100 movements a minute are not at all unusual. To speak broadly, rapid breathing may be caused by an elevation in the body temperature, by an interference with the blood aeration and by thoracic or abdominal pain.

As the increase in frequency may be unattended by any apparent effort, or true dyspnœa, it is well to make a rule of counting the respirations in every case in which the diagnosis is at all doubtful.

Diminished frequency, the movements being reduced to 16, 12, or even 8 in the minute, is noted in certain brain affections, as in chronic hydrocephalus, and in the later stages of tubercular meningitis. In such cases the rhythm may be greatly altered— a *tidal* form being assumed, in which the breathing ebbs and flows, beginning with an act which is scarcely perceptible or audible, gradually growing deeper until a full, noisy respiration is made, and then slowly subsiding into a period of absolute quiet, variable in its duration. This is termed Cheyne-Stokes' respiration.

Another form of breathing, in which the alteration is mainly in the rhythm, is termed *expiratory* respiration. In the normal act, inspiration is immediately succeeded by expiration, and between the latter and the next inspiration there is a period of silence or rest. Expiratory respiration, on the contrary, is characterized by the pause coming between inspiration and expiration, the expiratory effort, always very marked, being immediately succeeded by the inspiratory. This alteration occurs most frequently in young children, and is an evidence of dangerous pulmonary embarrassment.

Perfectly healthy children breathe through the nose, and so softly that it is necessary to place the ear close to the face to hear the breezy sound of the ingoing and outgoing air. A dry, hissing sound, or a moist sound of snuffling indicates partial obstruction of the nasal passages; oral respiration, complete occlusion. Difficult breathing with prolonged inspiration—*inspiratory dyspnœa*—shows an impediment to the entrance of air into the lungs and indicates laryngeal obstruction, due, most commonly, to spasm or to the formation of false membrane. In such cases the inspiratory act is also attended by a loud, piping, or rasping sound. Labored breathing with prolonged wheezing respiration—*expiratory dyspnœa**—occurs when the escape of air is impeded. The causative lesion is to be found, not in the larynx, but in the lungs. It may be a bronchial catarrh with excessive secretion, an emphysematous condition of the air vesicles, or asthma. In both forms of dyspnœa the movements are slow as well as difficult, and a combination of the two forms is met with in cases of marked laryngeal stenosis.

Yawning, one of the modifications of the respiratory act, if it

---

* I prefer to limit the term *dyspnœa* to difficult or labored respiration, and not to extend it so as to include simple accelerated breathing.

recur frequently, denotes great failure of the vital powers and is an unfavorable prognostic element.

THE PULSE.—To obtain any reliable data from the pulse it must be felt while the patient is perfectly quiet. The best time is during sleep, but if the child cannot be caught in this condition, advantage may be taken of its placidity while nursing at the breast, feeding from a bottle, or amused by a toy or book. With very young infants it is sometimes impossible to feel the beat of the radial artery, and it is necessary to ascertain the frequency of the pulse by directly ausculting the heart. After the second month palpation of the pulse at the wrist in the ordinary way presents no difficulties.

The child's pulse differs from the adult's by being much more frequent, more irregular, and more irritable, and necessarily of smaller volume.

The frequency, or the number of beats per minute, varies with the age. The following is the average rate:—

From birth, to the 2d month, ... 160 to 130.
From the 2d to the 6th month, ... 130 to 120.
" " 6th " 12th " . . . . 120 to 110.
" " 1st " 3d year, ... 110 to 100.
" " 3d " 5th " . . . . 100 to 90.
" " 5th " 10th " . . . . 90 to 80.
" " 10th " 12th " . . . . 80 to 70.

These figures represent the pulse in a waking but passive state. During sleep the frequency is less. Thus between the second and ninth years there are about sixteen beats less per minute while asleep than when awake; between the ninth and twelfth years, eight less; and between the twelfth and fifteenth years only two less. Below the age of two years the disparity is even greater.

The irregularity of the pulse in childhood is confined to an alteration of the rhythm. It is most marked in infants and is greatest during sleep, when the pulse is slowest.

The feature of irritability, that is, the facility with which its frequency is increased by muscular activity and mental excitement, is greater in proportion to the youth of the child. A rise of 20, 30 or even 40 beats a minute is not uncommon in early infancy under the excitement of the slightest effort or disturbance.

On account of these wide variations in health, little symptomatic meaning need be attached to alterations of the rhythm and frequency while unassociated with other abnormal features. When so associated they become important in diagnosis.

Increased frequency is a constant attendant of the febrile state. The extent of the increase corresponds with the degree of elevation of the temperature, though the pulse curve always runs higher than the temperature curve. The more frequent the pulse the higher the fever, is the rule, but in estimating the prognostic value of the increase, the law of the fever in question must be taken into consideration. For example, in scarlatina a pulse of 160 is usual and not indicative of special gravity, whereas in measles the same degree of acceleration would be abnormal and show great danger.

Jaundice and parenchymatous nephritis are accompanied by a diminution in the rate.

Irregularity is met with in diseases of the brain and heart, and sometimes in nervous and anæmic children.

THE TEMPERATURE must be estimated before removing the clothing. No reliable result can be obtained without the use of an accurate clinical thermometer. The instrument is usually placed in the rectum* of the infant and young child; in the axilla of one old enough to understand the importance of keeping the arm in a proper attitude. It should remain in position at least five minutes.

During the first week of life the temperature fluctuates considerably. After that the puerile norm—98.5° to 99° F.—is established, but until the fourth or fifth month it is greatly influenced by healthy causes of variation; the fluctuations ranging between .9° and 3.6°. By the fifth month regular morning

---

* The rectal temperature is normally at least 1° higher than the axillary.

and evening oscillations begin to be noticeable, and certain definite laws are followed. There is a fall in the evening of 1° or 2°. The greatest fall occurs between 7 and 9 P.M. and the minimum is reached at, or before, 2 A.M. After 2 A.M. there is a gradual rise, the maximum being reached between 8 and 10 A.M. Throughout the day the oscillation is trifling. These variations are independent of eating and sleeping.

In disease there may be either a rise above, or a fall below, the normal standard.

Fever is always associated with an elevation of the temperature. Rapid and transient rises attend slight catarrhs and passing indigestions; prolonged rises, inflammatory and essential fevers. The *degree* of elevation marks the type of the pyrexia. This is moderate when the mercury stands at 102°, severe at 104° or 105°, and very grave above 107°. The *duration* of the elevation and the peculiar *range* of the oscillations—for there are oscillations in disease as well as in health—determine the nature of the fever. The febrile oscillations differ from the healthy in that the lowest marking is noticed in the morning, the highest in the evening. *Variations* in the typical range of any given fever are important prognostic omens—a sudden fall of the temperature, together with improvement in the general symptoms, indicates the beginning of convalescence—a similar fall, with an increase of the general symptoms, is a precursor of death. When the morning temperature is equal to that of the preceding evening, there is great danger; if higher, greater danger still. Marked remission in continued fevers is generally a forerunner of convalescence.

Abnormal depression of temperature is occasioned by hemorrhage and by the loss of fluids in cholera infantum or enterocolitis. It is also met with in anæmia, in atrophy from insufficient nourishment, in diseases of the heart and lungs attended by imperfect blood aeration, and it constantly attends collapse and the death agony. A maintained temperature of 97° is dangerous in children, and for every degree of reduction below this point, the risk to life is more than proportionately increased.

While the physician must use the thermometer, to insure accuracy, he can, by placing the hand on the skin, detect gross differences of temperature. Reductions are best appreciated by touching the nose and extremities, while increased heat is most readily felt at the back of the head and in the palms of the hands.

Having determined the character of the respiration, pulse and temperature, the next step in the physical examination is to strip the child, in order to ascertain his general development, the condition of his skin, and so on.

THE GENERAL DEVELOPMENT.—The healthy child under two years of age is plump of body and round of limb with well-developed fat cushions and firm flesh, and with the head and abdomen large in proportion to the rest of the frame. As age advances, the figure gradually assumes the characteristics of adolescence.

To be robust, the newly-born child must have a certain average size and weight. Subsequently, under normal circumstances, there is a regular rate of increase in both of these respects. At birth the length is about 16 inches. Growth is quickest in the first week of life. In the first year there is an increase of from 5 to $6\frac{1}{2}$ inches; in the second, from $2\frac{2}{3}$ to $3\frac{1}{3}$ inches; in the third, from $2\frac{1}{3}$ to $2\frac{2}{3}$ inches; in the fourth, about 2 inches; and from the fifth to the sixteenth years the annual growth amounts to from $1\frac{2}{3}$ to 2 inches. The average weight at birth is from 6 to 8 pounds. The daily increase in weight should range from $\frac{1}{4}$ to $\frac{3}{4}$ of an ounce.

With these data it is quite possible to estimate what should be the normal size and weight of a child at any age. Consequently, if, on being measured and weighed, he be found to fall short of the normal standard, it is proper to infer the existence of some fault in the nutritive processes. A conclusion still further borne out by a want of rotundity of outline and by flabbiness of the muscles.

The age at which the child sits erect, at which it walks, and at which the anterior fontanelle becomes ossified, are points closely connected with the subject of development and nutrition.

For some time after birth, the child, if noticed while sitting upon the lap, will be observed to hold the head and shoulders forward or to "stoop" a little; the spine, from the cervical region to the sacrum, forming a continuous curve with the convexity directed backward. Toward the end of the eighth month the position begins to become more erect, and in a few weeks is perfectly so, the spine assuming an almost perpendicular line. Any marked delay in this change indicates general debility.

At the end of the fourteenth month the child should be able to walk alone. The spine then assumes the S-like curve seen in healthy adults. A delay in walking may be due to systemic weakness or to infantile paralysis affecting one or both legs. If the walking be done on the toes chiefly, if the gait be limping, and especially if knee-pain be complained of, and manipulation of the limbs causes suffering, the chances are that hip-joint disease is commencing.

The anterior fontanelle should be ossified or completely closed at some period between the fifteenth and twentieth months. The closure is much retarded in rickets, which is preëminently a disease of mal-nutrition. Hydrocephalus has a like effect. In a state of health, the opening, while still membranous, is level with the cranial bones or very slightly depressed. Conditions of systemic exhaustion cause marked sinking, and this depression is one of the best indications of the necessity of stimulation. Bulging of the fontanelle is a symptom of chronic hydrocephalus.

CONDITIONS OF THE SKIN.—The normal color of the integument, and the alterations produced by disease, have already been studied. The other characters possessed by the skin of a healthy child are, a velvety smoothness and softness, a scarcely perceptible moisture, and a great degree of elasticity.

Disease causes modifications in texture, in moisture, and in elasticity, and leads to the appearance of various eruptions and to œdema.

Mucous disease is attended with a dry, harsh skin, which is muddy in color, and covered, especially on the extensor surfaces of the arms and legs, by a more or less thick layer of exfoliating epidermis. Chronic abdominal affections, particularly tuberculosis of the intestines and mesenteric glands, lead to harshness, acridity, scurfiness, and a wrinkled appearance of the skin covering the abdomen and thorax, with enlargement of the superficial abdominal veins.

Protracted diarrhœa, and still more, vomiting combined with diarrhœa, cause absorption of the subcutaneous fat and wasting of the muscles. The skin becomes too large for the body, is dry, harsh, discolored, and so inelastic that it falls into wrinkles over the joints when the limbs are moved, and if pinched up retains the fold for a long time. The condition of general atrophy popularly known as "marasmus," presents these features most strikingly.

Dryness is a concomitant of the febrile state; excessive moisture, of prostration of the vital forces and collapse.

Eruptions appear upon the integument, in the skin diseases proper, in the exanthemata, in constitutional syphilis and in certain digestive disorders.

Œdema of the subcutaneous connective tissue may be due to affections of the heart, liver or kidneys. The cardiac variety usually shows itself first in the feet; the renal, in the eyelids; the hepatic, in the feet and legs, secondarily to ascites.

While examining the surface it is well to look for enlargement of the superficial lymphatic glands and swelling of the joints. The former occurs in scrofula and syphilis; the latter in rheumatism.

EXAMINATION OF THE ABDOMEN.—To examine this portion of the body the child, still stripped, must be placed on its back, upon the bed or nurse's lap. Quiet is most important, since struggling and crying are attended by such contraction of the abdominal muscles and rigidity of the walls that little can then be learned of the condition of the contained organs. The methods of investigation are those ordinarily employed in physical examination. Palpation or percussion should never be made with cold hands.

The abdomen of a healthy child is somewhat prominent,

uniformly soft, yielding, and painless to the touch, and to percussion gives a tympanitic sound, varying in tone according to the region percussed. The tympanitic note being lowest in pitch over the epigastric and left hypochondriac regions, the seat of the stomach; highest, over the umbilical region, the position of the small intestine.

In disease *inspection* reveals any disproportion in the size or form of the abdomen, the state of the integuments, of the superficial veins, and of the umbilicus. *Palpation* shows the temperature, pliability, moisture and tension of the walls, and the presence or absence of tenderness, of fluctuation, and of enlargement of the mesenteric glands, and other solid viscera. *Percussion* serves to demonstrate the nature of enlargements, whether due to accumulation of gas or liquid, or to solid growths. By it, also, the outline and size of the liver and spleen may be determined.

Distention of the abdomen is, in the vast majority of instances, due to flatulence. In children reduced by chronic disease the bowels are usually deranged, the food is badly digested, and the gases set free by the decomposition of the starchy foods accumulate, owing to feebleness of the intestinal walls, and give rise to much swelling and discomfort. Over such an abdomen the skin feels tense, the umbilicus is level or slightly prominent, there is no tenderness on pressure, and percussion is markedly tympanitic.

This simple cause of enlargement must be remembered, for a distended abdomen in a wasting child is often falsely attributed to caseation of the mesenteric glands. The latter disease is uncommon at any age, extremely rare under three years, and, moreover, is by no means uniformly attended by distention. On the contrary, unless the glandular disease be excessive, retraction is the rule. When distention does exist it depends upon associated intestinal disorder, and is merely an accidental complication. The only pathognomonic sign is the detection of the tumor caused by the enlarged glands. This is situated in the umbilical region, and is firm, lobulated and slightly tender to the touch. It is most readily detected by gently grasping the abdomen on either side with the hands and slowly bringing the fingers together toward the median line. Percussion over the tumor yields a dull, tympanitic sound. Whenever there is associated flatulence it is difficult or impossible to detect the tumor.

Drum-like distention, with great tenderness, and muffled tympanitic percussion note occur in general peritonitis.

Uniform distention, again, may be due to ascites depending upon simple or tubercular peritonitis, kidney disease, or less commonly, disease of the liver. The abdomen is barrel-shaped, painless to the touch, and there is extended fluctuation. Percussion is dull over the position of the fluid, but in nearly every instance there is an area of tympany which changes its position; moving always to the upper part of the abdomen, in reference to the posture of the patient. This variation is most important in the diagnosis.

Localized distention may be traced to gaseous accumulation, to enlargement of the liver and spleen, to fæcal accumulation, to circumscribed peritonitis, and to distention of the bladder.

Collections of gas are always tympanitic on percussion.

The extent of liver dulness is to be estimated by percussion. If the organ extend below the rib margin, the edge can usually be felt by laying the palm of the *warmed* hand flat upon the abdomen and making gentle pressure downward with the ends of the fingers.

An enlarged spleen may be felt by placing the fingers of the right hand on the back, directly below the twelfth rib and outside of the lumbar muscles; the fingers of the left, on the abdomen, directly opposite; then bringing the hands toward one another. If the hands have been rightly applied, and nothing is felt, the spleen may be considered to be normal in size. The fact that both the liver and spleen, though still unenlarged, may be more readily felt than natural when pressed downward by the diaphragm, must not be overlooked.

A fæcal accumulation is distinguished by the absence of tenderness, by the oblong shape of the tumor, by the situation in the region of the transverse or descending colon, to which its long axis corresponds, and by its shape being capable of some modification by pressure. Percussion over such a mass is dull.

Distention of the bladder gives rise to a bulging tumor in the hypogastric region, which is elastic to the touch and dull on percussion.

A shrunken or scaphoid condition of the abdomen is met with in serious brain affections, notably tubercular meningitis, also in cholera infantum, follicular enteritis and dysentery.

Tenderness to pressure indicates inflammatory lesion of the intestines. The presence or absence of this sign in an infant can be determined by forcing the attention, by bringing it before a strong light, for instance, and then making pressure on the abdomen. If crying be produced, there is tenderness, if not, the reverse.

EXAMINATION OF THE CHEST.—The stethoscope and pleximeter are unnecessary in examining the lungs. In the case of the heart, the former may be occasionally required, to localize murmurs. When used, it is better to give the instrument to the child to handle and become familiar with, before application. The thoracic end must never be adjusted without being warmed. The quieter the patient the more complete and satisfactory will be the results of the exploration. Unfortunately, though, it is too often necessary for one to do the best possible in the midst of cries and struggling. However, by skilfully seizing opportune moments, much reliable information may be gained. Aid is also derived from the fact that in serious lung affections, as croupous pneumonia, the child is quiet from choice, crying interfering with the respiratory act, upon which his attention is concentrated.

The steps of the examination are, first, inspection; second, auscultation; third, palpation; and fourth, percussion. The reason for making the order different from that practiced in adults, is to place the most disturbing element last. Mensuration and succussion are infrequently resorted to in children. If required, they are best postponed until the end of the examination.

Inspection.—The sitting posture, the child being stripped and in a good light, is the best for this process. Note is to be taken of the shape of the chest, the character of the breathing, and the position of the apex beat of the heart.

In the newborn baby, the chest is nearly circular in shape, the antero-posterior diameter being almost as great as the lateral. Later, it gradually becomes elliptical, the lateral diameter in time considerably exceeding the antero-posterior. The intercostal spaces are poorly marked, and the scapulæ lie so close that their outline is scarcely perceptible. The circular shape of the chest allows of little lateral expansion, and for this reason the respiration is chiefly abdominal in type. Together with the movement of the abdominal walls, every act of inspiration is attended by a certain amount of recession of the lower part of the chest walls, the yielding ribs being forced inward by the pressure of the external air before they can be sufficiently supported by the expanding lung. The rise and fall of the cardiac apex can be seen—except when there is a great accumulation of fat—a short distance below and to the right of the left nipple.

Disease may alter all of these conditions. The tuberculous diathesis is characterized by a small chest, and one which has either the *alar* or the *flat* shape. In rickets the thorax becomes irregularly triangular in outline. Emphysema causes a barrel-shaped chest, with stooping shoulders and round back. Pleuritis with large effusion produces bulging of the affected side, and sometimes prominence of the intercostal spaces. After absorption has taken place there may be marked retraction, sinking of the interspaces, falling of the shoulders, and curvature of the spine toward the healthy side.

Cessation of the costal respiratory movements indicates inflammation of the lung or pleura, or a large pleuritic effusion. Cessation of the abdominal play, inflammation of the peritoneum or of the intestines; excessive ascites and gaseous accumulations produce the same effect.

Rachitic softening of the ribs, and those diseases of the lungs and air passages which offer a direct obstacle to the entrance of air, are associated with a great increase in the normal recession of the lower portion of the chest on inspiration. In certain cases, a deep furrow appears across the chest, marking the upper borders of the abdominal viscera. The depth of the furrow indicates the degree of softening and of obstruction to the ingoing air.

The position of the apex beat is altered by cardiac diseases, by pleuritis, and occasionally by gaseous distention of the stomach. When the left ventricle is enlarged, it is shifted downward and to the left. Transmitted epigastric pulsation shows enlargement of the right ventricle. An extended impulse is not necessarily a sign of disease, since the chest-walls are so elastic in childhood that the normal impact of the apex is apt to affect a wide area. The effusion of pleurisy pushes the heart to the right or left, while the retraction, after absorption or evacuation, draws it in one or other direction. The apex is pushed upward and to the left in gastric flatulence. Emphysema, by pushing the heart away from the thoracic wall, diminishes or entirely hides the impulse.

Auscultation.—With infants, the back of the chest is most conveniently ausculted when the child is held in the nurse's left arm, with his breast against hers, his chin resting upon her left shoulder, his left arm around her neck, and his head kept in position by her disengaged hand. The front, when reclining on the back on a pillow. The sides, when sitting upright on the lap, first one arm and then the other being lifted up to allow the observer's ear to be applied. Older children may be made to take the same position as adults.

It is not sufficient to auscult the posterior aspect of the thorax alone, as is stated by some authors. The whole chest should be examined, particularly in doubtful cases. The signs of croupous pneumonia are most frequently discoverable at one or other base, posteriorly; the friction sound of pleuritis at the junction of the middle and lower third of the chest, laterally; and the signs of emphysema at the apices, anteriorly. Therefore,

THE INVESTIGATION OF DISEASE. 53

unless the exploration be thorough, important lesions may be overlooked.

In healthy infants the inspiratory act in ordinary breathing is superficial, and the respiratory murmur, as a consequence, feeble. If, however, a deep inspiration be taken, a frequent occurrence under excitement and during crying, the murmur becomes loud, or assumes the character that Laennec termed *puerile* breathing. After the age of two years this form of respiration is habitual.

Puerile breathing is characterized by its *intensity*, a property depending upon the thinness and elasticity of the chest-walls in childhood. There is no alteration in *rhythm*, the inspiratory element of the murmur being directly followed by the expiratory, and this in turn by an interval of silence; neither is there any change in the *pitch* or *duration* of the expiratory sound, which remains lower and shorter than that of inspiration. In other words, puerile respiration is simply a very intense vesicular respiration. The normal respiratory murmur is then feebler in infants, and louder in children over two years old, than in adults.

The breathing is loudest over the anterior, lateral and posterior inferior regions of the thorax. Faintest over the scapulæ and the præcordial area. Sometimes the expiratory element is wanting. This absence occurs most frequently in young children, and is most noticeable over the lower posterior portions of the lungs. In the inter-scapular region, the ear, being directly over the larger bronchi, readily detects a deviation from the vesicular quality. Here the inspiratory murmur is loud, harsh and somewhat tubular in character. There is a slight pause between it and the expiratory murmur, and the latter is longer in duration and higher in pitch. There is, in fact, an approach to the bronchial type of breathing, which may always be heard in its purity by listening over the trachea.

Sometimes a difference in the breathing can be detected over the apices anteriorly. On the left side the vesicular quality is purer, on the right, the intensity is greater. The difference is most decided in the expiratory element, which, also, may be slightly prolonged on the right when compared with the left side. These modifications are due principally to the larger size and more horizontal course of the right primary bronchus. They are perfectly compatible with a normal state of the lungs. Should, however, the conditions at the apices be reversed, and the intensity and prolongation of the expiratory sound be greater on the left side, the commencement of phthisis is indicated.

If the child speaks, cries, or coughs while the ear is applied to the chest, a muffled rumbling sound, the normal vocal resonance, will be heard. At the same time, vibration of the walls, the vocal fremitus, can be felt.

To develop the respiratory sounds it is often necessary to instruct the patient how to breathe, and if an infant is being examined, to take advantage of the deep inspirations that precede coughing, and occur during crying.

The cardiac sounds are readily heard when the ear is placed on the præcordia. In young infants the examination is somewhat difficult, on account of the rapid and excitable action of the heart, but after the first year, the circulation becoming slower and more regular, there is little trouble in distinguishing the sounds, and even slight alterations produced in them by disease. The first sound is longer and graver than the second, and the rhythm is ordinarily quite regular. In health the sounds may be heard under both clavicles for a short distance to the right of the sternum, and sometimes over the whole anterior surface of the chest. After muscular effort or during agitation, the heart sounds may be audible over the posterior aspect of the chest, but they are most distinct in this position when the lower lobe of either lung is consolidated by pneumonic exudation. The latter point is often of great value in distinguishing doubtful cases of pneumonia from pleural effusion.

Palpation.—In practicing palpation the palmar surface of the well-warmed hand must be applied to the naked chest. This method of exploration is useful as a means of determining the number of respiratory movements, the degree of expansion of the thoracic walls, the position of the cardiac apex beat, the presence or absence of painful regions and of pleural or bronchial fremitus, the existence of fluctuation in the intercostal spaces, and the character of vocal fremitus. For the last purpose, though, it is hardly worth while to make a separate step in the examination, for the vocal vibrations can be readily distinguished by the ear when applied to the chest in auscultation.

Percussion.—In percussing the different surfaces of the chest, the child must be placed in the same positions as for auscultation. When contrasting the two sides, percussion should be made in identical regions, and during the same period of the respiratory movement. Babies when constrained or when disturbed, hold their breath in the intervals of crying, and as they always do so at the end of an inspiration, this is a favorable time to seize for the comparative examination. The percussion strokes must be lighter than in the adult, but in other respects the operation in no wise differs.

In health the resonance will be found to correspond closely with the respiratory murmur. Thus, in infants under one year, the respiratory murmur being feeble, percussion is rather insonorous. Even at this age the case is different, when a deep breath is taken, and so soon as puerile respiration becomes established the resonance is uniformly intense. With the exception of this greater intensity, the sound is exactly similar to that obtainable in adults. It is always attended, too, by a sensation of elasticity, appreciated by the finger used as the pleximeter.

Different portions of the thorax possess, normally, different degrees of sonorousness.

In front, the right side is markedly resonant from the clavicle down to the fifth interspace, or the upper border of the sixth rib in the mammary line, where the liver dulness begins. On the left side the resonance is equally intense, but it is encroached upon by the gastric tympany, which extends upward as high as the seventh or sixth rib, as well as by the area of cardiac dulness. The latter forms an irregular triangle, of which one side is represented by a vertical line passing down the middle of the sternum, from the level of the fourth to the sixth rib; the other, by an oblique line touching the upper extremity of the first, and extending outward to the left, and downward, to terminate at the point of the apex beat; and the base, by a line drawn from the central point of the lower edge of the sternum (the inferior extremity of the first line), along the sixth costal cartilage, to the apex of the heart. Diminished resonance and elasticity are at once noticeable when the percussion passes from the lung to this area, though the præcordial dulness is never so decidedly marked in children as it is in adults.

Laterally, both supra-axillary regions are very resonant. The upper portions of the infra-axillary regions are a degree less resonant, and the lower portions are dull on account of the presence of the liver on the right and the spleen on the left side. The superior border of the liver dulness is found in the seventh interspace or at the eighth rib; that of the spleen, at the upper edge of the ninth rib. Gastric tympany may supplant the pulmonary resonance over the left infra-axillary region.

Posteriorly, there is little resonance in the scapular region, particularly the supra-spinous portions. Over the interscapular space the sound improves, but it is less resonant than anteriorly or laterally. Over the infra-scapular regions the resonance is but little less pure than in front, until the tenth rib is reached on the right side and the liver dulness is again met with. On the left side the resonance extends to the very base, the posterior splenic dulness being detected with difficulty. The right base is, therefore, naturally less resonant than the left, and this difference is especially marked during expiration, the liver rising higher at that time.

Affections of the lungs produce various alterations in the percussion sound. The chief of these are the substitution of tympany, of dulness, and of flatness for the normal resonance, and of increased resistance to the finger for elasticity. Cardiac diseases cause changes in both the extent and the shape of the area of præcordial dulness.

EXAMINATION OF THE MOUTH AND FAUCES.—This portion of the examination is most apt to cause crying and struggling, but it must never be omitted. In infants, gentle pressure of the fingers upon the chin is sufficient to cause wide opening of the mouth. An older child will frequently open the mouth when requested, but if he refuse, the finger, the handle of a spoon, or some other smooth, flat instrument may be inserted in the mouth, and downward pressure made upon the tongue, when the jaws will be widely separated. In some cases, when the child is old enough to do as bid, the fauces can be seen by directing the mouth to be opened wide and the tongue to be alternately protruded and retracted, or a prolonged sound of "*Ah*" to be made. With the refractory, and always with infants, the tongue has to be held down by a spoon handle or tongue depressor. If there be resistance, the patient must be taken on the lap of the nurse, who holds his back against her breast, directs his face toward a bright light, and controls the movements of his hands and feet.

The healthy oral mucous membrane has a deep pink color, and is smooth, moist and warm to the touch. The color is deeper on the lips and cheeks, lighter on the gums. The latter, up to the sixth month, as a rule, have a moderately sharp edge. Subsequently, the edge begins to broaden and soften, and the color of the investing mucous membrane deepens to a vivid red, and becomes hot, as the teeth begin to force their way through. The first, or *milk teeth*, so called from their color, are twenty in number, all told, ten to each jaw; they make their appearance in the following order, the corresponding teeth appearing a little earlier in the upper jaw * :—

The two lower central incisors, from	4 to 7	months after birth.
" four upper " "	8 to 10	" " "
" two lower lateral incisors and the four anterior molars, from	12 to 15	" " "
" four canines, "	18 to 24	" " "
" " posterior molars, "	20 to 30	" " "

* Upon this point, however, there is little uniformity.

58 DISEASES OF DIGESTIVE ORGANS IN CHILDREN.

The order of eruption of the permanent teeth is as follows.—

    The two central incisors of lower jaw, from the 6th to 8th year.
" " " " upper " " " 7th to 8th "
" four lateral "   " " 8th to 9th "
" " first bicuspids,   " " 9th to 10th "
" " canines,   " " 10th to 11th "
" " second bicuspids,   " " 12th to 13th "

These replace the temporary teeth; those which are developed *de novo* appear thus:—

    The four first molars, from the 6th to 7th year.
" " second " ." " 12th to 13th "
" " third "   " " 17th to 21st "

There are, therefore, twelve more permanent teeth, making thirty-two in all, sixteen in each jaw.

The tongue should be freely movable. It is pink in color, and the dorsum, or upper surface, marked in the centre by a slight longitudinal depression, has a velvety appearance, and is soft, moist, and warm to the finger. The velvety nap is due to the numberless hair-like processes of the filiform papillæ. There are also scattered over the surface, but most closely at the tip, a number of eminences, the size of a small pin's head, circular in outline, and deeper pink than the general surface—the fungiform papillæ. While far back, defining the papillary layer, are the circumvallate papillæ, numbering about twelve, and arranged in a V-shaped row. These have the form of an inverted cone, surrounded by an annular elevation.

The hard palate is roughened anteriorly by transverse ridges. The soft palate is smooth and its mucous membrane is paler than that of the rest of the mouth. The fauces, on the contrary, are redder. In the triangular recess between the half arches of the palate the tonsils can always be seen. They should be about the size and shape of almond kernels, and they present a number of circular openings, the orifices of pouches, into which the follicles open. The uvula is short and tongue-shaped. The posterior wall of the pharynx should be red, smooth and moist.

THE INVESTIGATION OF DISEASE. 59

Disease produces a great variety of changes in the mouth, tongue and fauces. Fever makes the mouth hot and dry, and causes the tongue to be frosted or coated. Affections of the gastro-intestinal tract are always attended by coating of the tongue, and the various appearances of this coating are of important diagnostic and therapeutic significance. Inflammation of the mouth itself, reddens the mucous membrane, makes it hot and tender to the touch, increases its moisture, alters the surface of the tongue and leads to the formation of aphthæ, to ulceration, and even to gangrene.

The eruptions of scarlet fever, measles, varicella and varioloid make their appearance first on the mucous membrane of the palate and fauces.

Irregular dentition indicates faulty nutrition; delayed dentition, rickets; and certain peculiarities in the formation of the permanent teeth, constitutional syphilis.

Finally, the conclusive evidences of the existence of diphtheria, of croup, and of the various tonsillar affections, are found in the fauces.

PART II.— THE GENERAL MANAGEMENT OF CHILDREN.

It is the duty of the child's physician not only to remove disease, but also to manage convalescence and every-day life in such a way that the little subjects confided to his care may be led to complete recovery, and kept in as perfect health as possible. To accomplish these objects, the ability to direct intelligently the daily regimen is much more important than a mere knowledge of drugs and of the principles of therapeutics.

The daily regimen embraces several factors; these are feeding, bathing, clothing, sleep and exercise, and under such headings the subject will be briefly outlined, for little more is possible, in the present chapter.

1. **Feeding.**

Age bears so close a relation to the choice of food and the method of feeding, that it will greatly simplify the study of these questions to consider them from the standpoint of the two stages of a child's life, namely, infancy, or the period extending from birth to the age of two and a half years; and childhood, the time elapsing between completion of the first dentition and puberty.

An INFANT may be fed in either one of three ways—1st, from the mother's breast; 2d, from the breast of a wet-nurse; and 3d, from a bottle, the latter being the method known as artificial or hand-feeding.

1st. *Feeding from the maternal breast.* There can be no doubt that this, being the natural, is at the same time the proper method of nourishing the human infant; and fortunate is the babe that, in our day of advanced civilization and city-living, can draw from the breast of a robust mother an abundant supply of pure, health-giving, tissue-building food.

It follows, therefore, that every woman who is free from certain contra-indicating diseases, to be mentioned later, should nourish her child solely from her breast up to the age of eight months, and partially to the end of the first year, or, failing in either limit, so long as possible.

The infant should be put to the breast as soon as the mother has recovered somewhat from the fatigue of labor—some four or eight hours after birth. Of course no milk can be drawn at this early date, but the babe gets a small quantity of thin, watery fluid, called colostrum, which affords sufficient nourishment, and at the same time, from its laxative properties, clears away the greenish or black, viscid material that collects in the infant's intestinal canal during intra-uterine life. This procedure, too, is of great advantage to the mother, for it insures proper contraction of the womb, draws out the nipples, and encourages the formation of milk.

As the secretion of milk is never fully established until the third day after labor, it stands to reason that no food other than the colostrum is required before that time. Hence, the practice of filling the infant's stomach with gruel, sugar and water, and other sweetened mixtures, is more than useless, for it diminishes the activity of sucking and the consequent stimulation of milk production. Put the child to the breast every two hours while the mother is awake, and there need be no fear of starvation.

After the third day, should the breasts not yield a supply of milk, a little pure cow's milk diluted with double its quantity of water and sweetened with sugar of milk, may be given every fourth hour, the babe being put to the breast in the meanwhile. So soon as the flow begins, however, the artificial feeding is to be discontinued.

Usually on the fourth day milk is secreted and regular lactation commences. Many untrained mothers make a failure of nursing because they know nothing of the manner of giving suck; of the length of time the child should be kept at the breast; of the proper time for, and interval between feeding, and of the importance of regularity. Upon these points the physician should give minute instructions.

When giving the breast, the infant must be held partly on its side, on the right or left arm, according to the gland about to be drawn, while the mother must bend her body forward, so that the nipple may fall easily into the child's mouth, and steady the breast with the first and second finger of the disengaged hand, placed above and below the nipple. In case the milk runs too freely—a condition very apt to excite vomiting—the flow is easily regulated by gentle pressure with the supporting fingers. Each of the breasts should be drawn alternately, the contents of one being usually sufficient for a meal; and a healthy child may be allowed to nurse until satisfied, when he will stop of his own accord, drop the nipple and fall asleep with milk still flowing over his lips.

During the first six weeks the breast is required every second hour, from 5 A. M. until 11 P. M. At night the infant should be put in a crib by the mother's bed, or in an adjoining room, under the care of a competent nurse, and there remain quietly until the morning feeding. This secures the mother six hours of uninterrupted repose, a matter of great importance to her general health and consequent capacity for prolonged lactation. As to the infant, he may rebel at first, and wake and cry, so that it is necessary to quiet him with a little milk and water administered from a bottle; but often after a few days, and certainly at the end of a week or two, the good habit of sleeping at night is formed, and there is no further trouble.

Regularity in meal hours is even of more importance in early than in adult life, on account of the natural feebleness of digestion. To secure this, it is only necessary to have a little perseverance, for infants are such creatures of habit that a short training brings them into the way of expecting food only at certain times, and, when healthy, they wake to suck the breast with almost the precision of the clock. While insisting upon this rule, one must recognize the fact that, although in the vast majority of instances a two-hours' interval is most suitable up to the second month, there is no absolute law as to the number of daily nursings. Some infants seem to need food less frequently, and it is best to respect their peculiarity and not force the breast upon them so long as they sleep well, do not fret when awake, and thrive generally. Others, again, may require it oftener, every hour and a half, perhaps, and once or twice at night. In these exceptional cases an appropriate schedule can only be made by close observation of individual characteristics.

A common and most ruinous mistake is to resort to constant feeding as a means of pacifying crying. Babies certainly do cry from hunger, but just as frequently the crying results from colic, or from the discomfort and pain of indigestion. Every mother should be able to recognize the difference. The cry from hunger usually begins after a sound sleep. It is not peevish, and stops at the sight of the breast, when the infant rouses himself, presents an expression of pleasure, clinches his hands and flexes his limbs. The cry of colic is violent and paroxysmal; the face is livid and wears an expression of suffering; the abdomen is distended and hard; the hands and feet are cold; the legs are drawn up or kicked violently about; and an explosion of wind from the mouth or bowels ends the attack. A peevish cry, hot skin and sour breath attend indigestion.

It stands without saying that the cry of hunger must be relieved by giving food; but this is the very worst thing to do under other circumstances, for it both breaks up good habits and produces serious mischief. The pain of colic and the discomfort of indigestion are chiefly due to the accumulation of flatus resulting from the fermentation of food. Mothers soon learn, and unfortunately infants too, that the breast milk temporarily relieves suffering. This it does in the same way as any other warm liquid; but, unlike a simple fluid, milk only adds more material to the already fermenting contents of the gastro-intestinal canal, and every nursing is soon followed by more pain, until between crying and sucking and sucking and crying, the infant's life is passed in misery, if not cut short altogether.

Instead of continuous feeding, the plan for relief is to decrease the quantity of food by increasing the intervals between nursing and by abridging the time of lying at the breast, while medicines are employed to strike at the root of the evil.

After the sixth week the interval between nursing may be slowly increased until, by the fourth month, it reaches three hours. During this period, also, the time of lying at the breast may be gradually lengthened, for the quantity of milk secreted and the child's appetite and capacity for food are all augmented as the days pass by. At the end of the sixth month, feeding every fourth hour suits some children well, but as a rule the three-hour interval must be adhered to from the fourth month to the end of lactation.

Many authorities recommend additional artificial feeding, alternating with nursing, after the sixth or eighth month. Such a plan is perfectly proper, if the babe ceases to gain strength and flesh while on the breast. If otherwise, the maxim of not interfering with any course that is doing well is as applicable here as elsewhere, and the breast may be relied upon entirely until the time comes for weaning.

Should additional nutriment be required, the food must be selected with due reference to age and prepared in the same manner as in regular hand-feeding.

The date of weaning cannot be fixed for all cases, since it must depend upon two conditions,—the health of the mother and the development of the child. When the former continues to be robust and the child steadily grows and gains flesh, lactation can be prolonged until the tenth or twelfth month. If persevered in longer, the mother's strength begins to fail, her milk is lessened in quantity or becomes poor in quality, the child's nutrition suffers, and he grows pale, thin and flabby, and may develop the disease known as rickets.

Change in the manner of feeding may be accomplished gradually or suddenly. In gradual weaning, about four weeks are required to prepare for the absolute withdrawal of the breast. For instance, if suck be given every three hours, from 5 A. M.

until 11 P. M., or seven times a day, there should be, during the first week of preparation, one artificial feeding and six nursings daily; during the second, two and five; during the third, four and three; during the fourth, six and one. Then the breast must be entirely withheld. Carefully prepared milk-food, administered from a bottle, is the best substitute. At the age of ten months a mixture that ordinarily agrees well is:—

  Cream, ... f ℥ ss.
  Milk, ... f ℥ iv.
  Sugar of milk, ... ℨ j.
  Water, ... f ℥ iss.

This is to be poured into a perfectly clean bottle, warmed in a water bath, and taken through a clean, plain rubber tip. Should the quantity (six fluidounces) be insufficient to satisfy the child's appetite, all the ingredients except the cream may be increased until the mixture measures eight or even twelve fluidounces, according to the demand.

When such accidents as fever, disordered digestion, with vomiting and diarrhœa, or the actual cutting of one or more teeth occur during the period of preparation, the number of artificial feedings must be reduced, or the breast resumed until the disturbance be passed; then the course may be begun again and carried to its completion.

Usually there is little trouble in weaning infants in this way. Sometimes they become fretful under the change and may refuse food entirely for a day or more; but a little determination on the part of the mother and the cravings of hunger will soon overcome this difficulty.

Occasionally the child refuses to suck milk from a bottle or to drink it from a cup or spoon, in fact seems to object to any form of liquid food except that drawn from the mother, while at the same time he is eager for bread or other solid food. Under these circumstances prepare for each meal a moderate portion of either rice pudding or junket. After these have been taken for a day or two, add to each meal a little milk, reducing the amount of pudding or junket; stir the whole together and feed from a spoon; next day still further reduce the solid and increase the liquid, and so proceed until finally a taste for milk is cultivated.

Sudden weaning is not advisable unless, while the breast is being presented, there is an absolute refusal to take artificial food from either a bottle or a spoon. This is most apt to occur when food has been given too frequently, and when the breast has been used as a means to quiet crying. Sudden weaning is also to be recommended when the mother's health becomes so affected as to render any further sucking a positive peril to the child's life; attacks of erysipelas or of smallpox are instances in point.

The physician is often forced to decide upon the advisability of premature weaning. His decision must be made cautiously and after thorough investigation of two propositions, namely: *a*, the effect of further lactation upon the health of the mother, and *b*, the requirements of the child.

a. Lactation being a physiological process is not a drain upon the systemic strength so long as the functions of nutrition are actively performed, but under other circumstances it very frequently becomes so. Premature weaning is necessary when the mother is attacked by any acute disease threatening dangerous temporary prostration, such as typhoid or typhus fever. A change must also be made if pulmonary consumption be developed, or, being already present, rapidly advances under the drain of milk secretion. Ordinarily, however, the general condition that leads to withdrawal of the breasts is one of simple loss of strength and flesh on the part of the mother.

Undoubtedly these indications often warrant the procedure, but every one who has seen much of children's practice must have met with many cases in which the advice to wean has been given carelessly and unnecessarily, and in which the child might have had its natural food had proper attention been given to the health of the mother.

If a woman be worn out by household cares; if she wear herself out by a round of dinners, balls or shopping, or if she expose herself to injurious atmospheric conditions and eats improper food, she grows weak and thin whether she be nursing or not; and a woman heedless of her health will probably care little whether she suckles her child or gives it up to a wet-nurse or to the bottle.

In addition to making nursing the important duty of her life for the time being, a mother must be as free from household cares as possible. Mental and physical fatigue is to be avoided, sufficient exercise must be taken to maintain a healthy appetite and digestion, and abundant time devoted to rest and sleep. Beyond securing a plentiful supply of plain and easily digestible food, with a judicious portion of meat, vegetables, and fruit, it is unnecessary to give special attention to the diet.

Should the secretion of milk be scanty, it may often be increased by the free use of animal broths, chocolate, gruel, or milk, and sometimes the moderate employment of stimulants, in the form of ale and porter, may be necessary. Such tonics as malt extract, ferrated elixir of cinchona, bitter wine of iron, and the preparation known as "beef, wine and iron," are useful when there is anæmia, or when the general failure of strength cannot be overcome by food and attention to hygienic rules.

The ordinary local conditions indicating the necessity of premature weaning, on the mother's account, are fissures of the nipple and mammary abscess.

Fissure being usually a unilateral condition, it is only necessary to retire the affected side from duty and nourish the child alternately from the unaffected gland and from the bottle until healing takes place, the disabled breast being pumped in the meantime to keep up secretory activity. Should both sides be affected, weaning may be imperative, on account of the extreme pain produced by sucking, though, even under these circumstances, an effort must be made to maintain the flow of milk by regular pumping. Sometimes women are able to struggle through the attack by taking advantage of the relief and protection afforded by a nipple-shield.

Fissures of the nipple may be preceded by various diseases of the delicate skin of the part. They result, also, from want of cleanliness or from keeping the nipple too moist, as when constant sucking is allowed or when there is a continual flow of milk. They may be prevented by proper attention to the nipple before confinement. During the latter months of pregnancy the clothing covering the breast must be loose, and the wearing of a wire tea-strainer over the nipple to prevent pressure has been recommended by one authority. Each day for three months before labor, the nipples should be washed thoroughly with hot water in the evening and anointed with cocoa-butter in the morning. At the same time, should the nipples be small or retracted, the woman must be taught to use her thumb and finger to draw them out. This process is not only an advantage in giving proper size and shape, but brings the skin into good condition without hardening it. The application of alcoholic and astringent lotions is not to be recommended. They tend to harden the tissue, which should be soft and pliable rather than tanned, and render the nipples liable to crack.

When a fissure exists, it is best first to see whether or not nursing can be continued by means of a nipple-shield. Should the child refuse this, a good plan is to fill the shield with warm milk and invert it over the nipple. The infant then draws the fluid at once and without difficulty, and will often continue sucking until the breast milk follows. After nursing and removing the shield, the nipple must be dried thoroughly with absorbent cotton, and the following lotion applied with a camel's-hair brush:—

    ℞. Acid. Boracic, ... gr. xx.
       Mucilag. Acaciae, ... f ℥ j.   M.

b. On the part of the infant, there are several indications for anticipating the time of withdrawing the mother's breast. It must be done if the occurrence of pregnancy or the recurrence of menstruation render the milk unwholesome; if the mother contract a dangerous contagious disease, as smallpox, scarlet fever, or erysipelas; if the mammary glands become inflamed; if the breast does not afford sufficient nourishment and artificial food be refused; and, finally, if dentition be markedly delayed and the premonitory symptoms of rickets appear. As to the amount of nourishment, it must be remembered that the breast milk may be of good quality, but so diminished in quantity that it is insufficient; or, while abundant in quantity, so poor in quality that it does not meet the demands of nutrition. Even without a minute examination of the milk, it is possible to form a good idea of which condition is present from the behavior of the infant in the act of sucking. If the milk be good in quality but deficient in quantity, the babe, when put to the breast, seizes the nipple as if famished, and draws upon it vigorously for a time, and then drops it with a scream of rage. On the contrary, should there be an abundant supply of poor milk, the nipple is grasped languidly, the child lies a long time at the breast and falls asleep there. Consideration of the final indication opens the question of the propriety of regulating weaning by the progress of dentition. This is certainly a good guide, but not in the way implied in the old precept, that the child must not be taken from the breast until evolution of the stomach and eye teeth. Insufficient food is one of the chief causes of rickets, and rickets more than any other disease delays dentition; consequently, should the teeth not pierce the gum in time, the inference is for other food rather than a continuance of the faulty maternal supply.

Upon deciding to anticipate the time of weaning, the next point to consider is whether the infant shall be brought up by hand or by a wet-nurse.

2d. *Feeding by a wet-nurse.* The advantage of feeding from the breast of a wet-nurse is that the mother's milk is substituted by the milk of another woman; in other words, that natural feeding is continued—a matter of moment in all cases, and of inestimable importance with delicate children. The disadvantage consists in the difficulty of finding, in a woman belonging to the class from which wet-nurses come, all the moral and physical characters essential to a good substitute, and in the fact that a stranger is introduced into the household, often to deceive and annoy the family, and on the slightest provocation to leave her charge to fate or to the tender mercies of another of her kind. For these reasons it is preferable, in the majority of instances, to trust to careful bottle-feeding. Nevertheless, as some children must have human milk if their lives are to be saved, the rules for selecting a wet-nurse must be understood.

The woman chosen must be strong and robust, but rather spare than fat. Her bill of health must be perfectly free from hereditary tendency to mental or physical disease and from taint of syphilis, consumption or scrofula. She must be cheerful, good-natured, active, careful, and temperate in habits. Her age should be between twenty and thirty years; she should understand the care of an infant and the manner of giving suck; her child ought to be nearly the same age as the infant to be adopted, and she must be able to afford an abundant supply of good milk.

The last quality can be estimated by inspecting the breasts, by examining some of the milk drawn by a pump, and by ascertaining the condition of the woman's own child. The breasts of a good nurse are not necessarily large, but are firm to the touch and pyriform in shape, with well-developed, prominent nipples, and with the skin distinctly marbled with large blue veins. The milk, which ought to flow readily on pressure or on suction, should be opaque and dull white in color, have a specific gravity of 1.031, an alkaline reaction, and show, when placed under the microscope, a number of minute, equal-sized, fat globules. Its quantity may be ascertained by weighing the child before and after sucking, the normal gain being from three to six ounces. There is, however, no better or more readily applied test of the quality of a nurse than the size, weight, and general development of her own child; and if it be weak and ill-nourished, no amount of fitness in other respects can warrant her engagement.

Even when a woman is found fulfilling in her single person all the required conditions—a rare thing, indeed—it does not necessarily follow that her milk will suit the babe to be suckled.

Then changes and new trials must be made until the desired end be attained.

The diet of a wet-nurse and the manner of weaning, must be governed by the rules already given for maternal guidance.

Personally, I have had such good results from carefully regulated bottle-feeding, that I have almost given up the employment of wet-nurses, preferring to regulate the artificial food myself rather than allow an ignorant woman to supplement surreptitiously her deficient supply of breast milk by an unskilfully proportioned food,—an event of not uncommon occurrence.

3d. *Artificial feeding.* In my experience, there are few American women, especially in the well-to-do classes, who do not look upon the duty of nursing their babies as a pleasant one; but there are many who are completely unable to do so, and a vast number in whom the secretion of milk fails after a few weeks or months of lactation. They must, therefore, through no fault of their own, resort to a wet-nurse or to artificial feeding. Usually, they select the latter method, with results that vary in direct proportion to the care and intelligence displayed in carrying it out.

There is no artificial food equal to the milk of a robust woman. The fluid, however, secreted from the glands of a feeble or unhealthy mother, though often sufficient in quantity to fill the suckling's stomach and satisfy the cravings of hunger, does not contain enough pabulum to meet the demands of nutrition. In such unfortunate cases, good cows' milk, properly prepared, is a better food than the bad breast milk. More care and trouble, though, are involved in bottle than in breast feeding. If the child has been nourished in the natural way—*i. e.*, breast-fed—even for a few weeks, or when the powers of digestion are inherently active, the task is far easier to accomplish. In these cases the stomach and intestinal canal, inactive in fœtal life, are trained to their new duties under normal conditions, and so prepared for the digestion of properly selected artificial food. On the contrary, if digestion be naturally feeble, or if the infant must be bottle-fed from the first, great difficulty may be expected, and most skilful handling is necessary.

To insure success in hand-feeding, it must be remembered that an infant is not nourished alone by the food he swallows, but by that portion of it he digests and assimilates. The best diet, therefore, is one so adapted to age and digestive power that everything eaten will be digested and absorbed. But as children differ as much in constitution as in feature, it is impossible to formulate exactly a food that will be applicable to every case, or one that needs no change from month to month of progressing growth. As age and strength increase, there is a corresponding development of the gastro-intestinal functions and a demand for more and stronger food. On the other hand, should the system be accidentally reduced by disease, the digestion, sympathizing in the general debility, temporarily loses its normal activity and assumes that of an earlier age. In such a case more nourishment is certainly needed to build up the failing strength, but it is to be supplied by giving such food as can be completely assimilated, and not by forcing down strong food merely because it is *strong;* for the latter, when not vomited, passes through the bowels undigested, and the little creature starves to death in the midst of plenty, or dies from the ill effects of the constant presence of fermenting food in the alimentary canal. On these accounts many changes in diet, as to quality and quantity, must be anticipated and made.

Important matters, therefore, to be studied in detail are: *a*, the selection of a proper substitute for the breast milk; *b*, the quantity to be given; *c*, the method of preparation; *d*, the mode of administration; and, *e*, the means of preservation.

a. Healthy breast milk must be taken as the type of infants' food, and the nearer an artificial substance can be made to approach it in chemical composition and physical properties, the more perfect it is.

Normal breast milk has a specific gravity of 1.031. It is a persistently alkaline fluid, having a somewhat animal, usually disagreeble, and very rarely sweetish taste. It is bluish-white in color and thin and watery in consistence.

According to Leeds' very thorough analysis, it contains:—

Water,	86.766 per cent.
Total solids,	13.234 "
Total solids not fat,	9.221 "
Fat,	4.013 "
Milk sugar,	6.997 "
Albuminoids,	2.058 "
Ash,	0.21 "

It contains, then, nitrogenous material, carbohydrates, salts and water—all the elements essential to repair tissue waste, to supply new material for growth and to maintain body heat, or, in other words, to constitute a perfect aliment; and these, too, are so proportioned in the combination as to most easily and completely meet the demands.

It must not be supposed, however, that the elements are uniformly present in the same proportion. On the contrary, the fluid varies both at different periods of lactation and in different individuals.

This fact is the most striking feature of the above observer's work, which shows that the most changeable constituent is the albumen, varying from a maximum of 4.86 per cent. to a minimum of 0.85; the next are the fats and salts, the maximum being about three times the minimum, and the least the sugar. The latter, in fact, varies but little from a standard of about 7 per cent. The function of albumen is nutritive; that of milk sugar calorifacient; hence the point seems to be that nature, while allowing a wide range of oscillation in the rapidity of tissue building, carefully provides an available fuel for the constant maintenance of animal heat; the supply of caloric due to cerebral impulses and self-originated locomotion being extremely small in early infancy.

In seeking a substitute for human milk, one naturally turns to the domestic animals for the source of supply. Between the milk of the ass, cow, goat and ewe there is little choice, so far as composition is concerned, though, perhaps, asses' milk resembles that of women a little more closely than the others; nevertheless, cows' milk is usually selected, because, being plentiful, it is easily obtained and cheap.

Cows' milk * (market milk) has a specific gravity of 1.029, is richer looking—that is, whiter and more opaque than human milk, and is slightly acid in reaction unless perfectly fresh from pasture-fed animals, when it may be neutral or alkaline, and contains—

Water,	87.7	per cent.
Total solids,	12.3	"
Total solids not fat,	8.48	"
Fat,	3.75	"
Milk sugar,	4.42	"
Albuminoids,	3.42	"
Ash,	0.64	"

Comparing this analysis with that previously given, it is readily seen that the two fluids differ in specific gravity and reac-

---

* The characters of cow's milk may be determined with sufficient accuracy in the following way :—

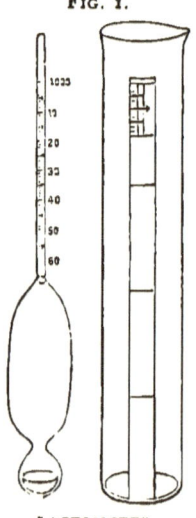

FIG. 1.

LACTOMETER.

Provide a urinometer, such as shown in Fig. 1, and which can be obtained at any drug shop. To obtain the specific gravity, fill the beaker to such a point with milk that it will float the specific gravity glass, and read the degree of density from the scale at a level with the surface of the milk. The chemical reaction is found by inserting a piece of blue litmus paper, which should turn slightly red a few moments after being wet. In applying this test small pieces of litmus paper should be examined under and in the milk, as exposure to air may redden paper dipped in milk though the fluid itself may not be acid. To ascertain the proportion of cream, cut a narrow strip of paper four inches long, and divide the upper half-inch, by cross-markings, into twelve equal parts; paste this on the beaker with the marked portion uppermost, and the lower edge coming accurately to the bottom of the beaker; then pour in enough milk to come just to the top of the paper, and place the whole aside for twenty-four hours. During this time the cream rises and appears as a yellow layer at the top; this layer should have the depth of ten or twelve spaces.

tion, and that cows' milk contains more nitrogenous material, but less fat and much less sugar than woman's milk.

The nitrogenous material differs in quality as well as in quantity. König, in a number of analyses that closely correspond with those of Leeds, divides the nitrogenous constituent into three groups; namely, caseine, albumen and albuminoids, basing the division upon the different effects of coagulating agents.

Upon this point Leeds remarks: " Whilst by present modes of analysis the separation of the so-called caseine from the so-called albumen is not accurately performed, yet the results are approximately correct (König's), and have a very great value in pointing out the most important of all the differences between the two secretions, which is, that the fraction of the total albuminoids in cows' milk which is coagulable by acids is far greater (perhaps four times) than the non-coagulable part.

"In woman's milk, on the contrary, the reverse is true, and the non-coagulable part much exceeds (perhaps by more than twice) the coagulable portion."

This difference is readily tested by adding rennet to the two fluids. In the case of cows' milk the caseine is coagulated into large, firm masses, while with human milk a light, loose curd is formed. In the stomach the acid gastric juice has the same effect, producing in the first instance a coagulum most difficult to digest; in the other, one readily attacked and broken down by the gastro-intestinal solvents.

These chemical and physical properties of cows' milk can be altered by various methods of preparation, and unless this be done there are few instances in which it will not prove a poor substitute for the natural food.

Condensed milk is frequently recommended by physicians and largely used by the laity, on their own responsibility. It keeps better than cows' milk and is supposed to be more readily digested by young infants. The latter supposition is a mistaken one, and arises from the overlooked fact that condensed milk is always given dissolved in a large proportion of water, while cows' milk is too frequently used insufficiently diluted or otherwise improperly prepared. The author is convinced of the accuracy of this statement from a number of years' close study of the subject.

Condensed milk contains a large proportion of sugar, forms fat quickly, and thus makes large babies; sugar also counteracts the tendency to constipation—often a troublesome complaint in hand-feeding. These advantages are unquestioned, and, together with the ease of preparation, are those which place it so high in the esteem of monthly nurses. It is equally true, however, that, as a food, it does not contain enough nutrient material to supply the wants of a growing baby.

Again, more than half of the saccharine ingredient of this preparation is cane sugar, added for the purpose of preservation, and this material is very liable, when in excess, to ferment in the alimentary canal, giving rise to irritant products that impede digestion.

Infants fed upon condensed milk, though fat, are pale, lethargic and flabby; although large, they are far from strong; have little power to resist diseases; often cut their teeth late, and are very liable to drift into rickets. It must be remembered also that condensed milk, when long kept, or when packed in imperfect cans, not unfrequently undergoes decomposition, and thus becomes utterly unfit for use.

For a temporary change of diet, and as a substitute during travelling or under circumstances in which sound cows' milk cannot be obtained, it may be resorted to with advantage.

The farinaceous substances so often selected, especially by the poor, to replace breast-milk, are not only bad foods, but have both directly and indirectly a deleterious effect upon the processes of nutrition.

They are bad for two reasons. First, they differ materially in chemical composition from human milk. For example, in arrowroot, which is the favorite, the proportion of the tissue-building to the heat-producing element is as one to twenty, while in human milk it is about one to five. Secondly, the heat-producing principle, starch, must be converted into sugar before it can be absorbed. This change is accomplished in the body by the saliva and pancreatic juice,—secretions that are not fully established until the fourth month.

While the starch lies undigested in the gastro-intestinal canal, it is subject to fermentation, resulting in the formation of irritant products that rapidly induce catarrh of the mucous membrane; a condition directly interfering with the digestion and absorption of food. Again, perfect nutrition demands rapid waste and removal of effete tissues as well as repair of the same. This is effected by oxidation. Now sugars are known to have a much greater affinity for oxygen than albuminates, and when the diet consists of farinaceous material, the small amount of sugar formed and absorbed appropriates oxygen that otherwise would go toward the removal of waste, and so retards the necessary changes.

Farinaceous food, as such, is never permissible before the fourth month; earlier, it is only to be employed for its mechanical action, as an addition to milk preparations. This will be mentioned later.

The nutrient value of the cereals and their products as they exist in so-called "infants' foods," has been imperfectly determined. They are undoubtedly useful as mechanical attenuants, but it is very questionable whether any of them, unless prepared with milk, can permanently meet the demands of nutrition. At the same time it is quite probable that the soluble albuminoid substances obtained by Liebig's process have a food value of their own, making them more serviceable than the starches.

b. The quantity of food to be allowed each day varies with the appetite and age. Some infants habitually eat little, others much; as both thrive, the question of the correct amount in a given case must be answered by observation. So long as the child develops with normal rapidity and keeps well, he may be allowed to eat as much or as little as he wants; for, if food of proper strength be given at proper intervals, the instinctive cravings of hunger, since they represent the wants of the system, rarely lead to excess in either direction. Nevertheless it is well to have some guide.

During the first four weeks, infants generally require from twelve and-a-half to sixteen fluidounces of food ; in the second and third months, about twenty-four fluidounces, and from this time to the twelfth month from two to two and-a-half or even three pints. After the twelfth month the quantity depends upon whether additions be made to the diet, or milk food be used exclusively. When the daily amount reaches three pints, the limit of the capacity of the stomach is usually attained, and the greater demand for nutriment, as growth advances month by month, must be met by adding to the strength of the food rather than by increasing its bulk. These two factors, strength and quantity, are intimately associated throughout the whole period of infancy, and in the earlier months a mere increase in the latter is not always sufficient to maintain the balance of nutrition.

As a rule, infants are overfed, and this opens the very interesting question of the normal capacity of the stomach at different ages. Rotch has recently written an important paper upon the subject. He states that, by actual measurement, the stomach of an infant five days old holds 25 c.c., or six and-a-quarter fluidrachms, a quantity very far short of that usually forced upon the babe during the first week. Frowlowsky's investigations show that there is a very rapid increase in the capacity of the stomach during the first two months of life, while in the third, fourth and fifth months the increase is slight. Guided by these data, the quantity of food should be rapidly augmented during the first six or eight weeks of life and then held at the same quantity up to the fifth or sixth month. Another considerable increase is also demanded between the sixth and the tenth months.

While the author has been unable to verify the above measurements, and has, on the contrary, found no uniformity in the size of the stomach for given ages, yet the following table (Rotch) is a useful one, and corresponds closely with conclusions drawn from clinical experience.

THE GENERAL MANAGEMENT OF CHILDREN.

GENERAL RULES FOR FEEDING.

Age.	Intervals of Feeding.	Average Amount at Each Feeding.	Average Amount in 24 Hours.
First week.	2 hours.	1 ounce.	10 ounces.
One to six weeks.	2½ hours.	1½ to 2 ounces.	12 to 16 ounces.
Six to twelve weeks and possibly to fifth or sixth month.	3 hours.	3 to 4 ounces.	18 to 24 ounces.
At six months.	3 hours.	6 ounces.	36 ounces.
At ten months.	3 hours.	8 ounces.	40 ounces.

c. The object to be accomplished in the preparation of cows' milk is to make it resemble human milk as much as possible in chemical composition and physical properties. To do this, it is necessary to reduce the proportion of caseine, to increase the proportion of fat and sugar, and to overcome the tendency of the caseine to coagulate into large, firm masses upon entering the stomach.

Dilution with water is all that need be done to reduce the amount of caseine to the proper level; but as this diminishes the already insufficient fat and sugar, it is essential to add these materials to the mixture of milk and water. Fat is best added in the form of cream, and of the sugars, either pure white loaf sugar or sugar of milk may be used. The latter is greatly preferable, as it is little apt to ferment, and contains some of the salts of milk, which are of nutritive value.

Firm clotting may be prevented by the addition of an alkali or a small quantity of some thickening substance.

Lime water is the alkali usually selected. It acts by partially neutralizing the acid of the gastric juice, so that the caseine is coagulated gradually and in small masses, or passes, in great part, unchanged into the intestine, to be there digested by the alkaline secretions. As it contains only half a grain of lime to the fluid-ounce, the desired result cannot be attained, unless at least a third part of the milk mixture be lime water. The quantity often used—one or two teaspoonfuls to the bottle of food—has no effect beyond neutralizing the natural acidity of the milk itself. When lime water is constantly employed, it becomes quite an item of expense if procured from the drug shop; this outlay is unnecessary, for it can be made quite as well in the nursery. Take a piece of unslaked lime as large as a walnut, drop it into two quarts of filtered water contained in an earthen vessel, stir thoroughly, allow to settle, and use only from the top, replacing the water and stirring as consumed.

Instead of lime water, two to four grains of bicarbonate of sodium may be added to each bottle, or, better still, from five to fifteen drops of the saccharated solution of lime.

This solution is made in the following way:—

Take of—

    Slaked lime, ... 1 ounce.
    Refined sugar, in powder, ... 2 ounces.
    Distilled water, ... 1 pint.

    Mix the lime and sugar by trituration in a mortar. Transfer the mixture to a bottle containing the water, and having closed this with a cork, shake it occasionally for a few hours. Finally, separate the clear solution with a siphon and keep it in a stoppered bottle.

Thickening substances—attenuants, such as barley-water, gelatine, or one of the digestible prepared foods—act purely mechanically by getting, as it were, between the particles of caseine during coagulation, preventing their running together and forming a large, compact mass.

To prepare the former, put two teaspoonfuls of washed pearl barley, with a pint of cold filtered water, into a saucepan, boil slowly down to two-thirds and strain. The liquid obtained does not possess the disadvantages of farinaceous foods generally. To be efficient, it must be used as a diluent instead of, and in the same proportion as, water.

Gelatine is prepared in the following way: put a piece of plate gelatine, an inch square, into a half-tumblerful of cold water, and let it stand for three hours; then turn the whole into a teacup, place this in a saucepan half full of water and boil until the gelatine is dissolved. When cold this forms a jelly; from one to two teaspoonfuls may be added to each bottle of milk food.

When an "infant's food" is used to act mechanically, care should be taken to select one in which the starch has been converted into dextrine and grape sugar by the process of manufacture. The articles known as "Mellin's Food" and "Horlick's Food" can be relied upon. One teaspoonful of either dissolved in a tablespoonful of hot water and added to each portion of food, makes a very easily digested mixture.

For the successful management of children, the mother or nurse must not only be familiar with the theory of feeding, but must practically understand the methods of preparing food. To this end a schedule of the diet of an infant from birth upward, with a sketch of the modifications that have to be made most frequently, will serve as a useful guide.

Diet during the first week:—

    Cream, ... f ʒ ij.
    Whey, ... f ʒ iij.
    Water (hot), ... f ʒ iij.
    Milk sugar, ... gr. xx.

    For each portion; to be given every two hours from 5 A. M. to 11 P. M., and in some cases once or twice at night; amounting to twelve fluidounces of food per diem.

Diet from the second to the sixth week:—

    Milk, ... f ℥ ss.
    Cream, ... f ʒ ij.
    Milk sugar, ... gr. xx.
    Water, ... f ℥ j.

    For one portion; to be given every two hours from 5 A. M. to 11 P. M.; amounting to seventeen fluidounces of food per diem.

Diet from the sixth week to the end of the second month:—

Milk,	f ℥j, f ʒij.
Cream,	f ℥ss.
Milk sugar,	ʒss.
Water,	f ℥j, f ʒij.

For each portion; to be given every two hours; amounting to thirty fluidounces per diem.

Diet from the beginning of the third month to the sixth month:—

Milk,	f ℥iiss.
Cream,	f ℥ss.
Milk sugar,	ʒj.
Water,	f ℥j.

For each portion; to be given every two and a half hours, or thirty-two fluidounces per diem.

Diet during the sixth month; six meals daily from 6 or 7 A. M. to 9 or 10 P. M.

Morning and midday bottles each:—

Milk,	f ℥ivss.
Cream,	f ℥ss.
Mellin's Food,	ʒj.
Hot water,	f ℥j.

Dissolve the Mellin's Food in the hot water and add, with stirring, to the previously mixed milk and cream.

Other bottles each:—

Milk,	f ℥ivss.
Cream,	f ℥ss.
Milk sugar,	ʒj.
Water,	f ℥j.

This gives an equivalent of thirty-six fluidounces of food in a day.

In the seventh month the Mellin's Food may be increased to two teaspoonfuls and given three times daily.

Throughout the eighth and ninth months five meals a day will be sufficient.

First meal at 7 A. M. :—

Milk,	f ℥ viss.
Cream,	f ℥ ss.
Milk sugar,	ℨ j.
Water,	f ℥ j.

Second meal at 10.30 A. M. Milk, cream and water in the same proportion ; Mellin's Food, one tablespoonful.

Third meal at 2 P. M.—Same as second.

Fourth meal at 6 P. M.—Same as second.

Fifth meal at 10 P. M.—Same as first.

This gives forty fluidounces of food per diem.

Instead of Mellin's Food, a teaspoonful of "flour-ball" * may be added.

Two meals of flour-ball daily—the second and fourth—are all that can be digested. To prepare these, rub one teaspoonful of the powder with a tablespoonful of milk into a smooth paste, then add a second tablespoonful of milk, constantly rubbing until a cream-like mixture is obtained. Pour this into eight ounces of hot milk, stirring well, and it is then ready for use. The other meals should be composed of milk, cream, sugar of milk and water, as already given.

Mellin's Food and flour-ball may be substituted by oatmeal or barley, or any one of the infants' food in which the starch has been converted, by Liebig's process, into dextrine and grape sugar.

---

* To make flour-ball, take a pound of good wheat flour—unbolted, if possible—tie it up very tightly in a strong pudding-bag, place it in a saucepan of water and boil constantly for ten hours; when cold, remove the cloth, cut away the soft, outer covering of dough that has been formed, and reduce the hard-baked interior by grating. In the yellowish-white powder obtained, almost all the starch has been converted into dextrine by the process of cooking, and the proportion of the nitrogenous principle to the calorifacient is as one to five, nearly the same as human milk.

Diet for the tenth and eleventh months:—
First meal, 7 A. M. :—

Milk,	f ℥ viiiss.
Cream,	f ℥ ss.
Mellin's Food,	℥ ss.
(Or flour-ball or barley jelly),	ʒ ij.
Water (used only with Mellin's Food),	f ℥ j.

Second meal, 10.30 A. M.—A breakfast-cupful of warm milk (eight fluidounces).

Third meal, 2 P. M.—The yelk of an egg lightly boiled, with stale bread crumbs.

Fourth meal, 6 P. M.—Same as first.

Fifth meal, 10 P. M.—Same as second.

On alternate days the third meal may consist of a teacupful (six fluidounces) of beef tea* containing a few stale bread crumbs.

A further variation can be made by occasionally using mutton, chicken or veal broth instead of beef tea.

As much more difficulty is experienced in feeding infants during the first twelve months than during the second, it would be well to pause here to consider what had best be done in case the food described should disagree.

If, after feeding, vomiting occur, with the expulsion of large, firm clots of caseine, the effect of adding lime water or barley water must be tried.

For instance, at the age of six weeks, make each bottle of:—

Milk,	f ℥ j, f ʒ ij.
Cream,	f ℥ ss.
Milk sugar,	ʒ ss.
Lime water,	f ℥ j, f ʒ ij.

---

* Beef tea for an infant is made in the following way: Half a pound of fresh rump-steak, free from fat, is cut into small pieces and put, with one pint of cold water, into a covered, tin saucepan. This must stand by the side of the fire for four hours, then be allowed to simmer gently (never boil) for two hours, and, finally, be thoroughly skimmed to remove all grease.

Or of :—

Milk,	f ℥j, f ʒij.
Cream,	f ℥ss.
Milk sugar,	ʒss.
Barley water,	f ℥j, f ʒij.

Sometimes, particularly if there be diarrhœa, boiling makes the milk more tolerable, and in this condition it may be used instead of fresh milk in either of the above mixtures. Condensed milk, too, can be employed temporarily, making each portion of :—

Condensed milk,	ʒj.
Cream,	f ℥ss.
Hot water,	f ℥iiss.

Should further alteration be necessary, goats' or asses' milk may be substituted for cows' milk, the strong odor of the former and the laxative properties of the latter being removed by boiling. One ass is capable of nourishing three children for the first three months of life, two children for the fourth and fifth months, and one child after this period to the ninth month. The milk should be used warm from the udder.

"Strippings" is another good substitute for cows' milk. It is obtained by re-milking the cow after the ordinary daily supply has been drawn, and contains much cream and but little curd. Assimilable proportions of this are :—

Strippings,	f ℥j.
Water,	f ℥ij.

And if the small amount of caseine, in such a mixture, be still undigested :—

Strippings,	f ℥iss.
Barley water,	f ℥iss.

Another good food is that recommended by Dr. A. V. Meigs. It consists of a combination of two parts of the cream, containing from fourteen to sixteen per cent. of fat; one part average milk; two parts lime water, and three parts sugar water, the latter consisting of seventeen and three-fourths drachms of milk sugar to one pint of water. This makes an alkaline mixture with the percentage of its ingredients closely corresponding to human milk.

When, in spite of careful preparation, all of these foods give rise to indigestion with fever, and the expulsion, by vomiting and diarrhœa, of hard curds from the stomach and intestines, the expedient of predigesting the milk must be resorted to.

The best method is to peptonize the milk by pancreatin. That manufactured under the name of extractum pancreatis, by Fairchild Brother & Foster, of New York, has proved most efficient in my hands. To accomplish this artificial digestion, put into a clean quart bottle five grains of extractum pancreatis, fifteen grains of bicarbonate of sodium, and four fluidounces of cool, filtered water; shake thoroughly together, and add a pint of fresh, cool milk. Place the bottle in water, not so hot but that the whole hand can be held in it for a minute without discomfort, and keep the bottle there for exactly thirty minutes. At the end of that time put the bottle on ice to check further digestion and to keep the milk from spoiling. The fluid obtained, while somewhat less white in color than milk, does not differ from it in taste; if, however, an acid be added, the caseine, instead of being coagulated into large, firm curds, takes the form of minute, soft flakes, or readily broken-down feathery masses of small size. When the process is carried just to the point described, the caseine is only partly converted into peptone; but every succeeding moment of continued warmth lessens the amount of caseine until peptonization is complete. Then the liquid is grayish yellow in color; has a distinctly bitter taste, and shows no coagulation whatever on the addition of an acid. This artificial digestion, therefore, may be carried just as far as circumstances indicate, although it is ordinarily best to stop it short of complete conversion, as children object to the markedly bitter taste, and often, on account of it, absolutely refuse the food. Partial peptonization, too, is usually sufficient to adapt the milk to ready assimilation. To seize the proper moment for arresting the process, the person conducting it must be told to taste the milk from time to time, and as soon as the least bitterness is appreciable, to remove the bottle from the hot water and place it upon ice for cooling and use. Such milk may be sweetened with sugar of milk, and given pure or diluted with water. For an infant of six weeks each meal may consist of:—

> Peptonized milk, ... f ℥ iij.
> Milk sugar, ... ℨ ss.
> Water, ... f ℥ j.

To this, cream may be added when desirable, and by diminishing the quantity of water and increasing that of milk the strength of the food may be made greater at any time.

Although every precaution be taken, the last of a quantity of predigested food is very apt to grow bitter; and if the attendants will take the trouble, it is much better to peptonize every meal separately. This is readily done by obtaining a number of powders of pancreatin and bicarbonate of sodium, so proportioned that each packet shall contain the proper amount for one bottle of food.

For example:—

> ℞. Extract. Pancreatis, ... gr. ix.
> Sodii Bicarb., ... gr. xxiv.
> M. et ft. chart., No. xij.
>
> SIG.—Put one powder into a nursing bottle with two fluidounces of filtered water and two fluidounces of fresh sweet milk; shake together and keep warm in a water-bath for about half an hour before feeding; sweeten with half a teaspoonful of milk sugar.

The great advantages of partial peptonization are that the necessity for lime water, barley water and thickening substances to keep apart the curd is done away with, and that, when the digestive disturbance requiring a careful preparation of food is removed, an ordinary milk diet can be gradually resumed by regularly diminishing the time artificial digestion is allowed to progress. This changes the caseine in a less and less degree, until, finally, it is taken in its natural form.

Instead of this ordinary peptonizing process, I have for several years past employed the "Peptogenic milk powder," prepared by the chemists already referred to. This powder contains a digestive ferment, pancreatin; an alkali, bicarbonate of sodium, and a due proportion of milk sugar.

The mode of employment is as follows:—

Take of—

Milk,	f℥ij.
Water,	f℥ij.
Cream,	℥ss.
Peptogenic milk powder,	1 measure.*

This mixture is to be heated over a brisk flame to a point that can be comfortably sipped by the preparer (about 115° F.) and kept at this heat for six minutes. When properly prepared, the resultant, so-called "humanized milk," presents the albuminoids in a minutely coagulable and digestible form; has an alkaline reaction; contains the proper proportion of salts, milk sugar and fat, and has the appearance of human milk.

Leeds gives the following analysis of this prepared milk:—

Water,	86.2	per cent.
Fat,	4.5	"
Milk sugar,	7.	"
Albuminoids,	2.	"
Ash (salts),	0.3	"

This corresponds very closely with his average analysis of human milk.

In using this powder, too, one can readily return to a plain milk diet by gradually shortening the time of heating; in other words, by slowly diminishing predigestion.

Great and deserving stress has recently been placed upon a method of preparing, or rather preserving, cows' milk, known as "Sterilization."

As milk exists in the healthy cow's udder it is aseptic, *i. e.*,

---

* Measure provided with each can of powder.

free from any poisonous or dangerous ingredient, but during milking, and subsequent handling and transportation, particles of manure or various forms of dirt get into it and are apt to set up fermentation or other injurious change. To deprive these accidentally introduced organic impurities of their activity, or, in other words, to *sterilize*, it is necessary to subject the fluid to high heat under pressure.

Several admirable implements have been devised for conduct-

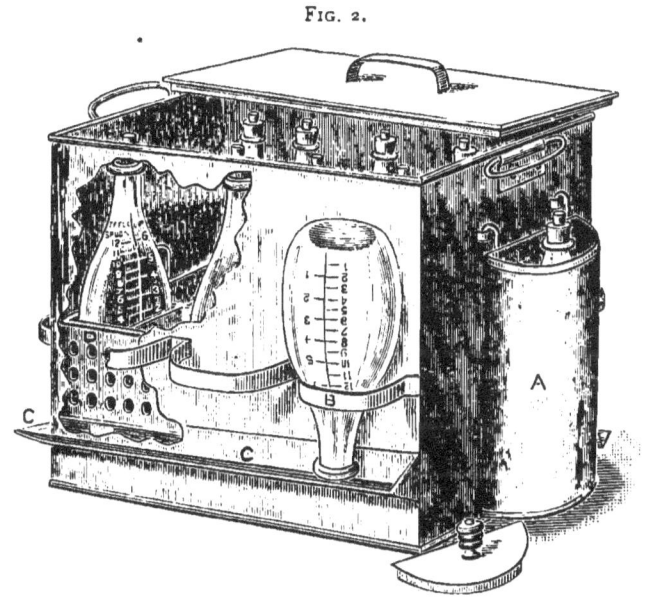

FIG. 2.

AUTHOR'S STERILIZER.

ing the process; one of the most simple, made after a design of my own, is shown in the accompanying figure.

This apparatus is made of tin, and consists of an oblong case provided with a well fitting cover, and having a movable perforated false bottom (D), which stands a short distance above the true one and has attached a framework capable of holding ten, six-ounce, nursing bottles. On the outside of the case is a row of supports (B) for holding inverted bottles while drying, and at the proper distance below these a gradually inclining gutter (C) for carrying off the drip. A movable water bottle (A) is hung to the side; in this each bottle of food may be heated at the time of administration.

The bottles are made of flint glass and according to the design described on page 96, the graduated markings being especially convenient for measurement and rendering the use of a separate measuring glass unnecessary, a matter of no little moment, as every implement that comes in contact with the milk in sterilization must be kept chemically clean. Ten bottles are used, so that the whole supply of milk intended for a day's consumption can be prepared at once. Each bottle is provided with a perforated rubber cork, which in turn is closed by a well-fitting glass stopper.

Sterilization should be performed in the morning as soon as possible after the milk has been served The process is as follows: First, see that the ten bottles are perfectly clean and dry; pour into each six fluidounces (12 tablespoonfuls) of milk; insert the perforated rubber corks, without the glass stoppers, however; remove the false bottom and place the bottles in the frame; pour into the case enough water to fill it to the height of about two inches; replace the false bottom carrying the bottles; adjust lid, and put the whole on the kitchen range. Allow the water to boil and, by occasionally removing the lid, ascertain that the expansion that immediately precedes boiling has taken place in the milk, then press the glass stoppers into the perforated corks, and thus hermetically close each bottle. After this, keep the apparatus on the fire and the water boiling for twenty minutes. Finally, remove the false bottom with the bottles; pour out the water, replace and carry the whole, covered with the lid, to the nursery.

When the hour of feeding arrives, put one of the bottles into the attached water bath and heat it to the proper point for administration. The milk may, of course, be diluted with filtered water, or receive the additions ordinarily made to adapt it to children of different ages. The tip used—and a tube must not be employed even here—should be thoroughly cleaned and immersed for a few moments in boiling water before it is attached to the bottle.

So soon as a bottle is emptied—and if the whole of its contents be not taken the remainder must be thrown away—it is washed in the ordinary manner with a solution of bicarbonate or salicylate of sodium (see p. 96) and placed in the rack (B) to drain and dry.

Milk sterilized by the above process will remain sound for several days, according to some authorities as many as eighteen * when the heating is continued for thirty minutes.

Sterilized milk is especially useful in travelling, when fresh milk cannot be obtained; for use in cities during the heat of summer, when milk is most apt to undergo injurious changes; for the feeding of delicate children, or for those suffering from disease of the stomach or intestinal canal.

A very good process has been inaugurated by some dairymen, in which the milk is sterilized on the farm directly after coming from the cow, and transported to the consumer in the original bottles. This procedure cannot be too highly recommended, provided the care is taken to preserve perfect cleanliness on the part of the original handlers, and to see that the process of sterilization is thoroughly carried out.

Sometimes milk, in every form and however carefully prepared, ferments soon after being swallowed and excites vomiting, or causes great flatulence and discomfort, while it affords little nourishment. With these cases the best plan is to withhold milk entirely for a time and try some other form of food. The following are good substitutes:—

Mellin's Food, ... ℨj.
Hot water, ... f℥iij.

For each portion; to be given every two hours at the age of six weeks.

---

* Since writing the above, this statement has been verified by my own experiments.

Veal broth (½ ℔ of meat to the pint), ... f ℥ iss.
Barley water, ... f ℥ iss.
For one portion.
Whey, ... f ℥ iss.
Barley water, ... f ℥ iss.
Milk sugar, ... ℥ ss.

A teaspoonful of the juice of raw beef every two hours will usually be retained when everything else is rejected.

Such foods are only to be used temporarily until the tendency to fermentation within the alimentary canal ceases; then milk may be gradually and cautiously resumed.

When infants approaching the end of the first year become affected with indigestion, it is often sufficient to reduce the strength and quantity of the food to a point compatible with digestive powers. For instance, at eight months the food may be reduced to that proper for a healthy child of six months, or even less. Here, too, predigestion of the food is very serviceable.

If a few grains of extractum pancreatis be added to a gobletful of thick, well-boiled starch gruel, at a temperature of 100° F., the gelatinous mucilage quickly grows thinner and soon is transformed into a fluid, the starch having been rendered soluble by the action of the pancreatin; by still longer contact, the hydrated starch is converted into dextrine and sugar. Advantage may be taken of this property to render the foods containing starch assimilable. Thus, to a mixture of barley jelly and milk, *e. g.* :—

Barley jelly, ... ℥ ij.
Milk sugar, ... ℥ j.
Warm milk, ... f ℥ viij.

Add three grains of extractum pancreatis, and five grains of bicarbonate of sodium, and keep warm for half an hour before administering.

The same process may be employed with food containing oatmeal, arrowroot or wheaten flour, with a view of converting the starchy ingredients into digestible elements without materially altering the taste.

THE GENERAL MANAGEMENT OF CHILDREN. 93

When the infant has arrived at an age to take meat broths, these too, when digestion is enfeebled, may be readily peptonized.

Returning to the regimen of the healthy infant, it will be found that after the first year far less change is required in the food from month to month.

Diet from the twelfth to the eighteenth month, five meals per day :—

First meal, 7 A.M.—A slice of stale bread, broken and soaked in a breakfast-cup (eight fluidounces) of new milk.

Second meal, 10 A.M.—A teacup of milk (six fluidounces) with a soda biscuit or thin slice of buttered bread.

Third meal, 2 P.M.—A teacup of beef tea (six fluidounces) with a slice of bread. One good tablespoonful of rice-and-milk pudding.

Fourth meal, 6 P.M.—Same as first.

Fifth meal, 10 P.M.—One tablespoonful of Mellin's Food with a breakfast-cupful of milk.

To alternate with this :—

First meal, 7 A.M.—The yelk of an egg lightly boiled, with bread crumbs ; a teacupful of new milk.

Second meal, 10 A.M.—A teacupful of milk with a thin slice of buttered bread.

Third meal, 2 P.M.—A mashed, baked potato, moistened with four tablespoonfuls of beef tea ; two good tablespoonfuls of junket.

Fourth meal, 6 P.M.—A breakfast-cupful of new milk with a slice of bread broken up and soaked in it.

Fifth meal, 10 P.M.—Same as second.

The fifth meal is often unnecessary, and sleep should never be disturbed for it; at the same time, should the child awake an hour or more before the first meal, he must break his fast upon a cup of warm milk, and not be allowed to go hungry until the set breakfast hour.

Diet from eighteen months to the end of two and one-half years, four meals a day :—

First meal, 7 A.M.—A breakfast-cupful of new milk ; the yelk of an egg lightly boiled ; two thin slices of bread and butter.

Second meal, 11 A.M.—A teacupful of milk with a soda biscuit.

Third meal, 2 P.M.—A breakfast-cupful of beef tea, mutton or chicken broth; a thin slice of stale bread; a saucer of rice-and-milk pudding.

Fourth meal, 6.30 P.M.—A breakfast-cupful of milk with bread and butter.

On alternate days:—

First meal, 7 A.M.—Two tablespoonfuls of thoroughly cooked oatmeal or wheaten grits with sugar and cream; a teacupful of new milk.

Second meal, 11 A.M.—A teacupful of milk with a slice of bread and butter.

Third meal, 2 P.M.—One tablespoonful of underdone mutton pounded to a paste; bread and butter, or mashed baked potato, moistened with good plain dish gravy; a saucer of junket.

Fourth meal, 6.30 P.M.—A breakfast-cupful of milk, a slice of soft milk toast, or a slice or two of bread and butter.

When sickness supervenes, all that is ordinarily necessary is a reduction of the diet to plain milk, or milk with Mellin's Food.

An important point, often neglected, is the matter of drink. Even the youngest infant requires water several times daily, and the demand increases with age. The water must be as pure as possible and should not be too cold. In the heat of summer, however, bits of ice and water moderately cooled by ice can be allowed without harm.

The foregoing schedule must, of course, be regarded only *as an average*. Many children can bear nothing but milk food up to the age of two or even three years, and, provided enough be taken, no fear for their nutrition need be entertained. If a child be thriving on milk, he is never to be forced to take additional food merely because a certain age has been reached; let the healthy appetite be the guide.

A young mother, in her solicitude to do her best, often finds great difficulty in adhering to simple rules in the diet of her child. Mrs. A., who has had great experience with children, having had some herself, tells her that the child would thrive far better if it ate such and such a thing, and did not keep to weak milk foods. Miss B. assures her that her cousin's last child grew much healthier after eating a chop with vegetables and pudding each day. Aunt C. comes with the announcement—which she breaks gently—that she knows the child is simply starving, and the ignorant nurse confirms the statement.

FIG. 3.

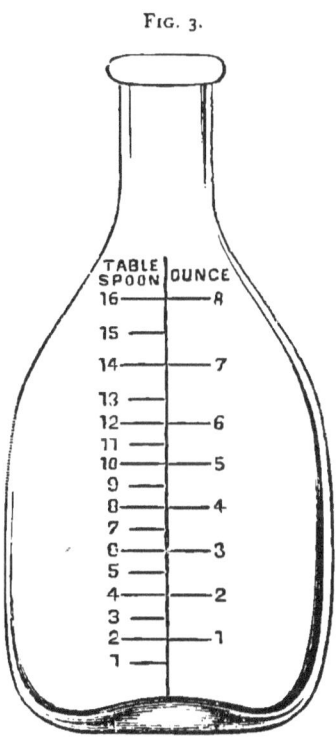

GRADUATED NURSING BOTTLE.

All their seemingly convincing theories are very upsetting to a mother who wants only to do what is right. She must bear in mind, however, that some children can eat anything and live; but she does not know how much better, more robust, and disease-resisting they would be, did they adhere to a simple diet. Let her remember that the so-called "weak milk foods" contain those nourishing qualities to which nature, in her wisdom, has limited the child's powers of digestion. Therefore, young mothers, let well enough alone.

d. Success in hand-feeding depends quite as much on the administration as upon the preparation of the food.

From birth up to such time as broth, bread, and eggs are added to the diet, all the food should be taken from a bottle. Even after this, as the bottle is a comfort and insures slow feeding, it may be allowed for milk preparations, until the child, of his own accord, tires of it. The only feeding apparatus to be admitted to the nursery is the simple bottle and tip. The bottle represented in Figure 3 is made, by my suggestion, by Mr. J. J. Ottinger, of Philadelphia. Its interior surface presents no angles for the collection of milk; it is easily cleaned, and the graduated scale is convenient for nursery use.

All complicated arrangements of rubber and glass tubing are not only an abomination, but a fruitful source of sickness and death. Rather than use them, it is far better to feed the infant with a spoon. In England, a bottle with a long rubber tube is almost universally employed. Should this be abandoned and a simple bottle and a rubber tip used, the objections of some authors to bottle-feeding would vanish.

The bottle shaped as above must be of transparent flint glass, so that the slightest foulness can be detected at a glance, and may vary in capacity from six to twelve fluidounces, according to the age of the child. Two should be on hand at a time, to be used alternately. Immediately after a meal the bottle must be thoroughly washed out with scalding water, filled with a solution of bicarbonate or salicylate of sodium—one teaspoonful of either to a pint of water—and thus allowed to stand until next required; then the soda solution being emptied, it must be thoroughly rinsed with cold water before receiving the food. The tips or nipples, of which there should also be two, must be composed of soft, flexible India-rubber, and a conical shape is to be preferred, as being more readily everted and cleaned; the opening at the point must be

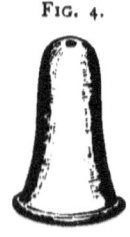

FIG. 4.

BOTTLE TIP.

free, but not large enough to permit the milk to flow in a stream without suction. At the end of each feeding the nipple must be removed at once from the bottle, cleansed externally by rubbing with a stiff brush wet with cold water, everted and treated in the same way, and then placed in cold water and allowed to stand in a cool place until again wanted.

While taking these precautions for perfect cleanliness, the nurse must satisfy herself of their efficacy by smelling both the bottle and the tip just before they are used, to be sure of the absence of any sour odor.

Next to cleanliness of the feeding apparatus, it is important to insist upon the separate preparation of each meal immediately before it is to be given. The practice of making, in the morning, the whole day's supply of food, though it saves trouble, is a most dangerous one. Changes almost invariably take place in the mixture, and by the close of the day it becomes unfit for consumption.

When the graduated bottle is not at hand, a common glass graduate, marked for fluidrachms and ounces and holding a pint, should be provided for the nursery. Some moments before mealtime, so as to avoid hurry, measure the different fluid ingredients of the food in this, one after the other; add the requisite quantity of milk sugar, and mix the whole thoroughly by stirring with a spoon, and pour into the feeding bottle. When the graduated bottle is employed, thorough shaking is sufficient. The food must now be heated to a temperature of about 95° F. This can be done by steeping the bottle in hot water, or by placing it in a water-bath over an alcohol lamp or gas jet. Finally, apply the tip and the meal is ready.

When feeding, the child must occupy a half-reclining position in the nurse's lap. The bottle should be held by the nurse, at first horizontally, but gradually more and more tilted up as it is emptied, the object being to keep the neck always full and prevent the drawing in and swallowing of air. Ample time, say five, ten or fifteen minutes, according to the quantity of food, should be allowed for the meal. It is best to withdraw the bottle occasionally for a brief rest, and after the meal is over, sucking from the empty bottle must not be allowed, even for a moment.

e. For children residing in cities, an honest dairyman must be found, who will serve sound milk and cream from country cows once every day in winter, and twice during the day in the heat of summer. The milk of ordinary stock cows is more suitable than that from Alderney or Durham breed, as the latter is too rich and, therefore, more difficult to digest. The mixed milk of a good herd is to be preferred to that from a single animal. It is less likely to be affected by peculiarities of feeding, and less liable to variation from alterations in health or different stages of lactation.

The care of the herd and of the milk is of great consequence. The cows should be healthy, and the milk of any animal that seems indisposed should not be mixed with that from perfectly healthy animals. The cows must not be fed upon swill or the refuse of breweries, glucose factories, or any other fermented food. They must not be allowed to drink stagnant water, and must not be heated or worried before being milked. The pasture must be free from noxious weeds, and the barn and yard must be kept clean. The udder should be washed, if dirty, before the milking. The milk must be at once thoroughly cooled. This is best accomplished by placing the can in a tank of cold spring water, or in ice water, the water being the same depth as the milk in the can. It is well to keep the water in the tank flowing; indeed, this is necessary unless ice water be used. The can should remain uncovered during the cooling and the milk should be gently stirred. The temperature should be reduced to 60° F. within an hour, and the can must remain in the cold water until the time for delivering.

In summer, when ready for delivery, the top should be placed in position and a cloth wet in cold water spread over the can, or refrigerator cans may be used. At no season should the milk be frozen, and at the same time no buyer should receive milk having a temperature over 65° F.

The milk and cream must be transported from the dairy in

perfectly clean vessels. To insure this it is best to provide two sets of small cans; one set to be thoroughly cleansed and aired while the other is taken away by the milkman to bring back the next supply. So soon as this arrives in the morning, or in the morning and evening in hot weather, the milk should be emptied into separate and absolutely clean earthenware or glass pitchers, and these put at once into a refrigerator reserved exclusively for them. This may stand in some convenient spot near the nursery, but not in it, and especially not in an adjoining bath room. With a good refrigerator there is no difficulty in keeping milk perfectly sweet for twenty-four hours in winter and for twelve hours in summer, except on intensely hot days; then it may be necessary to scald, lightly boil or sterilize the whole of the supply when received, in order to prevent change.

It is a well-known fact that milk is a fluid having active powers of absorption, and that it frequently acts as the medium of transmission of the contagion of such diseases as scarlatina, diphtheria and typhoid fever. Doctor V. C. Vaughan has also lately discovered in milk a special poison which he terms *tyrotoxicon* (cheese poison).

The clinical elements of interest in these discoveries is the close analogy between the symptoms produced by the experimental use of tyrotoxicon and those observed in cholera infantum —an analogy suggestive of the possibility of the latter disease being chiefly due to poisoned milk. This causal relation is scarcely more than a theory, though certain well-known features of the disease seem to bear it out. Thus, the affection occurs at a season when decomposition of milk takes place most rapidly; it occurs at places where absolutely fresh milk cannot be obtained; it prevails among classes of people whose surroundings are most favorable to fermentative changes; it is most fatal at an age when there is the greatest dependence upon milk as a food, when the gastro-intestinal mucous membrane is most susceptible to irritants, and when irritation and nervous fevers are most easily produced.

Drs. Newton and Wallace, of the New Jersey State Board of Health, have reported a number of cases of poisoning by milk that occurred in different hotels at Long Branch. These observers found that the affected milk was all obtained from one milkman, and that the cows furnishing it were milked at the unusual hours of midnight and noon. The noon milking was immediately placed in cans without being cooled, and "carted eight miles during the warmest part of the day in a very hot month." It was this milk that produced the poisonous effects, the morning's milk being always good. No statement is made as to the health of the cows or the nature of the poison, but there is a probability of its having been tyrotoxicon, and of this material or its ferment having been generated by the careless collection and transportation of the milk, combined with the high atmospheric temperature.

Childhood.—Children who have cut their milk teeth may be fed for a twelvemonth—namely, up to the age of three and a half years—in the following way :—

First meal, 7 A. M.—One or two tumblerfuls of milk, a saucer of thoroughly cooked oatmeal or wheaten grits, and a slice of bread and butter.

Second meal, 11 A. M. (if hungry).—A tumblerful of milk or a teacupful of beef tea with a biscuit.

Third meal, 2 P. M.—A slice of underdone roast beef or mutton or a bit of roast chicken or turkey, minced as fine as possible ; a baked potato thoroughly mashed with a fork and moistened with gravy; a slice of bread and butter; a saucer of junket or rice-and-milk pudding.

Fourth meal, 7 P. M.—A tumblerful of milk and one or two slices of well-moistened milk toast.

From three and a half years up the child must take his meals at the table with his parents, or with some reliable attendant who will see that he eats leisurely. The diet, while plain, must be varied. The following list will give an idea of the food to be selected :—

BREAKFAST.

EVERY DAY.	ONE DISH ONLY EACH DAY.	
Milk.	Fresh fish.	Eggs, plain omelette.
Porridge and cream.	Eggs, lightly boiled.	Chicken hash.
Bread and butter.	" poached.	Stewed kidney.
	" scrambled.	" liver.

Sound fruits may be allowed before and after the meal, according to taste, as oranges, grapes without pulp (seeds not to be swallowed), peaches, thoroughly ripe pears, cantaloupes and strawberries.

DINNER.

EVERY DAY.	TWO DISHES EACH DAY.	
Clear soup.	Potatoes, baked.	Hominy.
Meat, roasted or broiled, and cut into small pieces.	" mashed.	Macaroni, plain.
	Spinach.	Peas.
	Stewed celery.	String-beans, young.
Bread and butter.	Cauliflower.	Green corn, grated.

Junket, rice-and-milk or other light pudding, and occasionally ice cream, may be allowed for dessert.

SUPPER.

EVERY DAY.

Milk.
Milk toast or bread and butter.
Stewed fruit.

Fried food, highly-seasoned or made-up dishes are to be excluded, and no condiment but salt is to be used.

Eating, however little, between meals, must be absolutely avoided. Keep a young child from knowing the taste of cakes or bonbons, or, having learned it, let him feel that they are as unattainable as the thousand other things beyond his reach, and he soon ceases to ask for them. Even a piece of bread between meals should be forbidden. His appetite then remains natural, and he will eat proper food at his regular meal hours.

Filtered or spring water should be the only drink; tea, coffee, wine or beer being entirely forbidden.

As to the quantity, a healthy child may be permitted to satisfy his appetite at each meal, under the one condition that he eats slowly and masticates thoroughly.

In case of illness, the diet must be reduced in quantity and quality, according to the rules that are applicable to adults.

2. Bathing.

During the first two and a half years of life a child ought to be bathed once every day. The bath should be given at a regular time, and it is best to select some hour in the early morning, midway between two meals—ten o'clock, for instance. The tub should be placed near the fire or in a warm room in winter, and away from currents of air in summer. It should contain enough water to cover the child up to the neck when in a sitting posture, and the temperature must be about 95° F. Upon undressing the child, the first step is to wet his head; then he is to be plunged into the water and thoroughly washed with a soft rag or sponge, and pure, unscented castile soap. After remaining in the water from three to five minutes the surface must be well dried, and rubbed with a flannel cloth or soft towel; then the body must be enveloped in a light blanket and the infant either returned to his crib to sleep, or kept in the lap for ten or fifteen minutes, until thoroughly warm and rested, and finally dressed. If there be repugnance to the bath, the tub may be covered over with a blanket, and the child being placed upon it, may be slowly lowered into the water without seeing anything to excite his fears.

In very hot weather, in addition to the morning full bath, the body may be sponged twice daily, with water, at a temperature of 90° F.; this, contrary to what might be expected, has a greater and more permanent cooling effect than bathing with cold water.

After the third year, three baths a week are quite sufficient. An evening hour is now to be preferred, but the water must still be heated to 90°.

About the tenth year cooler baths can be begun, from 72° to 75° being the proper temperature. The cold sponge or cold plunge is not admissible as a daily routine until youth is well advanced.

The hot bath—95° to 100°—is employed for various purposes, notably for a derivative action; to cause diaphoresis, to relieve nervous irritability, and to promote sleep. Whether a full bath or merely a foot-bath be required, five minutes is a sufficient time for immersion; then, with or without drying, according to the degree of sweating desirable, the whole body, or only the feet and legs in case of a foot-bath, must be enveloped in a blanket, and the child put to bed. To render these baths more stimulating, from a teaspoonful to a tablespoonful of mustard flour may be added, and the child held in the water until the *arms* of the nurse begin to tingle.

It is important not to continue a hot bath too long, lest the primary stimulating effect be followed by depression. Cold baths, by shocking the system, first produce depression; but this is temporary, and is followed by reaction, during which the skin grows red, and the pulse becomes fuller and stronger. They have, therefore a general stimulant and tonic action, promoting nutrition and giving tone to the body. On account of the shock, the extent of which depends directly upon the coldness of the water, these baths must be used with caution, and are not to be employed in very young or feeble subjects.

When giving a cold bath, the child must be stripped in a warm room, and thoroughly rubbed with the palm of the hand until the whole body, especially the spinal region, is reddened; he must then stand in a tub containing enough hot water to cover the feet, and be rapidly sponged with the cold water. The temperature of the latter must never be below 60°, and the addition of half an ounce of sea-salt or a tablespoonful of concentrated sea water to the gallon, renders it more stimulating and insures a complete reaction. After the sponging, the surface must be thoroughly and quickly dried with a soft towel and shampooed with the open hand until aglow.

The cooled bath may be employed with advantage in extreme conditions of hyperpyrexia. The child is first immersed in water at 95°, and this is gradually lowered to 70° by the addition of cold water, the process occupying from fifteen to thirty minutes.

Various medicated baths are employed. Of these the most useful are :—

The Mustard Bath.

Take from two drachms to one ounce of powdered mustard; hot water, two to four gallons.

Derivative in form of foot-bath; stimulant as general bath.

Salt-Water Bath.

Take two ounces of rock salt, or Ditman's sea-salt, or concentrated sea-water (best); water (hot or cold, according to season), four gallons.

General bath, to be used every morning in chronic tuberculosis, scrofula, rickets and general debility. Bath to be followed by thorough rubbing of the surface, especially over the spine.

Bran Bath.

Take one pint of bran, tie up in a muslin bag, place in a quart of water, boil for an hour, squeeze bag thoroughly into the water; add to four gallons of warm water.

Useful in eczema and skin diseases.

Nitro-Muriatic Acid Bath.

Take muriatic acid, one fluidrachm; nitric acid, two fluidrachms; warm water, four gallons.

Serviceable in hepatic sluggishness. Make bath in a wooden tub. May be employed as a foot or general bath.

Mercurial Bath.

℞. Hydrarg. Chlorid. Corros., ... gr. v.
Alcohol, ... f ʒ ij.
Aq. Dest., ... f ʒ j.
M.

S.—Add to four gallons of water. Employed in syphilitic skin diseases.

Soda Bath.

Take half an ounce of bicarbonate of sodium; warm water, four gallons. Used in skin affections.

Astringent Bath.

Take one pound of oak bark, one quart of water, boil for half an hour, strain and add to four gallons of warm water.

3. Clothing.

Infants and young children have little power of resisting cold, and on this account require warm clothing. Too much cannot be said in condemnation of the fashion of allowing children to go, even while in the house, with bare legs and knees.

Every child is supplied with a certain amount of nerve force to be daily expended in the maintenance of the different functions of the body—respiration, circulation, digestion, calorification, etc. If an excessive proportion of this force be consumed in keeping up the heat of the body, as is the case when so much is left bare, the other functions, especially the digestive, must suffer in consequence.

During the oppressive heat of summer, the legs may be left uncovered; but throughout the rest of the year, the whole body must be encased in woolen underclothing. The thickness of this must vary, of course, with the season. Providing this be done, the outer clothing may be left to the taste of the mother; but all garments should fit loosely, that the functions of the different viscera may not be impeded by pressure.

The best pattern of a winter night-dress is a long, plain slip, with a drawing-string at the bottom, to prevent exposure of the feet and limbs, should the child kick off the bed-covering. This should be made of flannel, or, the more easily washed, Canton flannel. In summer, a loose muslin one may be put on, without the drawing-string. A flannel under-vest should always be worn at night, light gauze in summer and heavier wool in winter; care must be taken, however, to have one for night alone, discarding that worn in the daytime.

In infants under a year old, a broad flannel abdominal bandage, extending from the hips well up to the thorax, or, better still, a knitted worsted band shaped to fit the form, is very useful in keeping the abdominal organs warm, aiding digestion, and preventing pain.

All clothing should be changed sufficiently frequently to insure cleanliness.

Shoes must be large, well shaped and made of soft leather, with pliable soles, so as to allow the feet to grow freely.

When dressing a child for exercise in the open air in cold weather, the outer clothing must not be put on until just before leaving the house, and removed immediately on return.

It is important to protect the head from cold in winter by a close-fitting, thick cap; and from the direct rays of the sun in summer by a broad-brimmed, light straw hat.

Rubber shoes are necessary in wet weather to keep the feet warm and dry while walking out of doors.

4. Sleep.

For some time after birth, infants spend the intervals between being fed, washed and dressed in sleep, and thus pass fully eighteen out of the twenty-four hours. As age advances, the amount of sleep required becomes less, until at two years thirteen hours, and at three years eleven hours, are enough. Any marked diminution in the length of sleep or decided restlessness indicates disease, and demands attention from the physician. This matter, though, is, perhaps, more a question of training than any other item of nursery regimen, and many a mother, by want of judicious firmness, has rendered the early years of her child's life not only a burden to himself, but an annoyance to the entire household.

One cannot too soon begin to form the good habit of regularity in sleeping hours, and so far as circumstances will admit, the following rules may be enforced:—

From birth to the end of the sixth or eighth month, the infant must sleep from 11 P.M. to 5 A.M., and as many hours during the day as nature demands and the exigencies of feeding, washing and dressing will permit.

From eight months to the end of two and a half years, a morning nap should be taken, from 12 M. to 1.30 or 2 P.M., the child being undressed and put to bed. The night's rest must begin at 7 P.M. If a late meal be required, the child can be taken up at about ten o'clock, but if past the age for this, he may sleep undisturbed until he wakes of his own accord, some time between 6 and 8 A.M.

From two and a half to four years, an hour's sleep may or may not be taken in the morning, according to the disposition of the subject; but in every case the bed must be occupied from 7.30 P.M. to six or seven o'clock on the following morning.

After the fourth year, few children will sleep in the daytime; they are ready for bed by 8 P.M., and should be allowed to sleep for ten hours or more.

A later retiring hour than 9 P.M. ought not to be encouraged until after the twelfth or fifteenth year.

When feasible, different rooms should be used for the day nursery and the sleeping apartment. The latter should be large, airy, well ventilated, so situated as to be exposed for a certain period each day to the direct rays of the sun, and provided with an open fire-place—for wood, preferably—which serves for both heating and ventilating. It should contain a bed for the nurse and a crib for the child, and be without curtains, heavy hangings or superfluous furniture. A stationary washstand draining into a sewer is not to be permitted in the room, neither should it communicate with a bath-room. Soiled diapers or chamber utensils are to be removed at once, no matter what the time of night. The day nursery should have large windows, protected by blinds, and a southwestern exposure; all other requisites, with the exception of beds, are the same as in the sleeping room. It is very convenient to have the two chambers adjoining, but capable of entire separation by a door, so that one may be thoroughly aired without chilling the other. This arrangement, too, renders it practicable, by standing the door open and raising the windows in the day nursery, to keep the dormitory cool in hot weather without exposing the child to currents of air.

If an apartment has to be occupied during both the day and night, it must be vacated for half an hour or more in the evening and well aired before the child is put back to bed.

The temperature of the rooms must be as uniform as possible, the proper degree of heat being from 64° to 68° F.

The crib should have high sides, to prevent the child from falling out and injuring himself, and should be provided with springs and a soft hair mattress, protected by a gum cloth, placed under a double sheet. The bedclothes must be light in weight, while varying in warmth according to the weather; it is just as important to insist upon cleanliness here as in the clothing of the body.

5. Exercise.

A certain amount of muscular exercise is necessary for development and for the proper performance of the digestive functions. Infants, before they are able to stand, will use their muscles sufficiently if, when loosely clad, they are placed upon their backs in a bed and allowed to kick and turn about at pleasure. After the age of nine or ten months, a healthy child will begin to creep; at the end of a year, he will make efforts at standing, and from four to eight months later, will be able to walk by himself; children however, present great differences in this respect, and a delay of a few months must not be considered as abnormal. So soon as efforts at creeping are made, there need be no fear that insufficient exercise will be taken; the care should be rather to prevent over-fatigue.

Fresh air and sun-light are as necessary as muscular exercise. The child must be taken out of doors every day, weather permitting, after arriving at the proper age; this is four months for children born in the early fall and winter, and one month for those born in summer.

In cool weather, babies who are unable to walk should be taken out in a coach, or in the nurse's arms, for an hour in the morning and half an hour in the afternoon, while the sun is shining. In summer, they may pass the greater part of the waking hours in the open air, provided they be well protected from the direct rays of the sun.

Children old enough to walk may spend a longer time in the

air in winter, and may be out all day in summer. But until the fourth year, it is better to let them play about at will than take a long set walk.

Until well advanced in childhood, the house is the safest place in damp and rainy weather, when there is a strong east or north wind blowing, and when the thermometer stands below 15°.

MANAGEMENT OF WEAK AND IMMATURE INFANTS.—When premature expulsion of the fœtus cannot be checked, children are born in a condition of feebleness requiring particular care. Such children are under weight, breathe and eat imperfectly; have ill-formed organs and badly performed functions; their skin is soft and delicate, bright red in color, and so transparent that the superficial blood-vessels can often be seen, and their cry is feeble. Their muscles are inert, they hardly seem to contract, and the movements of the limbs are rare and without vigor. The infant, plunged in a sort of stupor, has not even strength enough to suck, the muscles of the cheeks and of the tongue and palate being apparently too weak to perform this act, and deglutition itself is often slow,—a grave symptom, since the regular accomplishment of this function alone renders life possible.

The employment of artificial heat and a well-regulated alimentation are the methods of combating this condition. Warmth and even temperature of the surrounding air are most important. The old method of accomplishing this was to envelop the infant's body and limbs, under the ordinary clothing, with a layer of cotton wadding, and place a fold of the same around the head. Two or three bottles filled with hot water were placed under the blankets of the bed, and renewed from time to time as they became cold. An effort was made to maintain the temperature of the chamber at 77° Fahr. All changes of clothing were made before a brisk fire, and two or three times every day massage or friction, either dry or with various stimulating embrocations, was practiced to strengthen the circulation. As an improvement upon this crude and very unsuccessful method, M. Tarnier has devised an apparatus called a "hatching-cradle."

It consists of a box made of wood, sixty-five centimetres long by fifty high and thirty-six wide, with sides twenty-five millimetres thick. The inside of the box is divided by a partial partition into two parts; this partition, which is horizontal, is placed about fifteen centimetres from the bottom. The lower story is intended for hot-water bottles.

The cut shows the apparatus.

There are two doors; one is a sliding door on the side of the box to push to either side for the purpose of introducing the hot-water bottles; the other is at one of the ends (at T in this figure);

FIG. 5.

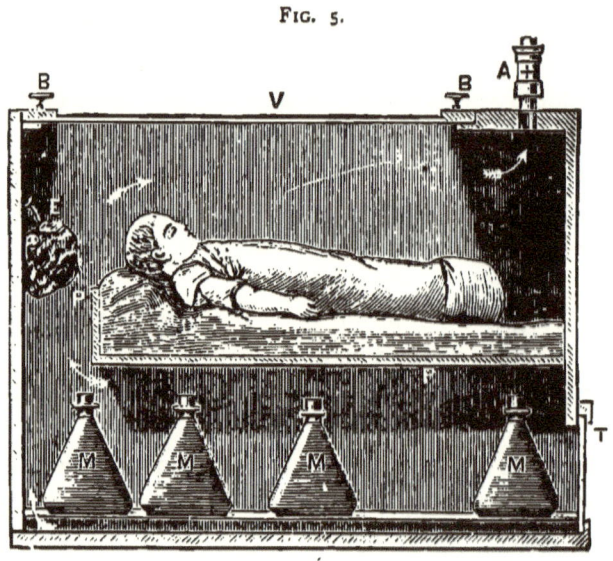

TARNIER'S "HATCHING-CRADLE."

it does not completely close the orifice, but allows air to enter. The upper part, for the baby, contains the bedding, and is covered with a glass top at V; it should close tightly and be held by two screws at BB. At A is an outlet for the air, to which a small ventilator can be attached. In the opening between the two chambers a wet sponge is placed to keep the air slightly moist, and here also a thermometer is placed to mark the temperature. The heat is supplied by earthenware jugs at M; they contain a pint of water each; four or five are required to keep the temperature at the proper point,—87–90° F. The chamber must be heated to this degree before the infant can be placed in it, and every one and a half or two hours one of the water bottles must be changed in order to maintain a constant temperature. The air passes in by the door, T, is heated by the bottles, and passing by the sponge, E, escapes at A; the movements of the small ventilator in the latter position is the index that the air is circulating. The infant must be dressed in swaddling clothes, as it has been observed that the temperature is always two or three degrees higher under the clothing than in the chamber itself. Every hour or two, according to the case, the little patient should be taken out to receive food and have its napkins changed. The shorter time occupied in these processes the better.

Auvard has suggested an improvement in Tarnier's hatching cradle. In his instrument a cylindrical reservoir of metal takes the place of the hot-water jars in the lower compartment of the couveuse. This reservoir is filled by means of a metallic funnel at one end of the box and communicating with the cylinder through a metallic tube.

The overflow of the cylinder is provided for by a curved metallic tube at the lower part of the cylinder, beneath the inlet through which the reservoir is filled.

The air enters by a register on one side of the couveuse instead of at the end, as in Tarnier's apparatus. The other portions of the apparatus are the same as Tarnier's.

The metallic cylinder is capable of holding ten litres of liquid (a litre is a little over a quart). To start the apparatus, about five litres of boiling water should be poured in, after which three litres may be poured in every hour. When ten litres are contained in the cylinder, the overflow-pipe carries off the excess. Auvard suggests having two vessels, capable of holding three litres each, keeping one under the escape-pipe and the other over the fire, reheating the water in the vessel filled by the escape-pipe and having it in readiness for the next changes. The two vessels may be thus used alternately, and but little time consumed in the heating of the apparatus as compared with that required in the use of Tarnier's invention.

To empty the cylinder, a rubber tube is attached to the escape-pipe, by which it is made to act as a siphon—a small quantity of water poured into the cylinder through the funnel being sufficient to start the liquid.

The length of time the child remains in a couveuse will vary from fifteen days to three weeks, a month, or even more. It should not be removed permanently until it has acquired sufficient vigor to live in the ordinary atmosphere of the apartment. To accustom the child to this atmosphere, it should, as it grows stronger, be removed for an hour at a time from the couveuse during the warmest part of the day.

It is best to continue the use of the apparatus at night for some time after the child becomes accustomed by day to removal from the couveuse, for the danger of chilling from changes in the atmosphere is greater at night.

Auvard recommends the use of the couveuse in all cases where the vitality of the child is enfeebled either by external causes, as cold, or internal causes, as prematurely congenital feebleness, cyanosis, or "blue disease," wasting, or other general maladies enfeebling to the newborn.

The excellent results obtained by these cradles is shown by the following statistics obtained from the Maternité, in Paris :—

WEIGHT OF CHILD.	NO. OF INFANTS.	NO. THAT LIVED.	NO. THAT DIED.
1000-1500 grammes.	40	12	28, or 70 per cent.
1501-2000 "	131	96	35, or 26.7 "
2001-2500 "	112	101	11, or 9.8 "

Before the introduction of the machine, infants died at the rate of 66 per cent.; since, the average proportion is 36.6 per cent.

The heated cradle has also been used with success in the treatment of sclerema, œdema and cyanosis attacking the newly born. From the very first day an attempt must be made to put these feeble infants to the breast; and if they be too weak to suck, the milk may be squeezed into the mouth, or first into a warm spoon and then given to the child. The mother's or nurse's milk, without dilution or addition, is the best food, though if this cannot be obtained asses' milk may be used. This must be mixed with equal quantities of warmed sugar and water —3 parts to 100. When the cows' milk is employed, the mixture should be one part to three of the same sugared water. M. Tarnier recommends the cows' milk to be prepared thus: The mixture of milk and sweetened water is placed in an air-tight pot, and this is placed in boiling water for half an hour. It is given to the child from a small spoon. When the infant is very small, six to eight grammes (f3ij) are enough for a meal; larger babies require from ten to fifteen grammes (f3iiss-f3iiiss). There should be at least twelve meals every twenty-four hours.

It often happens that the babe will drink badly and throw up half the liquid given. Under this deficient feeding the little sufferer gets rapidly worse, loses weight, and frequently has diarrhœa. In these cases "gavage" is resorted to. The apparatus is quite simple, being nothing more than a urethral catheter of red rubber (Nos. 14–16 French), at the open end of which a small glass funnel is adjusted. The infant upon whom gavage is to be practiced is placed on the knee, with its head slightly raised; the catheter, being wetted, is introduced as far as the base of the tongue, whence, by the instinctive efforts at deglutition, it is carried as far down as the œsophagus and into the stomach. The liquid food is next poured into the funnel, and by its weight soon finds its way into the stomach. After a few seconds the catheter must be removed, and here is the great point in the operation: it must be removed with a rapid motion and at once, for if it be withdrawn slowly all the food introduced will be vomited.

The number and quantity of meals thus given must vary with the age and strength of the infant. As a rule, eight grammes (f3ii) of food every hour will suffice when the subject is small, but there must be an increase as circumstances require. Mother's milk is the best in gavage, but other foods may be used if it be impossible to obtain it.

Should the gavage be too copious, the infant gains rapidly in weight and size. This increase, however, is due to œdema, and quickly disappears when a proper quantity of food is administered. When excessive feeding is continued, indigestion soon sets in, and the patient dies of gastritis or enteritis. As soon as the child gains strength this mode of feeding may be alternated with nursing, and gradually breast-feeding may be entirely substituted for it. Nevertheless, the least digestive disturbance indicates the necessity of a return to gavage.

Even when the child is old enough to nurse, should it be weak, it is useful, besides regularly giving the breast, to resort to gavage three or four times a day. This is what M. Tarnier calls *gavage de renfort*, as it keeps up the strength of the infant so that it can take the breast and digest well.

The absence of the sensation of hunger and of the necessary strength to suck are not contra-indications to this mode of feeding; and by it, together with the use of hatching machines, the actual period of vitality has approached the legal period, which in French law is six months of intra-uterine life.

LAVAGE.—Epstein, of Prague, has practiced lavage of the stomach in nursing children with good results. The apparatus employed consists of an elastic tube joined to a small glass tube, to the other extremity of which another piece of elastic tubing with a wide opening is adapted. Lavage may be practiced a few days after birth without the least danger to the infant. The instrument is inserted while the child is in the dorsal decubitus position, the trunk and arms being enveloped in napkins. The child's mouth is opened by exerting a slight pressure upon the chin, while the larynx is slightly pressed inward by the index finger of the right hand. The tube having been previously dipped in warm water, it is held as a pen, and the smaller extremity slowly introduced, advancing by the simple, repeated act of deglutition. The contact of the tube with the stomach causes contractions of the walls, thereby expelling a quantity of liquid through the tube, the broad end of which is depressed somewhat until the stomach is empty. The author employs distilled water with a little hydrocarbonate of soda, using from twenty to twenty-five cubic centimeters of the liquid for each lavage. The funnel-shaped end of the tube is raised to pour in the water and lowered to expel it. The washing may be repeated two or three times in succession until the liquid returns nearly clear.

Lavage is indicated : 1. In cases of repeated vomiting. 2. In cases where there is present an affection of the mouth which is capable of extending to the stomach. 3. In cases of eclampsia caused by indigestible substances. 4. In cases of poisoning.

After the lavage the child should remain perfectly quiet for fifteen or twenty minutes before nursing.

PART III.—MASSAGE IN PÆDIATRICS.

Systematic manipulation is of great value both as a means of preserving health and as a scientific method of treating certain diseases in children.

Mere rubbing or friction of the surface cannot be included under massage in its literal sense, still, it is a useful form of manipulation, and needs no special instruction, being possible to any intelligent, soft-handed mother or nurse.

Massage, on the contrary, is an art, and, like every other art, requires study and patient preparation for its successful practice. It is a powerful remedy too, and, like other agents of its class, as potent for evil as for good in unskilled hands. Therefore, to insure good results, a trained masseuse is necessary—and she must act under the direction of the physician.

Massage includes several processes of manipulation. Those given by Murrell, from whose excellent little work* I have taken much of the description of the different "movements," are *effleurage*, *pétrissage*, *friction* and *tapotement*.

EFFLEURAGE is a stroking movement made with the palm of the hand passing with more or less force over the surface of the body centripetally. The movements are made to follow as nearly as possible the direction of the muscle fibres, and for deep-seated tissues the knuckles can be used instead of the palm. This method is of minor value in itself but of great use when combined, as is the rule, with the procedures to be described.

PÉTRISSAGE consists essentially in picking up a portion of muscle or other tissue with both hands, or the fingers of one hand, and subjecting it to firm pressure, at the same time rolling

---

* "Massage as a Mode of Treatment." W. Murrell.

it between the fingers and the subjacent tissues. The hands must move simultaneously and in opposite directions, the skin must move with the hands to avoid giving pain, and the thumb and fingers must be kept wide apart in order to grasp a bulk of tissue, a whole muscle belly, for instance. The manipulation must be uniform, in a direction from the extremities toward the centre of the body, bearing in mind the arrangement of groups of superficial muscles and keeping well in the interstitia.

FRICTION, or *massage à frictions*, is performed with the tips of the fingers. It is a pressure movement rather than a rubbing. It is always associated with effleurage and, to be of any use, must be performed quickly and readily.

TAPOTEMENT is a percussion which may be made with the tips of the fingers, their palmar surfaces, the palm of the hand, the back of the half-closed hand, the ulnar or radial border of the hand, or with the hand flexed so as to contain, when brought in contact with the surface of the body, a cushion of air.

The hand of the masseuse must be perfectly clean and soft, and the finger nails short and smooth. The length and frequency of the sittings must vary with the individual case. Murrell is in favor of short and frequent *séances*, and also recommends *dry* massage, that is, without the use of oil, liniments or ointments; vaseline especially is to be avoided.

Our knowledge of the physiological action of massage is based upon experimental research and clinical experience. Experiments were made by Dr. Gopadze (quoted by Murrell) upon four medical students, who were kept in hospital and subjected to systematic manipulations for twenty minutes or more daily. The séance began with effleurage, followed by pétrissage, friction and tapotement, and ending with a second effleurage. The results were increased appetite and a notable gain in body weight. The axillary temperature fell, never more than .5°, for about thirty minutes after each massage; then it rose steadily, and an hour later was generally a degree higher than at the commencement of the operation. The respiratory movements were uniformly increased in frequency, depth and fulness. The pulse varied with the kind of "movement" used—light surface effleurage increased its frequency, while pétrissage made it slower.

Zabludowski, experimenting on himself and two servants for eighteen days, noted increased bodily and mental vigor and improved appetite and sleep.

Clinical experience shows that massage increases the activity of the circulation, reddens the skin and elevates the temperature in the part manipulated. It also increases the electrical contractility of muscular tissue, and stimulates the flow of lymph in the lymphatic vessels. Muscular stiffness and fatigue are relieved, nervous irritability is calmed, and restless and wakeful patients are soothed by it into refreshing sleep.

With these facts at hand, it is not difficult to see what a useful therapeutic agency we possess in skillfully employed massage. By its application we have the power to prevent the atrophy of muscles and to augment muscle tone, to build up such tissues as fat and blood, to improve nerve tone both directly by producing a better blood supply and indirectly by relieving irritability and giving rest and sleep, and, finally, to hasten the absorption of waste tissue and of morbid effusions. At the same time it must always be remembered that massage is a powerful remedy. A short séance with gentle movements may do good in infantile palsy, for example, but it does not follow that by doubling the time or force twice as much benefit will be derived. In fact, the reverse of the proposition is true; short, gentle massage maintains the size and tone of the muscles, while long, forcible manipulation causes them to atrophy quickly. The same truth runs through the whole question and must be observed.

Before entering upon the therapeutic application of massage proper, it will be well to revert to the process of simple rubbing, already mentioned. This is of much value as a general hygienic measure. Each day, after the bath, the skin having been thoroughly dried by a soft, warm towel, the whole surface should be gently rubbed with the palm of the hand, the process occupying about five minutes. This increases the capillary circulation, encourages thorough reaction, aids nutrition and adds vigor to the frame. Weakly children especially thrive under it. In older children, friction with a soft towel may be substituted for hand-rubbing, but this change should not be made before the fifth or sixth year.

Sometimes it is well to rub certain portions of the body more thoroughly than others. Thus in rickets the spine should receive especial attention, in indigestion and constipation the abdomen, in weak ankles the feet and legs, etc.; though even in these cases the general surface must receive a share.

Massage may be employed with advantage in the following diseases of childhood :—

(*a*) Chronic gastro-intestinal catarrh. In this condition the skin is harsh, and often so dry that a shower of epidermic scales falls on the removal of the underclothing, the muscle tone is faulty, general nutrition is impaired, and there is a determination of blood from the surface toward the mucous membranes. To get the skin active, and in this way balance the circulation, is an important step in the reëstablishment of normal digestion, secretion and excretion, the essentials of perfect nutrition. To accomplish this, a full, warm bath is administered every evening, just before bedtime, the patient remaining in the water for five minutes. Then the surface is thoroughly dried and half an ounce of olive oil is gently rubbed into the skin, the child enveloped in a light blanket and put to bed. After a little time diaphoresis begins. So soon as the sweating is free the skin is again dried and the night-dress put on in preparation for sleep. Next morning, at some convenient time after breakfast, the child is subjected to twenty minutes' massage (pétrissage with effleurage). The inunctions are continued until the skin becomes soft and active, and massage is employed daily until there is a decided improvement in the amount of flesh and general strength, a period generally of two or three weeks. Afterwards " movements" every third day will be sufficient to complete the cure.

In these cases massage not only aids the baths and inunctions in their general action, but directly and powerfully increases nutrition and muscle tone, and materially hastens an otherwise slow process of recovery.

(*b*) Constipation. Manipulation is a very efficient remedy in habitual constipation, and there are many cases that can be cured by it, combined with a properly regulated diet, without the use of drugs. Pétrissage of the colon is the best method, instructions being given to follow the natural course of the fæces through this portion of the gut; thus, beginning in the right iliac region to proceed upward to the right hypochondrium, to cross over to the left hypochondrium and then downward to the left iliac region. In this way the ascending transverse and descending colon are manipulated in order.

Five or ten minutes every morning, or every morning and evening in obstinate cases, constitute the proper duration and frequency of the applications. The pressure must be gentle, as delicate tissues are being dealt with.

In this condition I have not found the *dry* method so efficient as the combination of massage with the inunction of warm olive oil or a weak ammonia liniment. The addition of aloes to the liniment, a plan recommended by some authors, has never been necessary in my experience.

Sometimes tapotement with the flat hand, the hand partly closed forming a cushion, or with the margin of the hand, is necessary, but the course of the colon must always be followed. The therapeutic action of this mode of treatment is, undoubtedly, threefold : it increases the intestinal and other secretions; it increases the peristaltic action of the intestinal muscular fibres, and it mechanically forces accumulated fæcal matter toward the rectum.

(*c*) Colic. Every experienced mother knows how often flatus, the cause of colicky pain, is expelled from the stomach or intestines by gently rubbing the abdomen with the hand. Any approach to scientific manipulation is much more efficient, and two or three minutes' effleurage may be resorted to, as the urgency of the symptoms requires, with the most satisfactory effect. In this connection it must be remembered, also, that rubbing of the feet to increase the circulation is an important aid in relieving colic.

(*d*) General debility and anæmia. These conditions are much benefited by short, frequently repeated courses of massage. In the convalescence from many diseases, both acute and chronic, in which these states exist, manipulation improves general nutrition, and strength is rapidly gained.

(*e*) Infantile paralysis. Here massage of the paralyzed muscles brings more blood into them and maintains their nutrition until, in favorable cases, new cells in the cord take on the function of those which have been destroyed.

In essential paralysis the affected members are always cold, and the muscles contract feebly, if at all, under the influence of electricity. By systematic massage—pétrissage combined with effleurage and both performed centripetally—an improvement takes place with more or less rapidity. The first indication of this is an increase in the temperature of the parts, continuing for several hours after the rubbing. Then the electrical contractility of the muscles begins to return, and they respond to a current that at the commencement is entirely inoperative.

In recent cases the sittings should be of short duration and frequently repeated, five to ten minutes, three or four times daily. As improvement advances, the frequency may be reduced, and in chronic cases twice a day will be sufficient at any time.

Electricity is of great aid in the treatment, but it does not take the place of massage, for while it causes contraction and congestion of the muscles and hyperæmia of the skin, it does not have the same power of arresting rapid wasting. The constant current is to be employed. In the commencement the current must be mild, so as not to produce pain or emotional excitement, and often it is well to apply empty sponges for several sittings, to accustom the little patient to the novelty of the procedure without producing any sensation. The treatment may be begun about three weeks after the onset of the paralysis, earlier applications being attended by the risk of increasing spinal congestion.

Well wetted, large sponges should be used. The positive pole is kept stationary and placed close to the sacrum or lower part of the back when the legs are to be galvanized and to the back of the neck in case of the arms. The negative pole is slowly moved up and down over the surface of the affected limb, thus making and breaking the circuit gradually and without pain. The muscles that do not contract to Faradism are the ones to be influenced by galvanism; in other parts hyperæmia of the muscles and skin only is required.

Three or four electrical sittings a week are sufficient. They should be short at first, ten to fifteen minutes, and gradually increased in duration and force as tone and contractility return, care being taken never to over-fatigue the muscles.

(*f*) Chorea. So far as this branch of the management of chorea is concerned, it requires to be aided by proper diet and rest in bed. On the onset of an acute attack the patient is put to bed, given a full supply of good food and allowed to rest for two days without massage. Should the choreic movements be very violent, the sides of the bed are padded to prevent the child bruising himself, or, if too violent for this, to give security, he is slung in a hammock.

At the end of this time the regular treatment is initiated. The plan, a slight modification of that recommended by Goodhart, is as follows:—

The child—at seven years of age, for example—has, at 5.30 A. M., half a pint of warm milk; 7 A. M., half a pint of milk and three slices of bread and butter (each slice an ounce in weight); 9.45 A. M., a teaspoonful of Merck's dry malt in a little milk; 10 A. M., massage for fifteen minutes, followed by half a pint of warm milk; 12.30 P. M., a teacupful of rice pudding, half a pint of milk, green vegetables and mashed potatoes; 4.15 P. M., half a pint of warm milk, three slices of bread and butter and a lightly boiled egg; 7 P. M., malt as before; 7.30 P. M., massage for fifteen minutes, followed by half a pint of milk. At the end of ten days or a fortnight, the bread and butter is increased to four slices at 7 A. M. and 4.15 P. M.; a lean broiled chop is added to the mid-day meal, and an extra pint of milk is distributed over the twenty-four hours. After two or three weeks the patient may be allowed to sit up in bed, well supported by pillows, and may have a few toys to play with. It is a golden rule, however, never to hurry a patient with chorea out of bed. The muscular strength is more quickly recovered while at perfect rest, and too early exertion often causes a relapse. While carrying out this plan Goodhart employs no medicines, but in my experience recovery has been more rapid under the conjoint use of Fowler's solution, administered in daily increasing doses.

(*g*) Other nervous diseases in which massage is employed with success are pseudo-hypertrophic paralysis; facial paralysis; neurasthenia and spinal irritability occurring in girls about the approach of puberty, and that ill-defined and painful condition so often encountered in young subjects and known as "growing pains."

(*h*) Pleuritic effusions (serous); fibroid pleurisy; enlarged lymphatic glands, and stiffened rheumatic joints are all benefited by rubbing. In these special instances the manipulations are generally combined with the use of embrocations, though the curative effects cannot be attributed to the latter alone.

In concluding the subject of massage in childhood, it is a point of importance to mention that those cases in which the manipulation is immediately followed by a sensation of comfort or by refreshing sleep are most benefited by it. On the contrary, those cases that are stimulated, derive little benefit, and perhaps positive injury from rubbing. This I have especially noted in cases of general debility and anæmia, and my own experience has been confirmed by a number of practical observers in whose judgment I have the greatest confidence.

PART IV.—DISEASES OF THE DIGESTIVE ORGANS.

CHAPTER I.

AFFECTIONS OF THE MOUTH AND THROAT.

1. CATARRHAL STOMATITIS.

The ANATOMICAL LESION in this affection consists of a simple hyperæmia of the mucous membrane of the mouth, with its attendant redness, swelling, and altered secretion. This hyperæmia varies both in extent and degree. Sometimes it is limited to small, circumscribed points of the membrane, at others it extends over large patches, or involves the entire surface. In the latter cases it is most intense, the mucous glands of the lips and cheeks participate, becoming enlarged and prominent, and occasionally small herpetic patches appear.

The disease may be primary or secondary.

ETIOLOGY.—The causes of primary stomatitis are the ingestion of food or drinks which are acrid and irritating or too hot; the eruption of teeth; the presence of decaying teeth; want of cleanliness of the mouth; exposure to cold and wet; and the use of certain drugs, as mercury, iodine, antimony and arsenic. The secondary form occurs during the course of measles, scarlatina, typhoid fever, and disordered conditions of the stomach, particularly those attended by acid eructations. Catarrhal stomatitis is also met with in the earlier stages of more serious diseases of the mouth.

While not limited to any special age, the disease occurs most commonly during dentition, since at this period several of its causes are apt to be simultaneously operative.

SYMPTOMS.—These are mainly local. The lips are unnaturally full and red, and the skin at the angles of the mouth and on the chin may be excoriated by the dribbling saliva. The oral mucous membrane presents either a punctated, a patchy, or a diffuse redness. It is moderately swollen, and hot and tender to the touch. At first the mouth is dry, but soon the salivary flow is increased, the secretion becoming acid in reaction, and sometimes viscid and flocculent. The mucous glands of the cheeks and lips may project as yellowish-white or transparent nodules, yielding a drop of mucus on pressure. Infrequently, too, isolated collections of small vesicles develop and then quickly dry up, leaving scales behind them. The tongue is either red and smooth, with enlarged and reddened fungiform papillæ, or covered with a white frosting, through which the papillæ project in scarlet points. The last condition is most frequently seen when the stomatitis is secondary to gastric catarrh. The acts of sucking and eating are painful, and resistance is offered to inspection of the mouth. Cold drinks are craved.

Restlessness, irritability, slight heat of skin, anorexia—depending chiefly upon the local tenderness—and constipation, are the general symptoms of primary catarrhal stomatitis. In the secondary variety the general symptoms depend upon and vary with, the originating disease; the local features, however, remain the same.

The course of the disease depends upon the cause and the treatment adopted, though it is usually acute, rarely lasting longer than a week.

TREATMENT.—After attending to the removal of the exciting cause, if this be possible, the diet must be regulated. To sucklings, the breast or the carefully prepared bottle alone should be allowed, and milk guarded by lime-water must constitute the food of older children. If the act of sucking be so painful as to cause the infant to refuse the breast or bottle, it is necessary to give food, temporarily, from a spoon or glass.

The mouth should be thoroughly washed, at intervals of an hour, while the patient is awake, with a solution of borax or chlorate of potassium in rose water (gr. x to f℥j). After taking food, particularly, the mouth ought to be cleansed with cool water, and the lotion used. A little mass of absorbent cotton twisted around the end of a probe, or a soft rag folded around the index finger, are the best vehicles for carrying the lotion.

Regular evacuation of the bowels must be secured by saline laxatives. If the skin be hot and dry, liquor potassii citratis, in doses of a fluidrachm every two or three hours, for a child one year old, is indicated. When the tongue is heavily frosted, and the stomach disordered, recovery may be much hastened by using the following prescription:—

    ℞. Sodii Bicarbonatis, ... gr. xxiv.
        Pulv. Pepsinæ (Fairchild's), ... gr. xij.
        Pulv. Aromatici, ... gr. iij.
        M. et ft. chart. No. xij.
        S.—One powder four times daily, administered in milk or syrup, for a child between seven and twelve months old.

2. APHTHOUS STOMATITIS.

This is a much more common disease than uncomplicated catarrhal stomatitis, and is most frequently met with in children between the ages of six and fifteen months.

The ANATOMICAL LESIONS are hyperæmia of the mucous membrane of the mouth, and the formation of *aphthæ* or small, superficial, yellowish-white ulcers.

ETIOLOGY.—Any condition which reduces the general strength and interferes with nutrition may exert a predisposing influence. For instance, over-crowding; residence in damp, ill-ventilated houses or rooms; insufficient food and clothing; chronic diseases, especially of the digestive tract; scrofula and the tuberculous tendency.

The exciting agencies are, want of proper attention to the cleanliness of the mouth; foul nursing bottles; the administration of sour milk, or an excess of farinaceous food, and dentition. After the completion of the first dentition, an indulgence in pastry or candy is often followed by an attack of aphthæ, and certain children always suffer after eating some particular article of food, as honey, walnuts, or salted fish. All of these causes are active in the production of a catarrhal state of the stomach, which invariably precedes and attends the disease under consideration.

The disease is not contagious, though at times a sufficient number of cases occur simultaneously to constitute an epidemic.

SYMPTOMS.—For twenty-four hours prior to the appearance of aphthæ, there is fretfulness, increased thirst and poor appetite. Next, the mouth becomes hot, and a few hours later the ulcers appear, without any previous vesication.* The lips, swollen and vividly red, are held somewhat apart, and clear saliva drops from the mouth, excoriating the skin of the lower lip and chin. The oral mucous membrane is red, swollen, and hot, and presents the characteristic ulcers. These are usually discrete, and make their appearance first on the inside of the lower lip and the edges of the tongue, though they may, subsequently, extend to the cheeks, gums, soft palate, and even the tonsils. Their number varies from one to twenty, and their size, from that of a pin's point to a split pea. The ulcers, oval, round, or, more rarely, linear in shape, are slightly elevated above the surrounding surface, have deeply reddened edges and whitish or yellowish-white floors. They are excessively sensitive, and thus mechanically interfere with sucking, chewing, speaking, or other movement of the mouth. The edges of the tongue are clean and red, while its dorsum is covered with a thick, white coating.

Together with these local symptoms, there is restlessness, increased pulse rate, elevated surface temperature, dryness of the skin, thirst, anorexia, nausea with frequent eructations of acid liquid and occasional vomiting, and either constipation or di-

---

* For confirmation of this statement, see Vogel, " Ziemssen's Cyclopædia," Vol. vi, p. 779.

arrhœa. The loss of appetite is due both to the painful condition of the mouth and to the disordered state of the stomach.

When all the ulcers appear simultaneously, the disease runs its course in from four to seven days. The fibro-cellular exudation covering their floors then disappears, leaving the mucous membrane beneath intact but intensely red, though occasionally shallow, clean ulcers are left, which quickly heal. At the same time the local and constitutional symptoms rapidly subside. If, on the contrary, the ulcers develop in successive crops, as is sometimes the case, the duration may be prolonged for a fortnight or more.

There is another form of aphthous stomatitis, termed *confluent*, in which the aphthæ are very numerous, and tend to run together, forming large, irregular ulcers. The symptoms are proportionately severe. It occurs secondarily to grave constitutional diseases—namely, measles, variola, scarlet fever, diphtheria, typhoid fever, pneumonia and whooping cough.

The DIAGNOSIS of the ordinary, discrete form is unattended with difficulty. Thrush bears the closest superficial resemblance; but in this disease the creamy-white spots are slightly raised above the surface, being deposits upon the mucous membrane. There are no ulcerations, the color of the free membrane is rather purplish than scarlet, and finally the thrush fungus is discoverable by the microscope. The graver, confluent form is distinguished from ulcerative stomatitis, by the absence of fetor, and by the different seat and appearance of the lesions. The ulcers in the latter disease always begin at the margins of the gums, extend rapidly, and present grayish floors.

Aphthæ is usually a mild disorder, recovery taking place quickly and without difficulty. The confluent form, besides running a longer course, is more difficult to cure, on account of the general debility induced by the associated disease.

TREATMENT.—Since some disturbance of digestion is constantly at the bottom of the local trouble, attention to the feeding apparatus and to the diet is of great importance. Absolute cleanliness of both bottles and tips must be insisted upon, and if a complicated, patent arrangement of rubber and glass tubing has been used with the bottle, it must be at once discarded and a simple rubber tip substituted. Regular hours for meals—the frequency varying with the age of the child—are as essential as the selection of suitable food and its administration in proper quantities.

A child of six months should be fed every three hours, between 6 o'clock in the morning and 9 o'clock in the evening. A mixture such as the following—

Sound Milk, ... f ℥ iv.
Cream, ... f ℥ j.
Lime-water, ... f ℥ ij.
Sugar of Milk, ... one teaspoonful (gr. xl);

may be made immediately before the time of feeding; poured into an absolutely clean bottle, to which a clean tip is fitted, and the whole placed in a water-bath and heated to a temperature of about 100° F. This preparation is easily digested, contains enough lime-water to prevent rapid and firm clotting of the milk, and is not so great in quantity as to over-distend the delicate stomach and cause vomiting.

Children of two years of age and over should be placed on a simple diet. A breakfast, luncheon and supper of stale bread and milk guarded by lime-water (one part to three) and a midday dinner of broth and well-boiled rice.

The disease usually makes its appearance too long after the causative error of diet, to be stayed by the administration of an emetic. If, however, an overloaded stomach be indicated by fever, restlessness, and epigastric pain and distention, a dose of the wine or syrup of ipecacuanha* should be given. If the bowels be constipated a gentle laxative is required. Probably the best is calomel; for a child from six to twelve months old a powder containing half a grain of the mercurial and five grains of sugar may be placed dry upon the tongue in the evening, to be followed next morning by a small teaspoonful of magnesia in

---

* For a child of one year old the emetic dose of wine of ipecacuanha, is fifteen drops; of the syrup, half a teaspoonful, repeated if necessary.

milk or lemonade. If, on the contrary, there be diarrhœa, the bowels should be first cleared of irritating materials by a teaspoonful of castor oil, into which five drops of paregoric have been dropped, and the following prescription given:—

℞. Sodii Bicarbonatis, ... gr. xxiv.
Syr. Rhei Aromatici,
Syrupi, ... āā . . . . . . f ℥ ss.
Aq. Menthæ Piperitæ, ... q. s. ad f ℥ iij.
M.
S.—A teaspoonful every three hours.

The fever, when very moderate in degree, requires only a plentiful supply of cool water to drink, and a hot mustard bath in the evening. The strength of the latter should be one teaspoonful of strong mustard to as much water as will cover the child's legs and hips when in a sitting posture. The duration of the bath should be from five to ten minutes. If the skin be quite hot and dry, a saline diaphoretic is necessary. The best is liquor potassii citratis, in doses of one teaspoonful every two hours. Sometimes it is well to add one-quarter of a drop of tincture of aconite to each dose of the potash solution.

Locally, the best results will be obtained by lightly touching each ulcer with a point of lunar caustic. The pain incident to the application may be prevented by the previous application of a 4 per cent. solution of cocaine. In ordinary cases one such application suffices, in severe, it is necessary to repeat it daily for a week or more. In addition the mouth must be washed thoroughly and frequently, particularly after food is taken, with cool water, or with a solution of chlorate of potassium, as:—

℞. Potassii Chloratis, ... gr. xx.
Vini Opii, ... ♏v.
Glycerinæ, ... f ʒ j.
Aquæ Rosæ, ... q. s. ad f ℥ j.
M.

After the fever has subsided, a digestant will be required for a few days. Thus, half a teaspoonful of wine of pepsin three times daily may be ordered; or, if there be acidity with a coated tongue, the powder recommended, under the same circumstances, in catarrhal stomatitis.

The local treatment must be persevered in, until the ulcers have healed and the mucous membrane has returned to its normal condition.

3. ULCERATIVE STOMATITIS.

This affection of the mouth is quite common in childhood. It is usually seen in children between three and eight years of age ; is never met with before the commencement of dentition, and is not contagious though it sometimes occurs in almost epidemic profusion.

The ANATOMICAL LESIONS consist of parenchymatous inflammation of the gums, and often of the tongue and cheeks, with ulcerative destruction of the mucous membrane. Microscopical examination of the floors of the ulcers, reveals pus corpuscles, isolated blood corpuscles, and granulated cells, imbedded in an amorphous, finely granular mass which is filled with bacteria and micrococci. There is no trace of pseudo-membrane.

ETIOLOGY.—As the disease is not contagious there is probably no specific epidemic influence in its causation. When groups of cases, large enough to be classified as epidemics, do occur, they are generally limited to single houses or institutions, and may be traced to bad hygienic surroundings affecting alike all the inmates. Insufficient or bad food and residence in unhealthy, cold, damp, ill-ventilated houses constitute one set of causes.

Again, ulcerative stomatitis is very apt to follow in the wake of typhoid fever, scarlatina, measles, variola and dysentery, and since each of these primary diseases usually occurs as an epidemic, a similar tendency in the sequelæ is readily explained.

A certain amount of reduction of the constitutional vigor seems to be an essential precedent to the development of the disease. Sickly, rickety and scrofulous children are susceptible subjects, and when the gums are loose, soft and hyperæmic they are more readily affected than when firm and closely applied around the teeth. I have, however, seen it occur in the most healthy children. For instance, I lately saw it developed in two most robust children of 4 and 7 years, who had returned to the city after a summer in the mountains, during which there had not been a single day of illness, and, on the whole, two more robust specimens of healthy childhood could not well be found. This attack could, most probably, be attributed to the opening of a sewer within a block of their home.

The presence of decaying teeth, want of cleanliness of the mouth, and the careless administration of such medicines as mercury, lead and phosphorus, are exciting causes.

SYMPTOMS.—At first there is a sense of heat and pain in the mouth, and the breath grows offensive in odor. Next the gingival mucous membrane, immediately about and between the teeth, becomes red and swollen. The swelling rapidly increases, the points of the gum between the teeth standing out like flasks, and the whole margin becoming so soft and tender that it bleeds upon the lightest touch. In the course of twenty-four hours the edge of the gum, where it touches the teeth, changes from a bright red to a yellow or yellowish-gray color, and softens, breaking down into ulcers.

Ulceration generally commences on the external surface of the lower gum, and in the beginning appears as a more or less extended, narrow and indented gray band, following the line of the teeth. Later it may appear on the outer surface of the upper gum; on the internal surface of both the lower and upper gums; on the edges of the tongue, at points where the organ presses against the teeth; and finally on the cheeks. In the latter position it often happens that the ulceration corresponds exactly with that of one or both gingival borders, forming a single or double strip running parallel with the jaws.

The ulcers are depressed, have a ragged, dirty gray or brownish floor, and intensely red, swollen edges. The mucous membrane not involved shows the characteristics of catarrhal stomatitis.

When the disease is fully developed, the lips are tumid and red. They are held apart, and a stream of yellowish, sometimes bloody, always ill-smelling, acid and viscid saliva, constantly drips away, excoriating the skin over which it flows. If the mouth be kept closed, as it sometimes is, half an ounce or more of this fetid fluid gushes out whenever the lips are parted in speaking or in taking food. The submaxillary glands and the lymph glands of the neck are moderately enlarged, and there is often œdema of the face, limited or general, according to the extent of the ulceration.

The mouth is the seat of constant burning and pain; is hot and tender to the touch, and chewing causes great suffering. Between and upon the teeth there is a deposit of yellow unctuous material. The tongue, in addition to presenting the marginal ulcers, is swollen and heavily coated with a dirty, yellowish-white fur. The speech is thick; the breath has a characteristic heavy odor; there is loss of appetite, due principally to the pain produced by chewing and the contact of food with the ulcerated mucous membrane; thirst is moderately increased; and the bowels are normal, or inclined to constipation. The little sufferer is restless and sleeps badly. The pulse is feeble, and there are other evidences of general debility, but there is little febrile reaction, the temperature, even in well marked cases, rarely reaching a higher marking than 99.5° F. in the evening.

In severe and protracted attacks, the ulcers increase in breadth and depth, become covered with a gray or brownish pulp, and the teeth, deprived of the support of the gum, grow loose and are easily removed from the alveoli. Sometimes the periosteum of the jaw is destroyed, and more or less extensive necrosis results. Exceptionally, in very weak and badly nourished children, the stomatitis runs into actual gangrene or noma.

The symptoms, ordinarily, reach their height in from two to four days, and, under proper treatment, disappear in as many more, the ulcers cleaning off and healing without cicatrization. Severe or badly managed cases go from bad to worse for a time, and rarely recover under three or four weeks, during which the suffering is extreme. Those involving necrosis of the jaw, and those terminating in noma, run a still more protracted course.

DIAGNOSIS.—The appearances of the gums before ulceration, the position in which this process begins, the character of the individual ulcers, and the odor of the breath, furnish a train of symptoms distinguishing ulcerative stomatitis from any other affection of the mouth.

The PROGNOSIS, in the vast majority of instances, is most favorable. When necrosis of the jaw occurs the duration is greatly prolonged, but ultimate recovery is the rule. Intercurrent noma, on the contrary, often leads to the death of the child, and under the best of circumstances leaves its traces in permanent deformity of the face.

TREATMENT.—The first step is to improve the sanitary surroundings of the patient, or, if this be impossible, to remove him to healthy quarters. The importance of cleanliness, fresh air, and sunlight, are not to be lightly estimated.

The diet should be liquid, but nutritious. Apart from the fact that solid food will be refused on account of the pain caused by mastication, milk and animal broths are better suited to the somewhat enfeebled digestive powers, and should be relied upon entirely. Cool water ought to be allowed in sufficient quantities to satisfy the thirst.

Of drugs, chlorate of potassium is the most important, since it ranks almost as a specific for this disease. It may be given alone, simply dissolved in water, or combined with dilute muriatic acid, as in this prescription :—

   ℞. Potassii Chloratis, ... gr. xlviij.
    Acidi Muriatici dil., ... f ℨ j.
    Syrupi, ... f ℥ ss.
    Aquæ, ... q. s. ad . . f ℥ iij.
   M.

   S.—One teaspoonful, diluted, every two hours, for a child three years old.

In this combination the chlorate of potassium, being eliminated by the salivary glands, constantly comes in contact with, and acts as an alterative upon, the ulcers. The muriatic acid aids digestion, and acts as a tonic. If a more decided tonic effect be required, one-quarter to one-half of a grain of sulphate of quinia may be added to each dose.

Chlorate of potassium, too, constitutes the main element of the local treatment. Its action is somewhat improved by the addition of carbolic acid, as in the following wash:—

> ℞. Potassii Chloratis, ... gr. lxxx.
> Acidi Carbolici, ... gr. ij.
> Glycerinæ, ... f℥j.
> Aquæ, ... q. s. ad . . f℥viij.
> M.

A bit of absorbent cotton saturated with this wash should be thoroughly applied to all the ulcers at least once in every hour; or, at the same intervals, the child may take a quantity into the mouth, move the cheeks and tongue in such a way as to bring it in contact with the whole mucous surface, and then expel it. Should there be much pain, a four per cent. solution of cocaine may be applied to the ulcerated surfaces two or three times daily. I have also had good results from salicylate of sodium applied as a wash at intervals of two hours.

After the ulcers have healed, the specific treatment may be discontinued, and the patient placed upon a simple tonic, as ferrated elixir of cinchona, in doses of half a fluidrachm three times daily, until the health is perfectly restored.

As additions to this treatment, iron and stimulants will be required in severe and protracted cases. The tincture of the chloride is the best form of iron. It should be given in doses of three drops (♏iss) every two hours for a child three years old, and may be combined very well with the mixture of potash, acid and quinine (pp. 134–5). The best stimulant is whiskey, in doses of one-half to one teaspoonful, in milk or water, every three or four hours. Indolent ulcers may be stimulated to heal by touching them lightly with a solid stick of nitrate of silver, the parts being first anæsthetized by cocaine. Loosened teeth must always be allowed to remain in position, as they often become firm again after the termination of the disease.

When necrosis occurs, no change is necessary in the general

plan of treatment. Especial attention, however, must be paid to the cleanliness of the mouth, and poultices of flaxseed meal should be kept constantly applied to the cheek of the affected side. Surgical interference may become necessary.

4. GANGRENOUS STOMATITIS—NOMA.

This affection consists of a rapid gangrenous destruction of the cheek and adjacent parts, occasionally beginning on the lips, but usually near one corner of the mouth. It is generally asymmetrical, the left cheek being attacked in the majority of instances, but sometimes both cheeks are simultaneously involved.

ETIOLOGY.—Noma is, fortunately, an uncommon disease. Sucklings seem to be exempt from it, and most of the cases occur between the ages of two and twelve years. Girls are more liable to be attacked than boys. It is always of secondary origin, following severe maladies, such as measles, typhoid fever, gastrointestinal catarrh, ulcerative stomatitis, scarlet fever, smallpox, broncho-pneumonia, tuberculosis, protracted intermittent fever and whooping cough. This order also represents the etiological activity of the diseases mentioned. These, then, may be looked upon as predisposing causes, but despite the presence of any one of them, noma only occurs in those children who have been previously weak, ill-housed and ill-nourished.

There is no evidence to show that it is contagious, though it sometimes occurs as an endemic in overcrowded hospital wards and children's homes. These endemics may be explained in the same way as similar outbreaks of ulcerative stomatitis.

SYMPTOMS.—During convalescence from measles, or other of the diseases mentioned, a nodule, from a quarter to half an inch in diameter, appears spontaneously upon the child's cheek, in the neighborhood of the corner of the mouth. This can be easily detected from the outside, but it is best felt by opening the mouth and grasping the cheek between the thumb and forefinger. It is extremely hard, and very sensitive, especially at the periphery.

If the case be seen during the first few hours, the mucous membrane over the mass will be observed to be converted into a flat, ichorous bulla. Usually, however, this membrane is found hanging in ragged shreds from a black, gangrenous base. The skin over the induration is pale or mottled with purple spots, tense, and shiny as if oiled.

After twenty-four hours, the investing integument becomes bluish, the epidermis scales off, and a black eschar forms. This has a tendency to shrink, and in so doing, leaves a linear depression with ichor, which separates it from the healthy skin. Notwithstanding this line of demarcation, the tissue destruction rapidly extends, both in superficial area and depth. Soon the cheek is perforated, and a dirty, stinking, ichorous saliva, filled with shreds of broken-down tissue, flows out beside the eschar and over the cheek. At the same time the lips, chin, and uninvolved portions of the cheek become œdematous, the skin being tight and glistening, and the adjacent cervical glands enlarged.

At the very outset there are few constitutional symptoms. The child complains of little or no pain, persists in his amusements, has a good appetite, a temperature but slightly above the normal, and a pulse but moderately increased in frequency. As the eschar forms, the scene changes, symptoms of constitutional depression setting in. The face is pale and expressionless on the affected side, the skin cool and dry, the pulse feeble and frequent—sometimes counting 120 or 140 beats per minute—and there is œdema of the feet. The mind is apathetic, no complaints of pain are made, and at most a sense of discomfort is indicated by constant whimpering. The mouth is held partly open, the breath is fetid, the teeth and tongue are covered with sordes, and there is an abundant flow of bloody or dark-colored saliva. Severe hemorrhage never occurs, as the blood-vessels are closed at an early stage. The appetite is often retained, the thirst is intense, and the bowels are usually relaxed. In spite of the food taken, the strength rapidly declines, sometimes, though, it is wonderfully retained, the patient being able to sit up, and even leave his bed, until a few hours prior to death.

The air of the sick-room has a characteristic gangrenous odor. Perforation of the cheek occurs about the third day of the disease, and many cases die at this time. Others linger until the end of the first or second week. Under these circumstances the gangrene invades the lips as far as the median line, the corresponding ala of the nose, and the cheek as far as the lower eyelids, the tragus, and the inferior border of the lower jaw. Extending inward, the gums and periosteum of the jaws are destroyed; the bone becomes necrotic, and the teeth so loose that they can be readily pushed out by the finger, together with pieces of the alveoli. Finally the cheek is cast off in large, black sloughs, leaving huge openings, with black, ragged, and indurated edges, through which the blackened and necrosed bones and loosened teeth can be seen. The child's face is then unrecognizable; the symptoms of constitutional depression are greatly intensified; there is delirium, profuse diarrhœa, purulent and even gangrenous infiltration of the lungs, and occasionally, also, gangrene of the genitalia, in females. Death is the only result to be expected.

Exceptional cases do recover. In these, the gangrenous edges become clean and covered with granulations, the necrosed bone is thrown off, and after months, cicatrization takes place, with great disfigurement.

PATHOLOGY AND MORBID ANATOMY.—The fact that noma makes its appearance uniformly at one point, on the cheek, and is unilateral, suggests a localized, causative lesion. The most natural theory, that of embolism of a large arterial branch, due to weakness of the cardiac muscle or increased coagulability of the blood—effects of the primary disease—is untenable, because, with the given conditions, emboli ought, at least occasionally, to be found in other positions, which does not happen. It is necessary to look rather to the nerves; namely, the trifacial, the facial, or the vaso-motors. That the gangrene is due to a lesion of one of these, seems to be borne out by experiments. Thus, Magendie found that division of the trifacial in dogs caused destruction of the corresponding eyeball, and half of the tongue became dry, brown and fissured, the gums spongy and hemorrhagic, and the teeth loose. "In animals tenacious of life, the batrachians, for example, the soft portions of the face are cast off in shreds, just as in spontaneous gangrene. After three or four weeks only one-half of the face remains." *

The body of a child dead from noma has a gangrenous odor and decomposes quickly ; the skin is shriveled, and the face and the feet are œdematous. The gangrenous parts are converted into a blackish-brown mass, and the maxillary bones are naked, brownish in color and brittle. The nerves, when examined microscopically, are yellowish in color but unaltered in structure, and the blood vessels are thickened and filled with thrombi. In the uninvolved parts of the cheek there is a dense exudation, while the palate, tongue and tonsils are swollen and covered with black scales and crusts. The lungs are the seat of hemorrhagic infarctions, lobular or metastatic lobar pneumonia, and sometimes gangrene. The intestines are catarrhal. Evidences of the primary disease may also be present ; for example, the lesions of typhoid fever or dysentery. Noma of the genitalia, though rare, is occasionally encountered. I have seen several cases within the last three years. The local appearances and the clinical history as to causation and so on, correspond with what has already been stated. The possibility of such an occurrence should be borne in mind as a matter of interest.

DIAGNOSIS.—Noma is readily distinguished from other oral affections by its course, its peculiar and almost uniformly identical seat and its well marked local features.

Ulcerative stomatitis is the only other of the class at all likely to be confounded with it. This always begins with ulceration of the gingival margin, and when the cheek becomes involved, the ulcers situated there are linear in shape and have a grayish floor. There is no sloughing or gangrene of the mucous membrane. The cheek never presents a circumscribed induration, being at most simply œdematous. The skin shows no tension,

---

* Vogel, Ziemssen's Cyclopædia, Vol. vi, p. 812.

unctuous appearance or discoloration, and perforation of the soft parts never occurs. The breath is fetid but not gangrenous; salivation is less, and the saliva, though sometimes bloody, is not mixed with shreds of gangrenous tissue. The course is much less rapid, and the ulcers, while they extend in area, retain the same appearances throughout. Finally, the general symptoms are distinctive, the results of treatment are most satisfactory, and a fatal termination is extremely uncommon.

Malignant pustule closely resembles noma. The former, however, always begins on the exterior, involving the epidermis first and extending through the successive layers of skin to the deep structures.

The PROGNOSIS is most unfavorable. Vogel sets the mortality at 80 to 90 per cent.,* and out of one hundred and two cases that came under the observation of Steiner,† only four recovered. Death may occur at any time between the third and fourteenth day; a rapid course, however, is very much more frequent than the reverse. Even when recovery does take place, the patient is permanently disfigured by scars, or crippled by the development of ectropion, or of restricted movement of the jaw, in consequence of cicatricial contraction, or by the loss of teeth and portions of the maxillary bones. Such cases also drag through a very protracted convalescence.

TREATMENT is most unsatisfactory. Something can be done in the way of prophylaxis, by a proper management of the known predisposing diseases. Secure sound hygiene in the sick-room; give good nourishment, and avoid the abuse of mercurials and debilitating treatment generally.

If, notwithstanding these precautions, noma appears, it is of the first consequence to maintain the strength by the use of concentrated liquid food, tonics and stimulants. When perforation of the cheek takes place, the act of swallowing is mechanically interfered with. It is necessary then to resort to nutritious

---

* Ziemssen's Cyclopædia, Vol. vi, p. 814.
† Diseases of Children, p. 218.

enemata, suppositories of quinia, and even the rectal administration of stimulants.

The room in which the treatment is conducted must be large, airy, and so situated as to be exposed, for a part of the day, at least, to the sun's rays. In summer the windows should be kept constantly open, and in winter they must be raised for at least fifteen minutes several times daily, the patient being warmly covered in the meantime. The air of the chamber must also be kept as pure as possible by the use of disinfectants. For this purpose cloths saturated with a solution of chlorinated soda or with Platt's Chlorides, may be hung about the bed.

Early cauterization with the hot iron, with strong sulphuric or muriatic acid, or the solid stick of nitrate of silver, is recommended. All sloughs must be removed by scissors. The gangrenous spot should be frequently bathed with a strong solution of chlorate or permanganate of potassium, carbolic acid, or chlorinated lime. Pieces of lint soaked in one of these solutions, may, with advantage, be left in contact with the ulcer, if the child will tolerate a fixed dressing. In case of perforation, much of the wash will run into the mouth, and care must be taken to prevent its being swallowed. The mouth must be kept as clean as possible by repeated syringings with a solution of chlorate of potassium and carbolic acid, ten grains of the former and one grain of the latter to the fluidounce.

When recovery occurs, loss of tissue, and the deformities resulting from cicatricial contraction, may be, to some extent, remedied by plastic surgery.

5. PARASITIC STOMATITIS—THRUSH.

Thrush is characterized by the appearance of numerous, rapidly-growing, white, curd-like flakes upon the oral mucous membrane; the latter being in a more or less catarrhal condition, injected, swollen, hot and tender to the touch. The flakes are due to the development of a peculiar vegetable parasite, the

Oïdium Albicans. Thrush occurs both as a primary and a secondary affection.

ETIOLOGY.—The disease attacks sucklings, and is met with most frequently during the first three months of life. Children nursed at a healthy breast are never attacked, and rarely those belonging to the well-to-do classes, because of the attention given to cleanliness of the mouth and feeding apparatus.

Neglect of this fundamental principle, cleanliness, lies at the foundation of every case of primary thrush. Foulness of the mouth implies a condition in which the secretions and the food clinging to the mucous membrane are undergoing acid fermentation, a necessary precedent to the development of the fungus. Given this condition, thrush originates by contact with other cases, through the media of bottle tips, spoons, tumblers, or cups used in common, and is produced, *de novo*, by carelessly kept tips and bottles, and, more frequently still, by long nursing tubes which are always partially clogged, acid and ill smelling.

The secondary form has the same direct causes, but arises during the course of gastro-intestinal disorders, especially those resulting from a too free dietary or the over-use of farinaceous food. It may occur, also, as a complication of diseases that greatly impair the general nutrition, as the exanthemata, tuberculosis, scrofula, spinal caries, etc. The disease is most prevalent during the summer months.

The MORBID APPEARANCES are the same in both forms. Prior to the appearance of the flakes, the oral mucous membrane is purplish-red and sticky, and its secretion is acid in reaction. The latter shows, under the microscope, numerous spores, egg-shaped, sharply-outlined, and hanging together in twos and threes. Soon, isolated white points, as large as a pin's head, appear on the inside of the cheeks. These rapidly increase in extent and number; involve other parts of the mucous membrane, and often as early as the second or third day, large, white flakes are formed. Later still, the whole cavity of the mouth ; and, in some cases, even the pharynx and œsophagus are covered.

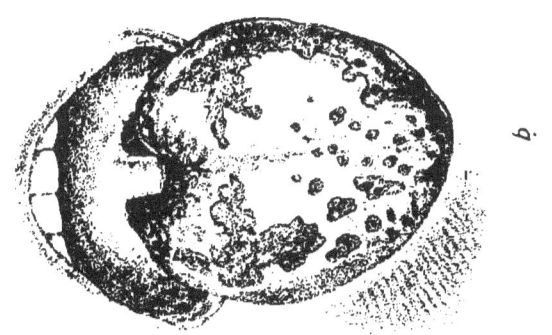

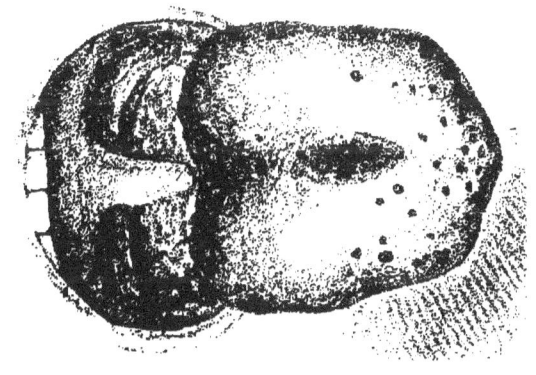

The patches, at first white, may become yellow, and sometimes brown, if bleeding occur from rough handling of the mucous membrane. Their surface is somewhat velvety, and they are soft, breaking down like curd under the finger. During the first few days they adhere firmly to the mucous membrane; afterwards they become quite loose, and can be wiped off readily, leaving the epithelial surface intact.

Microscopic examination of the fully formed patches reveals numerous irregularly developed fungoid filaments, with laterally branching arms and buds, interpersed with round or oval sporules, and imbedded in an amorphous, granular mass. A hardened section of a patch and the mucous membrane to which it adheres, shows, in addition to these characters, a partial loss of epithelium, and a tendency on the part of the filaments to penetrate into the mucous glands and between the cells of the deeper layers of the epithelium.

The fungus seems to grow only upon squamous epithelium, so that it is never found in the nasal cavities, the larynx or the trachea, and the presence of loose masses of it in the stomach may be regarded as accidental. On the other hand it may be formed upon the lower segment of the rectum, the female genitals, and on abraded surfaces about the mouth, chin and neck.

SYMPTOMS.—The primary form begins with heat, dryness, tenderness, slight swelling and uniform redness of the mucous membrane of the mouth. The redness is combined with a purple tinge which is most marked on the dorsum of the tongue. Here, too, prominence of the fungiform papillæ is noticeable. The child takes his food moderately well, but the meals are frequently interrupted on account of the pain caused by sucking. He is fretful and sleeps poorly. The bowels are moderately relaxed, the stools being liquid and yellow in color. In the course of twenty-four hours the thrush patches appear on the inside of the cheeks and then extend to the lips, tongue and palate. While extending, they increase in size, though they usually remain isolated and rarely overstep the limit of the posterior border of the soft palate. With the appearance of the patches, there is increased fretfulness, more pain on sucking, occasional vomiting and frequent evacuation of the bowels, the motions becoming green and acid. At some period, varying from six to twelve days from the beginning of the disease, the patches become loose and are removed by the act of sucking or in making applications to the mouth. The mucous membrane is left red but free from ulceration, and it soon returns to the normal condition. At the same time the general symptoms subside, and health is soon restored. Sometimes there are several crops of the fungus, but those coming last, being less firmly rooted than the first, are dislodged quickly and seldom prolong the course of the disease beyond two or three days.

In secondary thrush a history of previous gastro-intestinal or other disease will be obtained, together with an account of an immediately preceding diarrhœa and fever. Sometimes, however, the local symptoms are the first indications that the weak, badly-nourished child is ill. The preliminary catarrh of the mouth is very marked, the mucous membrane being intensely red and shining. The patches are thick, are apt to change from a white to a yellow or brown color, soon cover the whole oral cavity and frequently extend into the pharynx and down the œsophagus. They retain their attachment to the mucous membrane for a much longer period than in the idiopathic form. When they fall off they are quickly replaced by others, and a succession of crops is the rule up to the termination of the case in death. The mouth is hot, dry and tender to the touch, and throughout presents an acid reaction to chemical tests.

The appetite is gradually lost; there is vomiting, either occasional or so constant that every morsel of food taken into the stomach is rejected at once, and obstinate diarrhœa, the stools being numerous, liquid, green in color and acid. The abdomen is distended by flatus, and is tender to pressure, particularly in the epigastrium and right iliac region. Colic is a constant and annoying symptom. The pain is most severe just before or at the moment of an evacuation of the bowels.

The skin is hot and dry and the frequency of the pulse increased, a rate of 120, 140 or 160 beats per minute being not unusual.

The child sleeps badly, is restless and fretful, and when the pharynx is covered by the fungus, has a muffled, hoarse cry. The skin grows pale and inelastic, and the folds of the nates, the inner surface of the thighs and the heels are reddened, and eventually excoriated by the contact of the acid fæces. The strength and flesh are lost rapidly, the anterior fontanelle sinks, the eyeballs lie deep in their sockets and the nose and chin are pointed. Toward the latter end of the attack, which is rarely protracted more than a few weeks, the patient assumes the facies of a little, wrinkled old man. His skin is cool, and he lies in an apathetic condition on the bed or nurse's lap, with scarcely enough strength to whine over his suffering until death, from atrophy, ends the miserable life.

DIAGNOSIS.—Fragments of curdled milk adhering to the soft palate and cheeks, resemble very closely the thrush patches in their earlier stage. The normal condition of the mucous membrane, and the readiness with which the curds can be wiped away, constitute the distinctive characteristics.

Aphthous stomatitis bears a certain superficial likeness to thrush, but the differentiation is easily made by noting the fact that the yellowish-white spots of the former are depressed below the surface of the mucous membrane, being, in reality, the floors of ulcers, which in time are bounded by dark red borders.

Microscopic examination is always the crucial test, and the presence or absence of thallus-fibrils and spores decides the question as to the nature of any deposit in the mouth.

PROGNOSIS.—The primary form is a very trifling affection and almost uniformly ends in recovery. In the secondary form the result is very often unfavorable. This is especially apt to be the case when the disease occurs in a child who has been much weakened by a continued course of improper food. Here, the hope of improvement depends upon the rapidity and completeness with which new material for nutrition can be introduced into the system. Anything, therefore, that tends to prevent this introduction deprives the child of his only chance of recovery, and the existence of thrush implies a condition of the digestive tract extremely unfavorable to the assimilation of food. Attendant diarrhœa aids, too, in precipitating the fatal result.

The mere presence, then, of the thrush patches is not to be regarded with as much anxiety as the conditions accompanying their formation.

TREATMENT.—Much may be done to prevent the development of thrush by keeping the mouth clean. A strict rule should be made to wash out a child's mouth directly after each meal. This is best done by a large camel's-hair brush or a soft rag moistened with warm water. The bottles and tips must also be kept immaculately clean. An equally important precaution is to select a proper diet. The question of diet is, of course, a very comprehensive one, and no further consideration can be given it in this place than to state the general law. Babies under six months old, who are unfortunate enough to be deprived of their mother's milk, must be fed upon cows' milk so prepared that it may resemble as nearly as possible human milk. If farinaceous articles be used they must be employed with the object of rendering the cows' milk more digestible by separating the curd, and not as the staple of the food. The regularity and the length of the intervals between meals; the selection of the proper quantity of food, and the preparation of each portion immediately before it is given, are matters worthy of the most careful attention.*

Such measures, together with attention to general hygiene, constitute an important part of the curative treatment after the appearance of the fungus. In idiopathic or in mild cases all that is required in the way of general treatment will be an alkali

---

* See Part II.

combined with a digestant, as in the formula already given for catarrhal and aphthous stomatitis,* or if the stools be numerous, green and very acid :—

 ℞. Magnesii Carbonatis, ... ʒj.
  Syr. Rhei Aromatici, ... fʒij.
  Syrupi, ... f℥ss.
  Aq. Menthæ Piperitæ, ... q. s. ad f℥iij.
 M.
  Sig.—Teaspoonful every two or three hours, for a child three to six months old.

The local treatment consists in keeping the mouth perfectly clean. It should be thoroughly washed every hour at least, with a soft rag wrapped around the finger and wet with warm water. Immediately afterward, either one of the following lotions may be applied, upon a fresh piece of rag :—

 ℞. Sodii Boratis, ... gr. xxx.
  Glycerinæ, ... fʒj.
  Aquæ, ... q. s. ad f℥j.
 M.
 ℞. Sodii Hyposulphitis, ... gr. x.
  Aquæ Rosæ, ... f℥j.
 M.
 ℞. Acid. Carbol., ... gr. ij.
  Sodii Salicylat.,
  Sodii Boratis, ... āā . . . . gr. xxx.
  Glycerinæ, ... fʒij.
  Aquæ Rosæ, ... q. s. ad f℥j.
 M.

It is essential immediately to destroy the rag or other instrument used in cleansing the mouth or in carrying the lotion.

The same principles are applicable to the treatment of secondary thrush. Every means must be employed to arrest the vomiting and diarrhœa, to improve the digestive powers and maintain the strength. There is, however, but one promising remedy for this form of the disease, namely, the employment of a healthy wet-nurse.

---

* See page 126.

148 DISEASES OF DIGESTIVE ORGANS IN CHILDREN.

6. DENTITION.

ERUPTION OF THE TEMPORARY TEETH.

The eruption of the twenty milk teeth may, like other physiological processes, be unattended by noticeable symptoms, but in many instances it is accomplished with difficulty, giving rise to disturbances which, on the one hand, may be so trifling as simply to annoy the infant, or on the other, so serious as to endanger life.

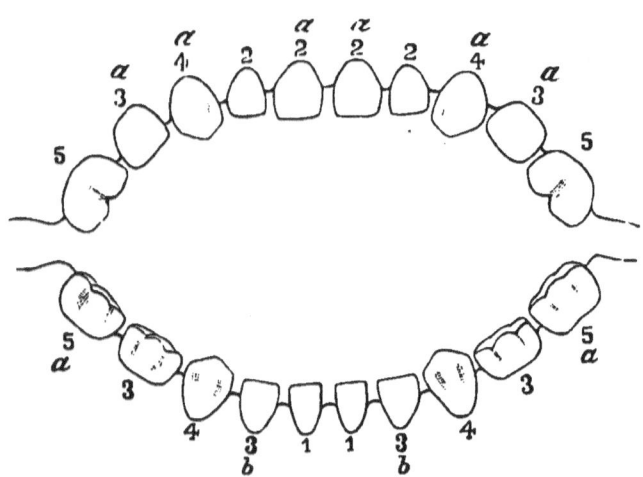

FIG. 6.

DIAGRAM SHOWING ERUPTION OF MILK TEETH.

1. 1. Between the 4th and 7th months. Pause of 3 to 9 weeks.
2. 2. 2. 2. Between the 8th and 10th months. Pause of 6 to 12 weeks.
3. 3. 3. 3. 3. 3. Between the 12th and 15th months. Pause until the 18th month.
4. 4. 4. 4. Between the 18th and 24th months. Pause of 2 to 3 months.
5. 5. 5. 5. Between the 20th and 30th months.

Normally the teeth are cut in groups, each effort being succeeded by a pause or period of rest. The above diagram and table show the grouping, the date of eruption, and the duration of the pauses. The numbers, 1 to 5, indicate the groups to which the individual teeth belong and their order of appearance, and the letters, *a* and *b*, the precedence of eruption in the different groups.

The dates given in the above table show the time within which the different teeth may be naturally expected. In regard to the period given for the eruption of the lower central incisors I would state that the fourth month, although an early, is not a very rare time for their appearance. For example, in the past winter alone, I have seen five cases in which these teeth pierced the gum at this age.

The pauses mentioned are, to say the least, most helpful, giving the infant's system an opportunity to rest after each effort, to recover from any coincident illness and to prepare for the next strain.

Even under normal conditions the edges of the gums lose their sharpness and become swollen, rounded and reddened as the teeth approach the surface. At the same time the saliva is increased in quantity, and the mouth is unnaturally warm and the seat of abnormal sensations, evidenced by the tendency to bite upon any object that comes to hand—in other words, there is a condition of mild catarrhal stomatitis. The consequent discomfort, though, is not sufficient to interfere with the child's appetite, good humor, or sleep, and when, after a few days, the margin of the tooth is free, all the local symptoms vanish.

Abnormal dentition is manifested either by departures from the laws of development already stated, or by actual difficulty in the process of cutting.

The standard rules for the eruption of the teeth may be departed from in three ways :—

1st. The appearance of the teeth may be premature. Children may be born with one or more of their teeth already cut. These are usually imperfect, and soon fall out, to be replaced at the proper age by well-formed milk teeth. Sometimes, however, they remain permanently, as in a case that came under my own observation. Natal teeth are always incisors. Instances of the lower central incisors being cut in the third month are not uncommon. Girls are more apt than boys to cut their teeth early, and, as an early dentition is likely to be an easy one, the occurrence is to be looked upon as fortunate.

2d. Dentition may be delayed. This deviation is more frequently seen and of more consequence than the first. Bottle-fed babies, as a class, are more tardy in cutting their teeth than those reared at the breast. With such, though healthy in every respect, a delay of one or two months is a common and not at all a serious event. On the contrary, whatever the method of feeding, if no teeth have appeared by the end of a year, it may be assumed that the child's general nutrition is faulty, or that rachitis is present. Delay does not necessarily imply difficulty in cutting the teeth, though the two conditions are often associated.

3d. The teeth may appear out of their regular order. Bottle-fed infants are most likely to show this irregularity, which is of some importance as an indication of general feebleness. In other instances, however, it is merely a family peculiarity, and as such, bears no special significance.

Difficult Dentition is attended by two classes of affections, viz. : local, and sympathetic or reflex.

The third and fourth groups of teeth are most prone to make trouble, and when the child is born at such a time of the year as to bring their eruption during the hot months, illness of some sort may be anticipated. This is often dangerous, sometimes fatal, hence the popular dread of the "second summer."

LOCAL AFFECTIONS.—Catarrhal stomatitis, already referred to as a physiological occurrence, frequently becomes greatly intensified, and is sometimes associated with enlargement of the submaxillary and cervical lymphatic glands. Aphthous and ulcerative stomatitis develop much more rarely.

Ulceration of the frænum linguæ often takes place after the cutting of the first lower incisors, being due to direct and continuous friction of the sharp, new teeth. Usually a single, flat, round ulcer, the size of a linseed, is formed. The base is yellow and lardaceous, the edges red and infiltrated. At times it is sufficiently painful to interfere with the movements of the tongue, though at others so indolent as to escape notice entirely. Such ulcers, if left to themselves, disappear in from eight to ten days. Their course may be much shortened by touching them lightly with a point of nitrate of silver, and then applying a solution of chlorate of potassium three or four times each day.

Another very common condition is an excessive increase in the secretion of the fluids of the mouth, attributable to irritation of the mucous and salivary glands. The constant driveling of this fluid reddens and excoriates the skin of the chin and anterior part of the neck. It also soaks the clothing; and the consequent chilling of the thorax both excites and tends to keep up a catarrhal state of the bronchial and mucous membrane. In this way may be explained the frequent concurrence of driveling and severe cough in teething children. The etiological association is further proven by the fact that if the chest be protected by a piece of oiled skin or other waterproof material, placed inside of the clothing, the cough either does not develop or quickly disappears if it has been present. If these two results are excepted, slobbering is not to be regarded as unfavorable, for such children rarely have alarming brain symptoms, or severe intestinal catarrh.

The SYMPATHETIC EFFECTS of difficult dentition show themselves in affections of the eyes and ears, gastro-intestinal disorders, skin eruptions, nervous affections, and fever.

Conjunctival blennorrhœa arises frequently during the eruption of the upper canines; hence the common name, "eye-teeth." This complication may be attributed to an extension of irritation to the antrum of Highmore, the nasal passages, and, finally, to the conjunctiva. The eyelids swell greatly and rapidly, there is difficulty in opening the eye, a free secretion of a stringy, translucent mucus, and pain and photophobia; the ball remains intact. It is distinguished from true blennorrhœa by frequently being unilateral, by its non-contagiousness, and by the absence of the characteristic, dark-yellow, thick, purulent discharge. Recovery quickly follows the appearance of the point of the tooth, and no treatment is required beyond cleanliness and the application of dry warmth.

Otitis is not uncommon. It is probably due to irritation conveyed from the inflamed gum to the otic ganglion, and thence deflected to the vessels supplying the tympanic membrane. In consequence, this membrane becomes congested, and there is great pain. When the irritation persists, suppurative inflammation is developed in the tympanic cavity, the drum is perforated, and pus flows from the external auditory meatus. The treatment belongs to the domain of aural surgery.

Simple diarrhœa, with yellow, pultaceous or somewhat greenish stools, is a very common occurrence. It is due either to an intestinal catarrh of mild grade, or to the swallowing of large quantities of saliva, the saline constituents of which act as laxatives. Under ordinary circumstances, attention to the diet is all that is necessary for its relief. If vomiting be associated or the stools be green, bicarbonate of sodium, carbonate of magnesium, or subcarbonate of bismuth, with an aromatic, may be ordered, and if there be griping it is well to clear out the bowels by a dose of castor oil. Opium and powerful astringents are to be avoided.

While the diarrhœa of teething is ordinarily of little moment it should receive careful treatment, for there can be no doubt that during dentition the mucous membrane of the digestive tract, sympathizing in the oral irritation, becomes more susceptible to such irritants as badly digested or improper food, and that a condition of simple catarrh is apt, under such influences, to pass into one of follicular enteritis, with secondary involvement of the mesenteric glands. This is soon followed by general atrophy (marasmus), frequently terminating in death. Hand-fed infants are particularly prone to be so affected. The symptoms are vomiting, the formation of thrush deposits in the mouth, anorexia, thirst, tympanitic distention of the abdomen, diarrhœa, and rapidly increasing emaciation. The very numerous fæcal evacuations are liquid, have a penetrating, cadaverous odor, and excoriate the anus and surrounding parts. This condition demands, and too often resists, the most careful treatment.*

---

* See sections on the diseases attended by diarrhœa.

When vomiting occurs as a complication, it may often be relieved, in sucklings, by shortening somewhat the time the child is allowed to lie at the breast, and thus preventing over-distention of the stomach. With hand-fed infants the same result may be attained by judiciously lessening the quantity of food given at each meal and reducing its strength, by the addition of lime-water. A teaspoonful of lime-water every half hour or hour will aid in checking the vomiting, and the following prescription is excellent:—

  ℞. Liq. Calcis,
    Aq. Cinnamomi, ... āā f℥j.
  M.
  S.—One teaspoonful every half hour or hour, for a child of
    seven months.

Several forms of cutaneous eruption attend difficult dentition. They usually appear in children with fair, delicate skins, show some hereditary and family tendency, and must be considered of vaso-motor origin.

The eruption may be present during the cutting of one set of teeth only, or may continue during the entire dentition. In the latter case it improves greatly or disappears in the pauses. The form usually remains unchanged throughout.

The varieties that may exist are:—

(*a*) *Urticaria.*—This consists of the appearance of a varying number of wheals, chiefly on the trunk and extensor surfaces of the limbs. The wheals are slightly prominent, flattened elevations of the cuticle, varying in size from that of a pea to a bean, somewhat paler than the normal skin, but surrounded by red areolæ gradually fading at the periphery into the healthy skin color. They develop suddenly, making their appearance especially when the child becomes warm in bed, or is overheated, and are the seat of some burning and intense itching. They last a few hours, and disappear rapidly, leaving the epidermis intact, except for accidental injuries occasioned by scratching.

Relief comes after a careful regulation of the diet. If the child be bottle-fed the milk must be well guarded with lime-water.

If older, and fed on mixed food, it is well to avoid farinaceous articles. The bowels, usually constipated, should be regulated by saline laxatives, and much benefit will result from the use of effervescing citrate of potassium, in doses of one-fourth of a teaspoonful (equal to gr. v), three times daily, for a child of eight months to a year old. The most soothing local application is:—

R. Sodii Hyposulphitis, ... ʒij-iv.
   Aq. Rosæ, ... fʒ xvj.
M.

This is to be sponged over the wheals at short intervals or applied by moistened compresses.

(*b*) *Strophulus.*—This presents itself in two forms, *strophulus intertinctus* or red gum, and *strophulus albidus* or white gum.

Red gum consists of an eruption of prominent, red papules, interspersed with small patches of erythema. The papules are scattered over the whole body, but are most abundant on the face, back of the hands and forearms; they vary considerably in size, though they are rarely larger than a pin's head; they are the seat of considerable itching, and after their disappearance there is trifling desquamation. It occurs in successive crops, each crop lasting from seven to fourteen days.

In white gum the papules are pearly-white instead of red, and each elevation is surrounded by a faint red areola. These papules are most numerous on the face, neck and breast; they are smaller than those of the former variety, but like them appear in crops, which remain about seven days.

With the exception of attention to digestion, which is often deranged, little treatment is required. Dusting with bismuth or lycopodium, or anointing with cold cream or simple cerate will relieve the itching.

(*c*) *Eczema.**—In teething children this affection may be found in any one of its various forms. It may extend over the entire surface or be limited to a small area; its usual position, however, is the face and scalp.

---

* This is the form of skin disease usually covered by the term "tooth rash."

Facial eczema begins with redness, induration and roughness of one or other cheek. There is also intense itching, and even at this early stage the child, if unhampered, will scratch the affected part until it becomes raw and bleeding. Soon minute vesicles or pustules form.

If the dermatitis be slight the eruption is vesicular. The vesicles may heal spontaneously, or being broken mechanically, they discharge a serous fluid which dries into thin scales or lamellated crusts. These fall off, leaving a reddened, moist, delicate surface, still intensely itchy, and soon to be covered by another crop of vesicles, which undergo the same changes.

If the inflammation of the skin be more severe, the eruption is pustular. The pustules on breaking, exude a material that dries into thick, yellowish-brown crusts, often extensive enough to cover the whole face, except the eyelids and nose, as with a mask. In hardening, cracking is apt to occur, with the exposure of the red and bleeding derm. When the crusts fall, or are scratched off, an intensely red, weeping surface is left, which is soon recovered by thinner, lighter-colored crusts, resembling those seen in the vesicular form. The neighboring lymph glands are always somewhat enlarged and tender. The process of crusting and cleaning off repeats itself again and again, frequently dragging through the entire dentition, though subject to improvement or even completely healing during the pauses.

Sometimes, especially toward its close, the eczema assumes the papular form.

The disease is attended by sensations of discomfort and tension about the face. The itching is continuous but subject to exacerbations in which it is most difficult to keep the child from tearing the skin with his nails. One often leaves a patient with his face covered by a thick crust, to find him at the next visit with clean, red and moist cheeks scarred by nail-scratches, from which the blood drops upon his clothing. The exacerbations occur chiefly at night. Under the circumstances it is but natural that the little sufferer should be restless and peevish, should sleep badly, and, from the latter cause, should gradually fail in health and strength. The appetite, too, is often diminished, the digestive powers impaired, and the bowels confined. There may be slight febrile reaction toward evening.

Eczema of the scalp—crusta lactea—involves at first only a small area, but has a marked tendency to spread. In some cases it appears in the form of minute, disseminated vesicles; these break and exude their serous contents, which harden into thin, scaly crusts. If these be removed the skin beneath will be found to be reddened and moist. In others the vesicles rapidly change into pustules, and when these are broken the puriform fluid dries into thick, yellowish-brown crusts. The underlying skin is red, swollen, and painful, and the source of a constant formation of pus, which in drying adds to the thickness of the preëxisting crusts. These mat the hairs together, become, day by day, thicker, harder on the external surface and darker in color, from the admixture of dust or of blood flowing from wounds caused by scratching. As the hair grows it lifts the crusts from the scalp and the pus finds its way out, running over and excoriating the skin of the forehead and neck. From the heat of the head and exposure to the air partial decomposition sets in, attended by a disgusting odor, and in carelessly kept children myriads of lice appear. In well-marked instances the crusts completely encase the scalp. The occipital and other lymphatic glands are usually enlarged, and occasionally, in strumous children, suppurate. Crusta lactea is no less obstinate than facial eczema. When it runs a protracted course the hair may fall out, but the loss is not permanent.

There can be no question as to the propriety of healing eczema of the face or scalp as quickly as possible. The idea that the cure of the rash leads to more serious mischief, as meningitis or hydrocephalus, is merely a remnant of the long abandoned doctrine that disease is due to the presence of an evil spirit, which if driven from one place will attack another. The objections of some self-important and misinformed parents to any local application will usually be overcome by the assurance that the eruption is not to be driven in, but cured as it comes out.

AFFECTIONS OF THE MOUTH AND THROAT. 157

Under the best of conditions, however, tooth rash is difficult to cure, and requires both general and local treatment.

The general treatment demands in the first place attention to the diet. Bottle-fed babies must have their milk well alkalinized with lime-water, and older children must avoid farinaceous and heavy food. Next, conditions of acid dyspepsia, which, by the way, frequently exist, must be corrected by alkalies. Then it is essential to keep the bowels freely moved; for this purpose the following mixture answers well:—

    ℞. Magnesii Carbonatis,
        Mannæ Opt., ... āā gr. xl.
        Ext. Sennæ Fld., ... ♏lxxx.
        Syr. Zingiberis, ... f ℥ ss.
        Aquæ, ... q. s. ad f ℥ j.
    M.
    S.—One teaspoonful once or twice daily, for a child seven months to a year old.

Finally, certain medicines may be used with advantage. The emulsion of cod-liver oil and lacto-phosphate of lime in debilitated, strumous or rachitic children; syrup of the iodide of iron in the anæmic; Fowler's solution when the eruption becomes chronic.

For the successful local treatment of facial eczema it is necessary to have many resources at command; to make frequent changes as the applications lose their beneficial effects, and to persevere in spite of discouragement. Little progress will be made if the child be allowed to scratch the face at will. During the waking hours this is to be prevented by careful watching and diversion on the part of the nurse. At night by muffling the hands in thick, soft cloths, or by wrapping a napkin around the body in such a manner as to confine the arms, or, better still, by fixing the lower ends of the night-gown sleeves to the diaper, with safety-pins. The latter arrangement allows of some movement of the hands but prevents their being lifted to the face. The child quickly becomes accustomed to the partial restraint.

As soon as redness, induration and roughness of the skin indicate the onset of the disease, a mucilage made from sassafras pith (never musty or sour), is to be mopped over the surface with a soft rag, at intervals of an hour, or less. After this has dried for five minutes, a thick layer of the ointment of the oxide of zinc should be gently applied with the finger. If the officinal zinc ointment increase the inflammation, as it sometimes does in young or delicate-skinned babies, it must be diluted one-half, or even three-quarters, with cold cream. The nose, eyes and lips should be kept clean by wiping them with a moistened cloth, but the affected portions of the face must never be washed. Both the mucilage and ointment greatly relieve the itching, and in mild cases are sufficient to effect a cure.

When vesicles or pustules become abundant, zinc ointment is still most serviceable, though it is no longer well to use the mucilage. When the former ceases to be of service, one of the following may be tried:—

℞. Cerat. Plumbi Sub-acetatis, ... ʒij.
    Ung. Aquæ Rosæ, ... ʒvj.
M.

℞. Bismuthi Sub-nitratis, ... ʒj–ij.
    Adipis, ... ʒvj.
M.

℞. Glycerinæ, ... fʒij.
    Adipis, ... ʒvj.
M.

℞. Olei Lini, ...
    Liquor. Calcis, ... āā fʒj.
M.

℞. Acidi Salicylici, ... ʒss.
    Adipis, ... ℥j.
M.

AFFECTIONS OF THE MOUTH AND THROAT. 159

The first three are more sedative than zinc ointment, the last two more stimulating.

Hebra's diachylon ointment, useful in very moist eczema, is prepared in the following way:—

℞. Olei Olivæ, ... f ℨj.
Lithargyri, ... ℨij.
Coque 1. a. in molle, dein adde.
Olei Lavandulæ, ... ♏viij.
M.
S.—Rub the ointment over the affected spot two or three times daily, or, better, apply upon linen compresses.

When the patient does not bear salves well, a paste of oxide of zinc with glycerin may be ordered, and sometimes a dusting powder is beneficial, as :—

℞. Zinci Oxidi, ... ℨj.
Amyli, ... ℨj.
M.
S.—Dust over the affected part frequently.

In such cases, too, lotions act well. For example, the officinal liquor plumbi sub-acetatis dilutus, or :—

℞. Zinci Sulphatis, ... gr.j.-ij.
Aquæ Rosæ, ... f ℨj.
M.

℞. Sodii Boratis, ... gr.v.-x.
Glycerinæ, ... f ℨj.
Aq. Rosæ, ... q. s. ad f ℨj.
M.

In chronic cases, especially where there is hypertrophic thickening of the corium, a tendency to the squamous form, and but little redness, wonderful results will be obtained by using either—

℞. Ung. Hydrargyri Ammoniati, ... ℨj-iv.
Ung. Aquæ Rosæ, ... q. s. ad ℨj.
M.
S.—Apply three or four times daily.

Or this lotion—

    ℞. Hydrargyri Chloridi Corrosivi, ... gr.j–ij.
        Aquæ, ... f℥j.
    M.
    S.—To be penciled over the affected surface two or three times daily.

A great point in the treatment is to continue the local applications, not only until the eruption has disappeared, but until the skin has become perfectly smooth, soft and altogether normal. Again, to begin the applications early, as the ease and quickness of cure is in proportion to the acuteness and scantiness of the eruption.

Crusta lactea may be cured with more or less rapidity by cropping of the hair, softening the crusts with olive oil, applying a poultice of flaxseed over night, and using an ointment containing mercury, as—

    ℞. Ung. Hydrargyri Nitratis, ... ʒj–ij.
        Ung. Aquæ Rosæ, ... q. s. ad ℥j.
    M.
    S.—Rub into scalp three times daily.

    ℞. Hydrargyri Chloridi Mitis, ... gr. xxx.
        Cosmoline, ... ℥j.
    M.

In some instances penciling with the solution of corrosive sublimate, already mentioned, gives better results.

When there is intense redness of the scalp, it is often best to begin with oxide of zinc ointment, reserving the mercurial preparations until the inflammation is lessened. When the disease becomes chronic, with thickening of the skin, loss of hair, and the assumption of the squamous form, more stimulating applications are indicated, for instance—

    ℞. Ung. Picis Liquidæ,
        Adipis, ... aā ʒiv.
    M.
    S.—Apply thrice daily.

R. Saponis Mollis, ... ℥iv.
Alcohol dil., ... f℥ij.
Spt. Lavandulæ, ... ♏xv.
M. et cola.
S.—Rub into the skin by means of a piece of flannel or a brush, to remove scales, etc.

Inunctions of cosmoline and vaseline are often most beneficial in eczema, and the latter is a good vehicle for the preparations of mercury, zinc, lead, etc.

The most dangerous of all the complications attending difficult dentition are the disturbances of the nervous system. These are due to a great increase in the normally excessive susceptibility of the infantile nervous system to reflex influences. They embrace slight spasms of isolated groups of muscles and general convulsions.

Slight spasms are very common, and are observable chiefly during sleep. They are revealed by upturned eyeballs, and half-open eyelids, exposing more or less of the whites of the eyes; by contraction of the muscles of the face, causing a smile, and by twitching movements of the fingers and limbs. These manifestations can be prevented and sound sleep secured, by a warm foot-bath at bedtime, with five or ten grains of bromide of potassium.

A general convulsion arises suddenly and unexpectedly, and generally begins with tonic spasm of the muscles. The head is thrown back, the spine arched forward, the limbs become rigid, and breathing is suspended. Soon clonic movements set in; the face, which is flushed, becomes distorted; foam, sometimes bloody, appears upon the lips; the limbs are jerked about; the trunk writhes; the respiration is unrhythmical and sighing, and consciousness is completely lost. After a time, as the convulsion ends, the face becomes pale, the lips bluish, and the skin moist. There may be but a single convulsion, lasting a few moments, or there may be a number, varying in duration, and following each other at longer or shorter intervals. In some cases, the child remains in a convulsed condition for several days, passing from one fit into another, with only brief intervals of imperfect calm. Often they pass off without leaving any traces, but may be followed by paralysis, strabismus or idiocy, and may terminate in death.

The treatment consists in removing pressure from the gums by the use of the lancet, and the employment of nerve sedatives, as chloral and bromide of potassium. Should it be impossible for the patient to swallow, these drugs may be administered by the rectum, in the following enema:—

  R. Chloral Hydrate, ... gr.xij.
    Potassii Bromidi, ... ʒss.
    Mucilag. Acaciæ, ... f℥j.
    Aquæ, ... q. s. ad f℥iij.
  M.
  S.—One tablespoonful for a dose.

The injections are to be repeated every half hour—at the age of one year—until the convulsive movements are checked, or three or four doses are given. If this quantity fail, it is better to omit the chloral for two hours, and then resume it as before; in the meanwhile continuing the bromide of potassium.

So-called dental paralysis is uncommon, and when it does occur can usually be traced to coincident anterior polio-myelitis. It is sudden in its onset. After a restless night the child wakes with one arm helpless, or with one arm and one leg powerless; more infrequently one arm and both legs, or both arms are affected. The means for relief are salt-water baths with frictions, attention to the general health, and the use of tonics.

Elevations in temperature, with other evidences of slight febrile reaction, are very common during teething. The pyrexia is of short duration, moderate in degree, and readily controlled by hot foot-baths, and diaphoretics, as tincture of aconite root in small doses, or citrate of potassium.

In both classes of affections arising from difficult dentition, the question of the propriety of lancing the gums often arises.

AFFECTIONS OF THE MOUTH AND THROAT. 163

Many authorities advise postponement of incision until the gum becomes swollen, tense and shining, and the edge of the tooth is perceptible to the touch, just beneath the mucous membrane, or until yellowness of the gum and fluctuation indicate the formation of pus about the approaching tooth. This rule applies merely to cases in which there is little difficulty. For in many instances the greatest discomfort and danger are present while the tooth is yet some distance from the edge of the gum, forcing its way through the deeper and denser tissues. Here the only safe course is to cut deep, and liberate the tooth, repeating the opera-

FIG. 7.*

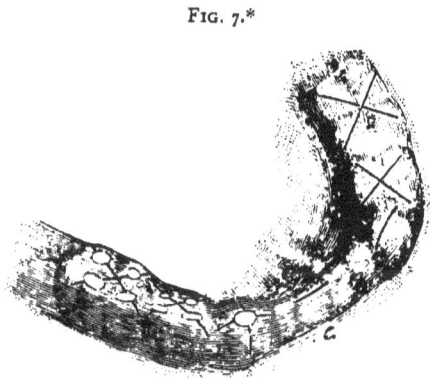

DIAGRAM OF LINES OF INCISION IN LANCING THE GUMS.

The above diagram plainly shows the lines of incision over the different teeth before eruption and after partial eruption.

tion if the original incisions heal—an event of little moment. My own practice in regard to lancing is guided entirely by circumstances. If there be fever, nervous irritability, sleeplessness, vomiting or diarrhœa during the progress of and dependent upon dentition, I invariably lance the gum,—provided the position of the tooth can be established by the touch—making the incision superficial or deep, according to the distance of the tooth from the surface. I feel confident that no one who has once

---

* From "Diseases Incident to First Dentition." James W. White, M. D., D. D. S.

attempted early lancing, and observed distressing and dangerous symptoms rapidly disappear, will ever hesitate a second time.

The form of incision is important. It must be linear in the case of the incisors and canines, and obliquely crucial in that of the molar teeth; the tissues must be divided until the edge of the lancet distinctly touches the tooth.

ERUPTION OF THE PERMANENT TEETH.

The eruption of the *milk teeth*, though physiological, as already stated, is so uniformly regarded as dangerous, that both physicians and parents congratulate themselves when infants under their charge pass through this process without trouble, or safely weather the various diseases that may arise during its course. The interval between the fourth and thirtieth months of an infant's life—the period of primary dentition—is an era of great and widely extended physical progress. The teeth are advancing; the follicular apparatus of the stomach and intestinal canal is undergoing development in preparation for the digestion and absorption of mixed food; the cerebro-spinal system is rapidly growing and functionally very active, and the organs and tissues of the whole body are in a state of active change. This period of normal transition must also be one in which there is great susceptibility to abnormal change, or disease, provided there be a causal influence at work. Such an influence may either originate outside of the body, as when there is exposure to cold, or to contagion, or come from within in the form of some perversion of a physiological process.

Difficult dentition stands prominent in the latter class. While the teeth are advancing, irritation of the gums very often produces stomatitis with fever, and the fever in turn leads to enfeebled digestion and impaired nutrition, conditions that open the way to catarrh of the mucous membrane of the bronchial tubes and gastro-intestinal tract, and to other intercurrent affections. Again the local irritation may be reflected through the widely extended connection of the dental nerves, and give rise to well-known disorders of distant organs, and tissues, for example, the brain, eyes, stomach, skin, and so on.

To appreciate this widely-extended nervous connection, it is only necessary to study Plate 1. This, which, by the way, is purely diagrammatic in character, illustrates the intimate anatomical relations existing between the trifacial, pneumogastric, and glosso-pharyngeal nerves, through the medium of the superior cervical ganglion of the sympathetic and its branches. This ganglion sends a branch directly to the jugular ganglion of the pneumogastric; another branch subdividing, sends one filament to join the ganglion of the root of the pneumogastric, while the other goes to the petrous ganglion of the glosso-pharyngeal, and, finally, branches pass between it and the ganglion of the trunk of the par vagum. Two ascending branches from the cervical ganglion pass to the internal carotid to form the carotid and cavernous plexuses; from these plexuses a filament passes to the under side of the ophthalmic branch of the fifth nerve, a second connects with the ophthalmic ganglion, while other communications with the ophthalmic nerve are formed by a branch passing between the ganglion and nasal branch of the ophthalmic, and the branch of the ophthalmic to the inferior oblique muscle. From the internal carotid plexus a filament reaches the spheno-palatine ganglion connected with the superior maxillary through the Vidian nerve. Filaments pass to the external carotid, forming a plexus from which those of the middle meningeal and facial arteries are derived. From the middle meningeal plexus a filament passes to the optic ganglion, which is connected with the inferior maxillary nerve, while from the facial plexus a branch reaches the submaxillary ganglion, which, in turn, is connected by several filaments with the gustatory branch of the trifacial nerve. Other communications exist between the Casserian ganglion and the cavernous plexus, the otic ganglion and the glosso-pharyngeal nerve, etc., etc.

Next let us trace the routes of transferred irritation.

1. **Stomach.**—Nerve supply of stomach from terminal branches of the right and left pneumogastric and from the sympathetic.

Irritation from the upper teeth travels, by means of the dental branches of the superior maxillary to the trunk of the superior maxillary division of the fifth pair, by the spheno-palatine branches of this nerve to Meckel's ganglion, and from this ganglion to the carotid plexus of the sympathetic. The carotid plexus being in connection with the superior cervical ganglion of the sympathetic, the irritation is transferred to the sympathetic branches which go to the stomach; and as it sends filaments to the jugular ganglion of the pneumogastric, the irritation reaches the gastric terminations of the vagus (or to Casserian ganglion, to carotid plexus, to superior cervical ganglion).

Irritation from the lower teeth passes by the superior dental branches of the inferior maxillary to the trunk of the nerve. This nerve, by its auriculo-temporal branch, is joined to the otic ganglion, and the irritation, after leaving this ganglion, reaches the meningeal plexus, is transferred to the superior cervical ganglion, and then reaches the stomach by the route above indicated.

Or, by trunk of nerve to Casserian ganglion, from this to carotid plexus and from this to superior cervical ganglion.

2. **Intestines.**—Same route as above to superior cervical ganglion, and from this by means of sympathetic to the intestines.

3. **Glands of Neck.**—An irritation from the teeth which reaches the superior cervical ganglion, passes by the outer branches of this ganglion to the four upper cervical spinal nerves. Also passes to middle cervical ganglion and by its outer branches to the fifth and sixth spinal nerves. Also passes to inferior cervical ganglion to seventh and eighth cervical nerves. These nerves are the chief source of supply for the cervical lymphatics.

4. **Salivary Glands.—To Parotid.**—An irritation from upper teeth is transferred to Meckel's ganglion by spheno-palatine branches and from this to carotid plexus, which plexus sends filaments direct to parotid gland.

An irritation from the lower teeth reaches the gland by means of the inferior dental branches of the inferior maxillary trunk of the inferior maxillary, auriculo-temporal branch and parotid branches.

Or, as before shown, by the superior cervical ganglion and the carotid plexus, which sends filaments to the gland.

To Submaxillary.—An irritation from the upper teeth passes by dental branches and trunk of superior maxillary to Casserian ganglion, from this to carotid plexus, from this to plexus on facial artery, and then to gland.

Or, from Casserian ganglion to gustatory branch of inferior maxillary, and by branches of this nerve to the gland.

Irritation from lower teeth by inferior dental branches and trunk of inferior maxillary to Casserian ganglion, and from this by gustatory and branches to submaxillary ganglion and gland.

Also, by mylohyoid branches of the inferior dental to gland.

To Sublingual.—From upper teeth by dental branches and superior maxillary to Casserian ganglion, and from this by gustatory and branches to gland. From lower teeth by dental branches and inferior maxillary to Casserian ganglion and from this by gustatory and branches to gland.

5. Lungs.—Nerve supply from anterior and posterior pulmonary plexuses formed by branches from sympathetic and pneumogastric reaches the superior cervical ganglion as indicated in No. 1, and is, as there shown, transferred to pneumogastric and sympathetic.

6. Eyes.—Irritation reaches carotid plexus from upper teeth as shown in No. 1. Filaments from carotid and cavernous plexuses to nerves of the eye. A filament to the under side of the ophthalmic division of the fifth pair. Another joins the ophthalmic ganglion, and branches pass between this ganglion, the nasal branch of the ophthalmic, and the branch of the ophthalmic which goes to the superior oblique muscle. From the carotid plexus come filaments to the abducens nerve. Irritation from lower teeth to superior cervical ganglion, as shown in No. 1. From here to carotid and cavernous plexuses, and to eye as above.

7. **Ears.**—Irritation from upper teeth to Meckel's ganglion as in No. 1. Meckel's ganglion by Vidian supplies part of mucous membrane of Eustachian tube, branches of it join the facial and the facial sends branches to the laxator tympani and stapedius.

Or, to superior cervical ganglion, as in No. 1. To the glossopharyngeal, to its petrous ganglion, and to Jacobson's nerve, which supplies the fenestra rotunda, fenestra ovale, and lining membrane of Eustachian tube and tympanum.

Or, from superior cervial ganglion to ganglion of pneumogastric, and by Arnold's nerve to external auditory meatus and membrana tympani.

Or, to superior cervial ganglion, to carotid plexus, and from this to tympanum.

From lower teeth by dental branches, as in No. 1, and inferior maxillary to otic ganglion, and from this a branch goes to tympanum.

Or, to otic ganglion, from this to meningeal plexus, from this to superior cervical ganglion, and from this to carotid plexus and tympanum.

Or, to Casserian ganglion, from this to carotid plexus, and from this to tympanum.

8. **Larynx.**—**Nerve Supply.**—Superior laryngeal branch of pneumogastric and recurrent laryngeal branch of pneumogastric and sympathetic reaches pneumogastric and sympathetic as in No. 1.

With these conditions fully recognized it is surprising that second dentition has not been accorded the position it deserves, as a cause of ill-health in later childhood. In second dentition the elements of local irritation with fever are quite as potent as before. There is, however, less activity of development or rapidity of change, if we except the radical alteration in the system occasioned by the approach of puberty. Therefore there must be less susceptibility to disease, and this fact, taken with the greater resisting power of advancing age, fully accounts for what every observer will find to be true, namely, that the disorders of this period are less frequent and, as a rule, less dangerous than those that attend the cutting of the milk teeth. Nevertheless the etiological relations of the eruption of the permanent teeth will fully repay a careful study.

The subject will be considered here in its relation to the time between the fifth year and the establishment of puberty, when childhood is over, and youth or maidenhood begins.

The permanent teeth are cut in the following order:—

(1) Four first molars, five to six years.
(2) Four central incisors, six to eight years.
(3) Four lateral incisors, seven to nine years.
(4) Four first bicuspids, nine to ten years.
(5) Four second bicuspids, ten to twelve years.
(6) Four canines, eleven to thirteen years.
(7) Four second molars, twelve to fourteen years.
(8) Four posterior molars (or "wisdom teeth," not entering into this study), seventeen to twenty-one years.

Of the twenty-eight teeth cut within the period already mentioned—the fifth to the fifteenth years—the first and seventh sets are developed *de novo*, and are more likely to give rise to trouble, particularly of the oral mucous membrane. The other sets take the place of corresponding milk teeth, and appear in very much the same order, the lower central incisors appearing before the upper, the upper lateral incisors before the lower, the upper bicuspids before the lower, etc.

Fig. 8 will aid in explaining the process.

As these teeth approach the surface, absorption begins in the alveoli and at the roots of the deciduous teeth, and this continues until the teeth are loosened and readily extracted, or if this be not done, until little is left but their crowns.

When the first and second molars approach the surface, the gums, just as in primary dentition, become red, swollen, rounded and tender. The salivary secretion is increased, the mouth is hot, the patient complains of aching in the gum, and, on account of tenderness, refuses food requiring mastication. With the other sets there is a gradual loosening of the superimposed temporary teeth, pain on mastication, redness and tumefaction of the gum, and augmented flow of saliva. As there is no impairment of the general health, these trifling symptoms must be regarded merely as manifestations of the progress of a physiological process. Such are the normal manifestations.

The most common disorders of second dentition, are (*a*) those of the mouth and throat; (*b*) of general nutrition; (*c*) of the stomach and intestinal canal; (*d*) of the cervical lymphatic glands; (*e*) of the eyes; (*f*) of the ears; (*g*) of the skin; (*h*) of the respiratory tract; (*i*) of the nervous system.

(*a*) Oral pain is often intense. It is lancinating or neuralgic

FIG. 8.

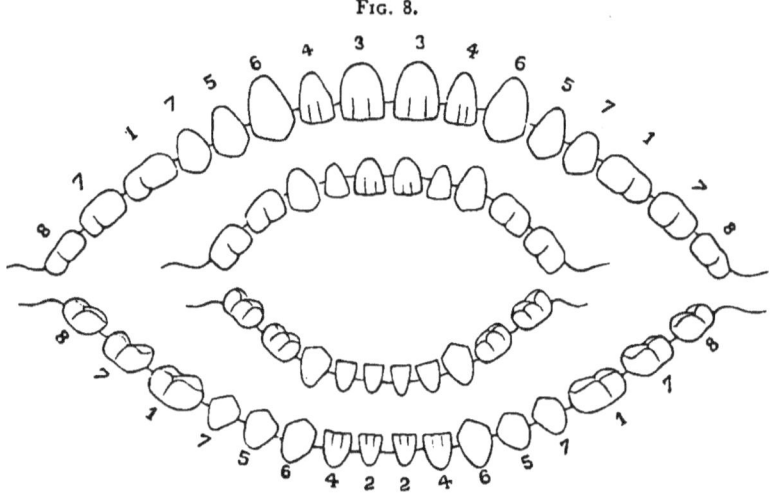

DIAGRAM SHOWING RELATION BETWEEN THE PERMANENT AND TEMPORARY TEETH.

The figures 1, 2, 3, etc., indicate the groups of teeth and the order of their appearance.

in character and may either be limited to the position of the advancing tooth or extend throughout the upper and lower jaws —the region supplied by the dental branches of the trifacial nerve. Sometimes the pain is referred to the eye, the ear, the face, or even to the forehead. Pain associated with tenderness most frequently attends the eruption of the first molars; then there is also redness and marked swelling of the gum, as in primary dentition.

The redness and swelling about an advancing tooth or around

a loosened milk tooth may, in debilitated or strumous subjects, extend to the mucous membrane of the whole mouth and give rise to catarrhal stomatitis. Again, as one of the first or second molars advances, the mucous membrane of the gum directly over the tooth breaks down and a circular ulcer is formed. This ulcer possesses all the characteristics of the marginal ulcer of ulcerative stomatitis, and is very liable, provided such favoring conditions as scrofula, overcrowding and bad hygiene exist—to run around the alveolar border, and extend to the inside of the cheek. A case has recently been under my care, in which all of the six-year molars were cut in this way, the resulting ulcerative stomatitis producing considerable discomfort, but yielding readily to treatment.

Superficial ulcers upon the edges and tip of the tongue are often encountered. These ulcers correspond in position and number to loosened and perhaps decaying deciduous teeth; are due to constant irritation of the mucous membrane; are the seat of moderate pain, and more or less interfere with the movements of the organ in mastication and speech. They vary in shape and size but are generally oval, with the greater diameter—rarely more than half an inch—extending in the direction of the axis of the tongue. Their bases are smooth, red and shining, and their edges red, indented, somewhat indurated and surrounded by a narrow band of white fur formed upon the neighboring healthy epithelium.

A boy, ten years old, was recently brought to consult me about the condition of his tongue. Nearly two months before the anterior deciduous molars had commenced to loosen. Soon after two ulcers appeared upon the tongue at points corresponding to the loose teeth in the lower jaw. These caused considerable discomfort and interfered with the movements of his tongue. Six weeks later the four loose teeth were extracted. At the time of his visit the points of the permanent teeth were distinctly visible. The ulcers, which presented the characteristics already described, were present, too, but they were much contracted and evidently in process of healing. This case is a clear illustration of the etiology of the condition, and of the rapid effect of the removal of the cause.

Loss, or perversion of taste, depending upon reflected irritation of the gustatory and glosso-pharyngeal nerves, has occasionally arisen in my experience. It is a feature that may be readily overlooked in childhood, and without doubt has never received due credit as a cause of the anorexia so often observed during second dentition.

Of throat affections, simple hypertrophy of the tonsils and follicular tonsillitis seem particularly apt to arise in late childhood. The extension of catarrhal inflammation from the mouth to the throat is certainly an element in the causation of the conditions, though anorexia, imperfect digestion and fever are more potent, as they lead to impaired nutrition and increased susceptibility to the action of cold and bad hygienic surroundings.

The treatment of this class of affections must vary with the symptoms presented. Should there be much inflammation and pain about a loose tooth, great relief can be obtained by painting the gum three or four times daily with a solution of—

   ℞. Cocaine Hydrochlorate, ... gr. iv.
     Glycerinæ, ... f ʒ ij.
     Aquæ, ... q. s., ad f ℥ j.
   M.
   S.—For local use.

When the first or second molars cause the trouble, free lancing with an oblique crucial incision is to be recommended. Much good can also be done in the way of softening the gums and lessening pain by a thorough application of—

   ℞. Zinci Chloridi, ... gr. j.
     Vin. Opii, ... f ʒ j.
     Glycerinæ, ... f ʒ ij.
     Aquæ Rosæ, ... q. s. ad f ℥ j.
   M.
   S.—Apply to tender gums with a brush or soft cloth thrice daily.

Such measures, too, will be more successful in relieving referred pains than any direct application to the place of reference.

Catarrhal and ulcerative stomatitis demand the usual methods of treatment.

Superficial ulcers on the tongue can often be healed by a daily application of a solution of nitrate of silver (ten grains to one fluidounce) and the frequent use of a borax or chlorate of potassium wash (fifteen grains of either to one fluidounce). Should there be much pain and discomfort, the solution of cocaine recommended above may be used two or three times daily. When the deciduous teeth are decayed, however, nothing short of their extraction will cure the ulceration.

Nothing can be done for loss or perversion of taste except removing loose teeth and freely lancing the gum over advancing molars.

Hypertrophy of the tonsils and follicular tonsillitis must be treated in the same way as when they occur independently of dentition, the question of the propriety of extraction and of lancing being always borne in mind.

(*b*) After safely passing through primary dentition children usually grow robust and enjoy good health, unless they be attacked by some one of the acute contagious diseases to which their age is liable. This state of affairs may happily endure throughout the remainder of childhood, but it is often supplanted, during the sixth and following years, by a condition best described as one of "general debility."

This ill-health is neither produced by disease of important viscera, as of the lungs, heart, kidneys and digestive organs, nor can bad hygiene be blamed in many cases. For the explanation one must look rather to an impairment of nutrition, resulting from the constitutional strain of cutting the second teeth, from the moderate fever associated with the process, and from the diminished consumption of food, attending oral discomfort and painful mastication. The severity of the symptoms depends somewhat upon the general vigor of the subject, though in my experience it bears little or no relation to the difficulty or ease of cutting the milk teeth.

Early in the sixth year, children so affected begin to lose

their rosy cheeks; the lips grow pale; the skin of the body becomes sallow and harsh; the hair dry and lustreless, and there is moderate loss of flesh with flabbiness of the muscles. The face wears an anxious expression and the temper is unstable; by day frequent complaints of weariness are made and little interest is taken in play, while at night sleep is restless and there is often slight fever, with a temperature rarely above 100° F. Pain and discomfort in the mouth are constant symptoms, and as these are increased by mastication, there is apparent anorexia. Examination of the mouth reveals redness, swelling and tenderness of the gums over advancing molars, or if these have been cut, around loose temporary teeth. The bowels are inclined to constipation and the urine limpid and voided in abundant quantity; the pulse is rather feeble though normal in frequency, and, as a rule, there is no cough nor other alteration in the respiratory function. Careful investigation shows an absence of lesion in the heart, lungs or kidneys, and of disease of the abdominal glands or digestive tract. As the teeth of the advancing group are cut, the symptoms disappear, to return with the approach of each succeeding group, but the anterior molars generally give rise to more marked disturbance than any of the teeth that replace the temporary set.

In addition to the ordinary risk of intercurrent disease existing in every case of general debility, the condition just described is very apt to be complicated by bronchitis or catarrh of the gastro-intestinal canal. Pyrexia, although it is comparatively slight in second dentition, accounts for this, for a feverish child is very susceptible to cold, and very liable to have his digestion disordered by food upon which he has previously thriven. The first cause, by driving the blood from the surface, produces bronchitis; the second, by direct and indirect irritation, leads to catarrh of the mucous membrane of the stomach and bowels.

This knowledge taken in connection with the course and history of the case and the condition of the mouth, should enable the observer to attribute the illness to its proper source rather than to any complicating affection, although the latter undoubtedly accentuates the symptoms, and may force itself into prominence. The negative results of physical exploration of the heart, lungs, and abdominal organs,—particularly the mesenteric glands—and of examination of the urine, are also important in establishing the correct relations of cause and effect.

Careful regulation of the diet and the administration of tonics, the methods of treatment that would naturally be suggested, are of little avail, unless oral pain and difficulty of mastication be relieved. Even then, it is often impossible to do more than maintain a moderate degree of health until advancing teeth are completely free.

Free lancing of the gums over molars, the application of cocaine to painful gums surrounding loose temporary teeth; the extraction of these when the substituting teeth are so advanced as to run no risk of impairing the arch of the jaw: regulation of the diet and hygiene, and the employment of tonics and laxatives are the measures to be recommended. The diet must be simple, non-farinaceous and nutritious; it is better to allow four small meals a day than three large ones. Of tonics a good formula is:—

R. Tr. Nucis Vomicæ, ... ℳxij.
Elix. Cinchon. Ferrat, ... f ʒ vj.
Syrupi, ... f ℥ ss.
Aquæ, ... q. s. ad f ℥ iij.
M.
S.—Two teaspoonfuls thrice daily at the age of six years.

Syrup of the iodide of iron, bitter wine of iron and an emulsion of cod-liver oil with lacto-phosphate of lime, are also very useful. The best laxatives are fluid extract of senna, which may be combined with the tonic mixtures; tincture of aloes and myrrh in small doses three times daily, compound licorice powder, glycerine suppositories or laxative tamarinds.*

Complicating bronchitis and catarrh of the gastro-intestinal canal demand active attention, and little else can be accomplished until they are relieved.

---

* See Habitual Constipation.

(c) Disorders of the digestive system, while unattended by such marked symptoms and rarely reaching the same degree of danger as in primary dentition, are among the most common of disturbances produced by the eruption of the second teeth. One cause of this is the intimate sympathy existing between the different portions of the digestive tract, and leading to reflection of irritation from point to point. Another and more active cause has already been indicated. It involves two conditions. First, general depression of vitality—from constitutional strain, fever, and so on—with a corresponding weakening of the function of digestion, so that food previously suitable and easily assimilated, becomes relatively too coarse and "strong," and being more or less imperfectly changed and absorbed, lies in the alimentary canal undergoing fermentation and decomposition with the formation of irritant acids and gases. Second, the well-recognized susceptibility of the gastro-intestinal mucous membrane to become inflamed, or to assume the catarrhal state, when subjected to the action of irritants. A susceptibility decided enough under the best of circumstances, but intensely marked if the general health and resisting power be below par.

Anorexia, vomiting, acute gastric catarrh, chronic gastro-intestinal catarrh, and diarrhœa are the more common of the digestive disorders attending second dentition. Loss of appetite when not due to perversion of taste, generally forms but one member of a group of symptoms depending upon catarrh of the stomach, and can be best studied under this head. The same may be said of vomiting. Of each of these symptoms it is also true, that they may be so prominent as to mask the associated features unless the observer be very careful.

Acute gastric catarrh,* in so far as it is related to second dentition, is most frequently encountered during the eruption of the six-year molars. The attack of so-called "biliousness" and "indigestion" may or may not be preceded by some indiscretion in diet.

---

* For description of this affection and its treatment see page 199.

It will fully repay the physician to inspect the mouth of every six-year old patient (and upward) suffering from acute indigestion, and after making the examination thoroughly to lance swollen gums over advancing molars, and apply soothing lotions to irritated gums about loosened temporary teeth, or order extraction if admissible. General treatment must be conducted on the same plan as when the attacks, as is often the case, occur independently of dentition.

Chronic gastro-intestinal catarrh owes its origin to second dentition more frequently than to any other cause save whooping-cough. In this condition—so aptly termed by Eustace Smith, "Mucous Diseases" *—there is a mucous flux from the whole internal surface of the alimentary canal, which mechanically interferes with the digestion and absorption of food and greatly impedes nutrition. It may be met with at any time during the eruption of the permanent teeth.

Diarrhœa is a very constant attendant upon second dentition. It is most apt to arise in the changeable weather of spring and autumn or in the heat of July, August or September. In this respect it resembles the diarrhœa so common with primary dentition, but unlike the latter, it shows little or no tendency to run into entero-colitis. Two forms are met with, namely catarrhal diarrhœa, and lientery.†

The general depression produced by the coming of the second teeth, would naturally favor the development of any constitutional tendency, and, having traced the connection in several instances, I have no doubt that not a few cases of tubercular ulceration of the bowels owe something to this etiological factor.

(*d*) Enlargement of the sub-maxillary gland, or one or more of the lymphatic glands of the neck, is a frequent occurrence during the approach of the first molars. Patients so affected may at the same time present other evidences of difficult dentition ; for example, they often show the symptoms of "general debility"

---

* For description of this condition and its treatment see Mucous Diseases.

† See Catarrhal Diarrhœa

already referred to, and often have mucous disease ; but they are quite as frequently perfectly healthy in constitution as strumous.

The swelling of the gland or glands is moderate in degree ; there is considerable hardness, moderate tenderness and pain, but the superimposed skin is movable and healthy. There is some tendency to chronic enlargement and induration, though little or none to suppuration, save in scrofulous subjects.

Should the gums be thoroughly lanced—and it is often necessary to sink the knife blade deep, to free the coming tooth—the glandular swelling soon subsides, though resolution can be hastened by painting with tincture of iodine, or using the following ointment :—

   ℞. Ichthyol . . . . ʒj.
   Lanoline . . . . ʒj.
   M.
   S.—Apply to the enlarged glands three times daily, with rubbing.

(*e* and *f*) Conjunctival blennorrhœa and otitis, sometimes noticed during primary dentition, have come within my observation, and, from the intimate connection between the nerves of the ear and eye and those of the teeth, it is more than probable that certain other disturbances of these two organs of special sense arise during second dentition, and depend upon dental irritation. Unfortunately so few cases of disease of these two organs come under my notice, and so little attention has been paid to this causal relation by specialists, that I have no data upon which to base conclusions.

(*g*) Herpes of the lips, eczema, and urticaria frequently appear during the eruption of the second teeth. They apparently depend upon gastro-intestinal disturbances and are relieved by measures directed to the cure of disorders of this tract, together with appropriate local treatment, and attention to the teeth and gums.

(*h*) Nasal catarrh, "teething cough," and an increased susceptibility to catarrh of the bronchial mucous membrane are the chief affections of the respiratory tract.

Nasal catarrh is generally sub-acute in type, and is attended by hypertrophy and redness of the Schneiderian membrane. There is a more or less copious discharge, which has a slight, heavy odor, may be composed of thin mucus or muco-pus, and which sometimes dries into thick crusts. This catarrh occasionally runs into ozæna in weak and strumous children.

Attention to the teeth, frequent syringing of the nostrils, with insufflation of boracic acid and bismuth (1 part to 2), the occasional application of a weak solution of nitrate of silver (gr. v-f₃j), judicious use of the electro-cautery, and the administration of tonics, constitute the best treatment for the ordinary form of catarrh. Ozæna calls for its special plan of management. "Teething cough" is due to reflex irritation of the pneumogastric nerve; it is identical with the "stomach cough" of mucous disease.

Bronchitis can never be said to be due directly to dentition. This cause acts only indirectly by reducing general vitality and the power of resisting disease, and thus rendering the delicate bronchial mucous membrane more than ordinarily susceptible to catarrhal inflammation from chilling of the surface and exposure to damp air. An attack of bronchitis may occur with the eruption of one group of permanent teeth, or the attack may be repeated with several successive sets.

I have in my mind now, the case of a delicate boy who, in spite of the utmost care on the part of his mother, had a most severe bronchitis during the eruption of the six-year molars, and a second persistent attack while the four central incisors were pushing their way through. Not two months before writing this, he again came under my care suffering from a third attack that promised equal obstinacy. On inspecting the mouth the four lateral temporary incisors were found to be quite loose and their permanent substitutes evidently advancing rapidly. Extraction of the loose teeth was ordered, together with a mild expectorant and a tonic mixture, and the catarrh soon subsided.

(*i*) Nervous disorders, both sensory and motor are encountered.

Headache is common. The pain is usually temporal and

unilateral. It may be seated however, in the occipital region or in different parts of the face, and sometimes shifts suddenly from the temporal to the occipital region. It is lancinating, more or less constant, with no distinct intermissions, and during its continuance there is restlessness, anorexia, a frequent, hard pulse, sweating, dilatation of the pupil on the affected side, and perhaps dimness of vision, diplopia and colored or uncolored spectra. One or more painful points can often be detected, and generally there is a hard, tender, moderately enlarged lymphatic gland in the submaxillary or cervical region.

These attacks are due to disordered vasomotor innervation, depending upon irritation of the sympathetic nerves and producing irregular contraction or spasm of the vessels—the temporal or occipital artery, as the case may be. The real source of irritation is to be found in the mouth. The mode of action may be twofold. First, from a swollen gum or carious tooth the lymphatic vessels readily convey irritating matter to a neighboring lymph gland, and the irritation here excited acts, in its turn, as a disturber of the sympathetic nerves which furnish the vaso-motor supply to the carotid artery and its branches. This theory of the production of migraine has the value of the support of T. Lauder Brunton.

The other method of production—and this I simply submit for consideration—is one of direct nervous connection; the submaxillary ganglion acting in the case of the lower teeth and the spheno-palatine in the upper, as the medium of transfer of irritation to the vaso-motor nerves.

Bromo-caffeine, gelsemium, saline laxatives and tonics may be employed to lessen the severity of the pain, but lancing or extraction are the only certain remedial measures.

The motor disturbances, while not quite so common as the sensory, are more varied.

Reflex spasm and paralysis of the eyelid have been noted.*
The former I have frequently seen, but the latter, so far as I can

---

* Brunton.

recall, has never fallen within my experience. For the reflex spasms no method of treatment availed until the dental irritation was removed; and the same statement may be hazarded, *a priori*, in regard to paralysis of the lid.

More extended paralysis also occurs. In this connection I can do no better than quote the words of Brunton, the correctness of which I can fully confirm. After speaking of paralysis of the eyelid this author states :—" Sometimes, however, paralysis occurs of a much more extensive character, in consequence of dental irritation, especially in children. Teething is recognized by Romberg and Henoch * as a frequent cause of paralysis appearing in children without any apparent cause. According to Fliess,† paralysis of this sort occurs more commonly during the period of the second dentition, whereas convulsions generally occur during the first. Its onset is sudden. The child is apparently in good health, but at night it sleeps restlessly, and is a little feverish. Next morning the arm, or more rarely, the leg, is paralyzed. The arm drops; it is warm but swollen, and of a reddish-blue color. It is quite immovable, but the child suffers little or no pain. Not infrequently paralysis is preceded by choreic movements. Sometimes recovery is rapid, but at other times the limb atrophies, and the paralysis may become associated with symptoms indicating more extensive disturbance of the spinal cord and brain, such as difficulty of breathing, asthma, palpitation, distortion of the face, and squint, ending in coma and death.

"It is only in very rare instances that we are able to gain any insight into the pathological anatomy of such cases, because they rarely prove fatal, and even when they do so the secondary changes are generally so considerable as to leave one in doubt as to the exact mode of commencement. This renders all the more valuable the case recorded by Fliess, in which a boy five years old, and apparently quite healthy, found his left arm completely paralyzed on awaking one morning after a restless night. The arm was red, but the boy suffered no pain, and played about

---

* Klinische Wahrnehmungen und Beobachtungen.

† Fliess, *Journ. der Kinderkr.*, 1849, July and August.

without paying much attention to the arm. The same day he fell from a wagon upon his head, and died in a few hours. Apart from the fracture of the skull which caused his death, the anatomical appearances which were found were congestion of the spinal cord, and great reddening and congestion of the meninges near the point of origin of the brachial nerves, where the veins were also much fuller than on the corresponding right side. There was no organic change perceptible, either in the spinal cord or in the brachial nerves. On the other hand, the turgescence of the veins extended from the shoulder and neck up to the face, and was very striking in the sub-maxillary region.

"This vascular congestion seems to point to vaso-motor disturbance of a somewhat similar kind to that which we have already noticed in connection with occipital headache, or with migraine accompanied by subjective appearances of either form or color." *

This form of paralysis certainly suggests acute anterior poliomyelitis, though the symptoms are not quite identical and the age is not that at which "infantile palsy" usually occurs.

Unfortunately, in all but the mildest cases, which get well quickly, when the paralysis appears the mischief is done and little benefit can be expected from attention to the teeth. The treatment must be conducted on the plan usually adopted in infantile paralysis.

As just stated dental irritation sometimes produces choreic movements as prodromata of paralysis, but it much oftener acts as the exciting cause of genuine chorea in nervous children. Approaching molars, or carious, loose milk teeth about to be shed, may be the source of irritation. The causal relation is proved by the fact that the chorea disappears or yields quickly to ordinary treatment when the new teeth pierce the gums or are freed by lancing, or when decaying teeth are removed. Epilepsy is another nervous affection which can occasionally be traced to the same source. In such cases one usually finds a history of repeated general convulsions during primary dentition.

---

* Brunton. "Disorders of Digestion," p. 93.

7. SIMPLE PHARYNGITIS.

Catarrh of the mucous membrane covering the soft palate, tonsils and pharynx—simple or erythematous pharyngitis—is a common occurrence in children who have reached the third or fourth year, though it is rarely met with before that age. It may either be primary or secondary in origin.

The ANATOMICAL LESION is hyperæmia of the affected mucous membrane. This is red, swollen, softened, granular, and at times œdematous.

ETIOLOGY.—The primary form is most prevalent during the winter and spring. Impaired health, from neglect, bad food, or insufficient clothing predispose to an attack; while sudden changes in temperature and exposure to wet and cold are the chief excitants. One attack is often followed by others. The disease is not contagious, but many cases often occur simultaneously. Secondary pharyngitis, which will not be studied here, constantly accompanies scarlet fever and measles, and often complicates bronchitis and pneumonia.

SYMPTOMS.—An attack of simple pharyngitis of ordinary gravity begins with fretfulness and lassitude; the child refuses food, and may vomit once or twice. Fever quickly follows, preceded by rigors, or in children nearing the age of puberty, by a single distinct chill. This fever is quite out of proportion to the local symptoms. The temperature in the course of a few hours rises to 102° or 104° F., and often higher; the pulse runs up to 130 or 140 beats per minute; the respiration is correspondingly rapid, though easy, the face is flushed and the skin dry. The voice becomes thick and husky, and there is a teasing, unproductive, hoarse cough, which may assume a brazen character toward evening. Older patients may complain of dryness and fulness of the throat, of a sensation leading to frequent efforts at deglutition, or of difficulty and pain in swallowing food; while infants manifest the latter conditions by refusing the breast or bottle. An entire absence of these subjective symptoms, however, is common.

On inspection, the mucous surface of the soft palate, uvula, tonsils and pharynx presents a reddened, tumefied, dry, granular appearance, and may be partially covered with flakes of whitish mucus or muco-pus. The tonsils are somewhat swollen, and at times the uvula is œdematous. The lymph glands about the angles of the jaw are slightly enlarged and tender to the touch.

On the second day the fever abates, the temperature often falling to the normal line, but there is an elevation on each succeeding evening until the end of the fourth or fifth day, when the attack begins to subside. In the meantime the local symptoms increase. Throughout, the child is peevish and restless, sleep is disturbed, the tongue is heavily coated, and there is loss of appetite, increased thirst, and a tendency to constipation.

In exceptional cases the disease is much more grave in type. These severe attacks begin with vomiting, excessive restlessness or drowsiness, occasionally convulsions, and always high fever, with a temperature reaching 106° or even more, and a rapid and bounding pulse. The affected mucous membrane becomes intensely red and covered with a muco-purulent secretion. All the ordinary symptoms are intensified, and in addition there may be mild delirium and a flushing of the entire cutaneous surface, suggesting the scarlatinal rash. These attacks vary in duration from three to eight days, and, notwithstanding the alarming character of the symptoms, usually terminate in recovery.

DIAGNOSIS.—It is quite possible to overlook the presence of pharyngitis on account of the frequent absence of symptoms calling attention to the throat. Thus the sudden onset of high fever with rapid pulse and respiration and dry cough would, in the absence of difficult deglutition and pain in the throat, suggest an attack of croupous pneumonia. If, under the same conditions, the pharyngitis be ushered in by vomiting, the fever might readily be referrred to a digestive disorder. Such errors are to be avoided only by making a rule to inspect carefully the throat in each doubtful case. A grave case, again, may in the beginning be taken for one of scarlet fever, the resemblance being increased by the uniform flushing of the surface. Distinction is to be found in the different course of the two diseases, and the non-appearance of certain characteristic symptoms of the exanthem.

Care must be taken not to confound the white or yellowish-white patches of mucus or muco-pus adhering to the inflamed surface with diphtheritic membrane. The former can be wiped away easily, leaving the mucous membrane intact.

TREATMENT.—If the case be seen on the first day, it is possible greatly to reduce the severity of the attack by giving the child a hot mustard foot-bath,* putting him to bed in a properly-warmed room, and by cautiously administering aconite, with some saline laxative, as a small teaspoonful of magnesia in a wineglassful of strong lemonade. Under such circumstances, tincture of aconite root may be given to a child of four years, in doses of a drop every fifteen minutes until four drops have been taken, and subsequently the same dose every hour until an effect is produced on the pulse, or the heat and dryness of the skin are lessened.

When the fever has been reduced in this way, or should the case not be seen until the second day, the following may be ordered:—

> ℞. Potassii Chloratis, ... gr. xlviij.
> Syrupi, ... f ℥ ss.
> Aquæ, ... q. s. ad f ℥ iij.
> M.
> S.—One teaspoonful every three hours, in water, for a child of four years.

If the fever returns as evening approaches, this mixture should be discontinued, and another foot-bath and a few doses of aconite given; or some simple diaphoretic may prove sufficient, as liquor potassii citratis, at intervals of an hour during the night.

Throughout the attack the diet should consist of milk and farinaceous articles prepared with milk, with a little meat broth as the fever subsides. A daily evacuation of the bowels must be secured, and the child must be kept in bed.

---

* The ordinary strength of such a bath for a child of three or four years is one tablespoonful of mustard-flour to two gallons of water.

Local treatment must not be neglected. If the child be able, he should gargle the throat every hour with a solution of chlorate of potassium, ten grains to the fluidounce. If too young to do this, the same solution should be applied to the throat at short intervals with a mop of absorbent cotton. Painting the throat daily with a solution of nitrate of silver (gr. v to f℥j) hastens the cure. At the same time it is well to redden the skin of the neck with some such liniment as:—

 ℞. Ol. Terebinthinæ, ... f℥j.
  Ol. Olivæ, ... f℥iij.
 M.
 S.—Apply twice daily.

Grave cases require no alteration of this plan. It is well, if there be great restlessness, to repeat the foot-bath, or even to give several full warm baths of ten minutes' duration. If there be intense inflammation of the pharynx the neck should be enveloped in a poultice, or in extreme cases a leech may be applied behind each angle of the lower jaw. Clogging of the throat by tenacious mucus demands an emetic.

When convalescence begins, the diet must be more liberal, and restoration to perfect health is hastened by administering a bitter tonic, as tincture of nux vomica, or compound tincture of gentian, in appropriate doses, three times daily.

8. SUPERFICIAL CATARRH OF THE TONSILS.

In this affection there is a simple hyperæmia of the mucous membrane covering the tonsils, accompanied by moderate swelling of the glands. It is produced by the same causes, and usually occurs as an element, merely, of general pharyngitis. In the exceptional cases in which it exists in an isolated form, the tonsils will be found reddened and moderately swollen, and several yellowish-white points, due to retained follicular secretion, will be seen on their surfaces. The local subjective, and the general symptoms are the same as those of pharyngitis, and they yield to the same measures of treatment.

9. FOLLICULAR TONSILLITIS.

In this disease there is, in addition to superficial hyperæmia, a catarrh of the lacunæ or follicles of the tonsils. According to the extent of the disease, several or all of the follicles become filled with a yellowish-white, curd-like material, consisting of epithelium and pus. When thin, this flows away; but, when thick, it is removed with difficulty, collects and distends the lacunæ, and may undergo desiccation, or even become calcified. The parenchyma of the tonsils becomes hyperæmic, and there is an infiltration of serum and a proliferation of the gland cells.

ETIOLOGY.—The affection is a common one after the fifth year. It is most apt to be met with in the winter and spring, but it may occur at any season. Exposure to wet and cold is usually considered to be the exciting cause, but an attack may quite as frequently be traced to over-eating, associated with excitement and fatigue. One attack predisposes to others, and I have seen many patients who are invariably affected after gorging themselves with rich food, pastry or candy. A combination of all of these causes—so well afforded by that worst of institutions, a child's party—invariably produces a crop of cases.

SYMPTOMS.—When due to over-eating, the attack usually sets in on the day succeeding the indulgence. It begins with headache, lassitude, pain in the back and legs, and more or less rigor. The tongue becomes frosted; there is thirst, anorexia and nausea, often followed by vomiting. Toward the evening of the first day the face becomes flushed, the skin hot and dry, and the pulse rapid. The bowels are sluggish, and the urine is scanty, high colored, and lateritious. On the morning of the second day the fever abates, but it returns in the afternoon, and this course is maintained for three or four days, when convalescence is established. In the meantime the anorexia and constipation continue, the patient sleeps badly; may even be slightly delirious at night, and, finally, is left so feeble that health is not restored for a week or more.

When the affection is due to exposure alone, there is less headache, and no nausea or vomiting. In other respects the course is similar, though the attack is followed by less prostration.

Whatever the cause, the local symptoms are the same. They consist of a sensation of dryness and heat in the throat, repeated efforts to clear the throat, difficult and painful deglutition, increased salivation, a nasal intonation of the voice, and a heavy, offensive breath. On inspection, a catarrhal condition of the palatine arches and pharynx is observed. The tonsils are enlarged, sometimes sufficiently so as almost to meet one another; their enveloping mucous membrane is reddened and swollen, and their surface is dotted with yellowish-white points, corresponding in number, shape and size with the follicles involved. These points are sometimes covered and concealed by muco-pus, and may be surrounded by shallow, circular erosions of the mucous membrane. On pressing the tonsils, ill-smelling masses of varying size and consistency may be pressed out. These are also expelled by hawking, or are forced out in deglutition and swallowed with the food. In whichever way removed, they leave the orifices of the follicles more widely open and gaping than in health. There is some tenderness on pressure beneath and behind the angles of the jaw.

The DIAGNOSIS is easily made from the appearance of the tonsils, and from the fact that gentle pressure with the finger will force out one or more masses of retained secretion—a pathognomonic sign. There is no doubt that these cases are by some classed as diphtheritic, though none but the most inexperienced could confound the numerous yellowish-white points, of irregular shape and size, depressed below or projecting beyond the well-defined lips of the follicles, and which, as already stated, can be often expelled by pressure on the tonsils, with diphtheritic membrane. Again, the difference between this affection and a patchy tonsillar diphtheria—a common enough disease—must strike any careful observer.

The PROGNOSIS is always favorable, except that one attack predisposes to others, which may lead to chronic hypertrophy of the tonsils.

TREATMENT.—If the attack be traced to over-eating, the administration of an emetic would naturally suggest itself as a preliminary. This, however, is rarely necessary, as the initial vomiting empties the stomach sufficiently. Usually, the first steps are to place the child's feet in a hot mustard bath, then put him to bed, and give, according to the age, one or two grains of calomel at once, or in broken doses if there be much nausea. If, on the first night, the fever be high, tincture of aconite should be resorted to; if more moderate, an effervescing draught, like the following, will suffice :—

R. Acidi Citrici, ... ʒjss
Aquæ, ... fʒiij.
M.
S.—Solution No. 1.

R. Potassii Bicarbonatis, ... ʒj.
Aquæ, ... fʒiij.
M.
S.—Solution No. 2.
A teaspoonful of each solution is to be poured into a tablespoon or glass and taken while effervescing.

This draught has the advantage of checking nausea at the same time that it reduces fever.

Small pieces of ice should be swallowed at short intervals to relieve thirst and lessen the inflammation of the tonsils, and the food must be restricted to small quantities of milk and limewater (3 to 1), or weak broths in case milk disagrees. On the second day, it is only necessary to look carefully after the diet, to allow nothing but milk and broths; keep the patient in bed, and give during the day the following:—

R. Pulv. Pepsinæ,
Sodii Bicarbonatis, ... āā ʒj.
M. et ft. chart. No. xij.
S.—One powder every three hours for a child of six years.

The effervescing mixture may still be used in the early night if the fever be high enough to require it.

Such measures should be continued until convalescence is

established, care being taken to keep the bowels regular with calomel in broken doses. Then the diet may be gradually increased and a bitter tonic given.

If the cause be exposure to wet and cold, the general treatment must be the same as for pharyngitis.

The local treatment embraces counter-irritation of the skin of the neck; touching the tonsils once daily with a solution of nitrate of silver, gr. v to f℥j; and frequent gargling with:—

    ℞. Potassii Chloratis, ... gr.lxxx.
        Acid. Carbolici, ... gr.ij.
        Glycerinæ, ... f℥j.
        Aquæ, ... q. s. ad f℥viij.
    M.
    S.—Use as a gargle every hour.

10. SUPPURATIVE TONSILLITIS.

Quinsy is a rare disease in childhood and is scarcely ever met with before the twelfth year. When it does occur, some family predisposition can generally be traced. One of the most common predisposing elements is the rheumatic diathesis. Fatigue and exposure are the exciting causes. It is most frequent during spring and autumn. One attack predisposes to others. It may arise as a primary affection or as a complication of scarlatinous, variolous, or diphtheritic anginas. One or both tonsils may be affected.

MORBID ANATOMY.—At first there is intense hyperæmia with serous infiltration of the glandular tissue, and the tonsils sometimes become swollen to more than double their size. The inflammation may now undergo resolution. Otherwise an infiltration of small cells takes place, into and between the follicles, into the inter-lacunar connective tissue, and in the capsule. Retrogression is still possible, or failing this a new formation of reticulated substance takes place, resulting in permanent hypertrophy; a frequent termination of repeated attacks in children. If the inflammation be very intense, an abscess forms, but suppuration is not the usual result of tonsillitis occurring before puberty. With these conditions there is always associated general pharyngitis and often follicular tonsillitis.

SYMPTOMS.—The disease begins with rigors or a distinct chill, followed by sneezing, epistaxis, headache, pain along the Eustachian tube, loss of appetite, and fever, with languor and muscular prostration during the day, and mild delirium at night. Soon the patient complains of dryness and burning in the throat, difficulty and pain in deglutition, and the voice becomes nasal. If the throat be inspected, the mucous membrane of the soft palate and pharynx is seen to be red and swollen, and one or both tonsils are reddened and enlarged, often presenting several whitish-yellow points of retained follicular secretion. If one tonsil only be affected, the œdematous uvula will be pushed to the opposite side—an important sign.

The symptoms gradually increase in severity. The temperature ranges from 99° or 100° F., in the morning, to 102° or 104° in the evening, and the pulse from 110 to 120; but the respiration, though snoring, is little increased in frequency. Pain and difficulty in deglutition grow worse; the voice assumes a peculiar, thick, nasal tone; the breath has a heavy odor; the salivary secretion is increased and dribbles from the mouth; the tongue is heavily furred, and the bowels are sluggish. The child's face wears an apathetic expression, is red or dusky in hue, and there is dulness of hearing. Talking is painful, and so also is any movement of the jaw. On this account it is difficult to obtain a view of the throat; but if such be had, the tonsils, when both are affected, are seen to be intensely congested, and so much swollen that they meet; or, when only one gland is involved, it often extends a third of an inch beyond the median line. The day is divided between the listless inaction of prostration and the uneasy tossing of discomfort, and the night, between the restlessness of fever and the wandering of delirium. What little sleep is obtained is interrupted by snoring.

The crisis usually occurs on the fifth day, although it may be postponed until the eighth. If the tonsillitis ends in resolution the fever rapidly subsides, disappearing entirely in twelve hours; the local symptoms simultaneously abate and convalescence is rapid. When the inflammation ends in the formation of new tissue and hypertrophy of the glands, the acute manifestations give place to a train of symptoms to be described in the next section. Finally, if suppuration take place, there is a chill followed by high fever. The abscess soon points toward the mucous surface of the gland, and, unless opened by lancing, is broken by an effort at deglutition or in an examination of the throat. The quantity of pus discharged is ordinarily small, and is swallowed, as a rule. After the opening of the abscess, the child passes at once from a condition of great distress to one of comparative comfort, and strength and health are soon regained.

The DIAGNOSIS of quinsy is unattended with difficulty, and the prognosis, so far as life is concerned, is always good, though the danger of chronic hypertrophy must not be forgotten.

TREATMENT.—If the patient can be seen when the peculiar tone of the voice, the pain in the line of the Eustachian tubes, and the deflection of the uvula indicate the beginning of tonsillitis, it is possible to abort, or, at least, greatly reduce the intensity of the inflammation. For this purpose he must be put to bed, and given a sufficient quantity of wine of ipecacuanha to empty the stomach. Then properly proportioned doses of tincture of aconite root must be administered every half hour until an effect is produced on the temperature and pulse, and small bits of ice must be swallowed at intervals of ten minutes. At the same time it is well to apply a sinapism to the side of the neck corresponding to the affected gland. Since the introduction of cocaine I have often succeeded in aborting tonsillitis by thoroughly mopping the affected parts three times daily with a four per cent. solution of this drug. Even in cases where this favorable result was not obtained, the cocaine so far allayed pain as to permit liquid food to be swallowed with ease. This is an invaluable aid in the treatment of severe quinsy occurring in feeble children.

When the case is not seen till later, the indications are to encourage resolution or hasten suppuration, and to maintain the strength. To fulfil the first, the neck should be enveloped in a poultice, the throat should be repeatedly gargled with warm water, and steam from an atomizer should be constantly inhaled. The strength is to be kept up by administering all the concentrated liquid food that it is possible for the patient to swallow and by using suppositories of quinine. The latter may be ordered in this way :—

  ℞. Quiniæ Bi-sulphatis, ... gr. xij.
    Ol. Theobromæ, ... ʒiij.
   M. et ft. supposit. No. xij.
   S.—Use every four hours for a child six years of age.

On account of the difficulty in swallowing it is well to avoid ordering any medicine by the mouth except a diaphoretic, such as the solution of the citrate of potassium, and an occasional dose of some saline laxative. When there is much restlessness or delirium at night, it is well to give bromide of potassium, in ten-grain doses, by the mouth or rectum.

If an abscess forms, a somewhat rough pressure of the finger against the involved tonsil will hasten its rupture, but incision is a better method of treatment and often lessens the duration of suffering by twenty-four hours or more.

After the crisis is past, the diet must be increased and a tonic ordered as :—

  ℞. Tr. Ferri Chloridi, ... fʒj.
    Quiniæ Sulphatis, ... gr. xij.
    Syrupi Zingiberis, ... f℥j.
    Aquæ, ... q. s. ad f℥iij.
   M.
   S.—One teaspoonful, in water, three times daily for a child
    six years old.

The subsidence of the tonsils to their normal size is hastened by painting them twice daily with—

  ℞. Acidi Tannici, ... ʒj.
    Glycerinæ, ... f℥j.
   M.

For prevention, gargles of cold water and astringents, applications of the glycerole of tannin, and measures to maintain a high standard of health and counteract any rheumatic tendency, should be employed.

11. HYPERTROPHY OF THE TONSILS.

Chronic enlargement of the tonsils is slow in its development, and must be considerable in degree before giving rise to definite symptoms. Consequently, the disease is rarely recognized before the third or fourth year of life, although its commencement in early infancy is quite possible. It is common between the seventh and twelfth years.

ETIOLOGY.—Repeated tonsillar inflammation and the irritation attending dentition are the ordinary exciting causes, but it may appear spontaneously in children who are out of health, strumous or syphilitic. As the symptoms are aggravated by any passing angina, more cases demand treatment during the winter and spring than at other seasons.

SYMPTOMS.—The first to attract attention is loud snoring during sleep, due to pressure upon the velum, and obstruction to the passage of air through the posterior nares. At the same time there is a decided nasal twang to the voice. Examination shows marked projection of both tonsils, or, more rarely, of one only; the follicular orifices are widely open and very distinct, and several of them may present the yellowish-white points of retained secretion. The investing mucous membrane is pale, as a rule, but it may be traversed by arborescent blood vessels. Such a degree of hypertrophy and the accompanying symptoms sometimes disappear spontaneously with the development of the mouth and vocal organs attendant upon puberty.

When the glands are so much enlarged that they touch in the mid-line of the throat, there are added to the other symptoms a constant hacking cough with labored respiration, and difficulty of hearing, due partially to pressure upon the orifices of the

Eustachian tubes, and partly to a state of habitual congestion kept up in the surrounding parts. The dyspnœa is much worse at night, and the little patient often starts from sleep in a state of terror. It may be so grave as to threaten life and necessitate tracheotomy.

When enlargement—so great as to almost completely obstruct the passage of air through the nose—has existed from an early age, noticeable anatomical changes take place. The nostrils become extremely small and compressed, while the superior dental arch retains the narrowness of infancy, not allowing room for the teeth, which, in consequence, overlap one another. The palate, also, becomes unusually high and arched. Furthermore, the obstacle to the free entrance of air prevents the lungs being readily filled in inspiration, so that a partial vacuum is formed between them and the chest-wall, to fill which the external air-pressure forces in the yielding parietes. The effect of external pressure is most marked where the resistance is least, namely, at the base of the thorax, and a constant and long-continued repetition of this leads to the production of a gutter of variable depth and three or four inches in width, extending laterally from the lower part of the sternum, and to a projection forward of this bone. Any tendency to pulmonary phthisis is increased by this deformity, and if tubercular disease be present, the impediment to the entrance of air, and the constant irritation of the air passages, maintain a condition most unfavorable to its arrest.

TREATMENT.—Moderate enlargement of the tonsils in a weakly child will sometimes disappear when puberty is passed, or as health is regained under a course of tonics. The best tonic is syrup of the iodide of iron, in doses of ten drops three times daily for a child of eight years of age. It is well to paint the tonsils once every day with one of the following astringents:—

℞. Tr. Ferri Chloridi, ... f℥j.
    Glycerinæ, ... q. s. ad f℥j.
M.

℞. Liq. Iodinii Comp., ... f℥ij.
    Glycerinæ, ... q. s. ad f℥j.
M.

When there is marked hypertrophy, the best and most rapid results (next to excision) are obtained by the careful use of the electro-cautery. Children of six or eight years readily submit to this treatment provided thorough cocaine anæsthesia be produced before each application of the heated wire. A gargle containing tannic acid must also be used four times daily, as :—

℞. Acidi Tannici, ... ʒss.
Glycerinæ, ... fʒss.
Aquæ, ... q. s. ad fʒviij.
M.

Syrup of the iodide of iron should be given three times daily, care being taken not to administer it at the time that the gargle is used. Cod-liver oil is also serviceable.

Together with this treatment enough nutritious food must be given to keep up the strength. This can be done with readiness, since, in spite of the size of the tonsils, there is usually no pain, and little difficulty in swallowing.

Excision must be practiced when there is excessive enlargement, provided the above measures of reduction have been thoroughly tested without avail, or if, at any time, there is dangerous interference with respiration. Constant or frequent cough, or the presence of any other symptom suggestive of phthisis, also demands an immediate operation.

If, after removal of a portion of the tonsils or their reduction by treatment, the chest is slow to regain its natural form, the use of light dumb-bells and carefully regulated gymnastics are of much service. Dupuytren's method of reducing the sternal prominence by placing the child's back against a wall, and pressing it firmly backward with the palm of the hand during each act-of expiration, is efficient, notwithstanding its apparent roughness.

12. RETROPHARYNGEAL ABSCESS.

Abscess behind the pharynx is an uncommon disease; so rare is it, indeed, that in many years' experience at the Children's Hospital I have seen but one case, and this, unfortunately, passed from observation before its termination.

Its occurrence is not limited to any age. It results from direct injury; from disease of the cervical vertebræ; as a sequel of fever; or, more frequently, arises idiopathically.

The symptoms are difficulty in swallowing and breathing, with a peculiar sound during the latter act. On lying down the respiratory embarrassment is increased, sometimes to such an extent as to threaten suffocation. There is, also, great stiffness of the neck, retraction and immobility of the head, and a diffuse swelling of the lateral cervical surfaces, often greater on one side than the other. If now the finger be carried over the root of the tongue, and down toward the pharynx, a firm or fluctuating swelling will be felt, more or less filling the pharyngeal canal, and projecting over the opening of the glottis. On inspecting the throat, the swelling can usually be seen, occupying one or other side or the middle of the pharynx, and pressing forward the uvula and soft palate. The investing mucous membrane may be normal or congested. Sometimes the mouth cannot be sufficiently opened to permit of inspection, and at others the abscess is seated so low in the pharynx that no tumor can be seen.

Duparcque enumerates three symptoms indicating the formation of an abscess behind the œsophagus, viz.: Severe pain, produced even by moderate pressure on the larynx and upper part of the trachea. The entire suspension of respiration by such pressure. Displacement of the larynx forward and to the right.

Fever and cerebral manifestations may or may not be present, and initial symptoms are far from being uniform, so that, unless an examination of the throat be made, the disease may be overlooked in its early stages. Ordinarily, however, the diagnosis can be made without difficulty.

Dr. West, describing the lesions in one of his cases, states: "Immediately on dividing the cervical fascia on the right side, a quantity of thick, yellow, healthy pus passed out. This matter had burrowed close to the œsophagus, to within a little more than an inch of the clavicle, and also in an oblique direction behind the œsophagus toward the left side, completely detaching it from its connections on the right side, though not on the left. It passed up behind the œsophagus and pharynx quite to the base of the skull, a few threads of cellular tissue bathed in pus being all that remained of their posterior attachments. The tonsils were not enlarged, and the glottis was neither red nor swollen, but quite natural."

The prognosis is very grave when the disease accompanies cervical caries; under other circumstances it is favorable. When untreated, the course is prolonged, as the abscess is slow to break spontaneously. Suffocation from the sudden discharge of pus is an exceptional event.

The treatment is simple. As soon as the abscess has formed, it must, when within easy reach, be punctured by a bistoury, the blade of which has been carefully wrapped with adhesive plaster to within a fourth of an inch of its point. If the abscess be situated low down, a trocar and canula is the safer instrument to employ.

For several days after the operation, occasional pressure must be made by the finger on the tumor, to ensure thorough evacuation of the pus. At the same time a general tonic and supporting treatment is advisable.

CHAPTER II.

AFFECTIONS OF THE STOMACH AND INTESTINES.

The fact that hyperæmia is the acknowledged condition of the gastro-intestinal mucous membrane during digestion, and the easily appreciated readiness with which this hyperæmia may pass from a normal to an abnormal degree under the influence of such apparently trifling irritants as food in excessive quantity or of improper quality, has led me to doubt the existence of what is usually termed "simple indigestion" or "functional dyspepsia." The doubt has been strengthened after years of special study of this class of affections in children, and I am now disposed to attribute all forms of disordered digestion to a distinct tissue lesion. This may be, and usually is, a simple catarrh; but it is none the less a lesion. The fact of its leaving no traces after death, when this event has occurred from other causes, is a poor argument, for no one expects to be able to detect the lesions of simple pharyngitis, for instance, under like circumstances.

In consequence of this belief, I have departed somewhat from the usual plan of classifying diseases of the digestive tract.

1. ACUTE GASTRIC CATARRH.

This is one of the most common ills of childhood, since, in addition to arising idiopathically, it attends every disease in which there is pyrexia, as well as many of those that are apyretic.

The IDIOPATHIC FORM may occur at any age, but is infrequent in breast-fed infants. Its origin under such circumstances is always traceable to some abnormal condition of the mother's milk. The ordinary predisposing causes are dentition, general feebleness of constitution, exposure, and imperfect hygiene. Exposure is also an excitant, but the chief of this class of causes is the administration of food that is either bad in quality or excessive in quantity. An attack, too, sometimes directly follows the use of emetic doses of antimony, sulphate of copper or ipecacuanha.

The ANATOMICAL LESION is hyperæmia of the mucous membrane of the stomach, producing an increased secretion of mucus, and a diminished flow of gastric juice.

SYMPTOMS.—An attack of what the nurse calls "indigestion" comes on in infants after a bottle of changed milk or a "taste" of some unusual food has been given; in older children after a mixed and indigestible meal, particularly when this has been attended by exposure and excitement. The child, after a few hours, becomes listless, has a hot, dry skin, loses appetite, is thirsty, sleeps restlessly, and, if old enough, complains of headache, abdominal discomfort and nausea. Then there is vomiting of sour-smelling, curdled milk, or of whatever food is in the stomach in a more or less imperfectly digested state. The first act of emesis is easy, but if repeated, as is often the case, there is painful retching, and nothing is expelled save a little bile-stained mucus. Soon the tongue becomes covered, except at the very tip and edges, which are red, with a thick white or yellowish-white fur, through which the fungiform papillæ protrude as bright scarlet points. The breath has a heavy or sour odor. There is some fever, the temperature ranging from one to three degrees above normal, and the pulse counting 110 or 120 per minute. There is moderate tenderness on pressure in the epigastric region. The bowels are confined, and the urine is lessened in quantity and lateritious. These symptoms continue from twenty-four to forty-eight hours.

The attack sometimes terminates suddenly, with several loose fæcal evacuations. In other cases the fever gradually subsides, the nausea and thirst diminish, the tongue cleans, and the appetite slowly returns, convalescence extending over a period of two or three days.

The DIAGNOSIS is readily established by the history of the causation, the character of the vomit, the state of the tongue, the moderate fever, the epigastric tenderness, and the course of the attack.

The PROGNOSIS is always favorable so far as recovery is concerned, but it must be remembered that one attack always increases the susceptibility to another.

TREATMENT.—Complete rest, on the nurse's lap for infants, and in bed for older children, is essential. During the first twelve or twenty-four hours there is no inclination for food, and if any be forced it is quickly rejected. Consequently it is better to avoid any attempt at feeding until the stomach becomes settled. Thirst is to be relieved by ice, swallowed in small bits at short intervals, and by frequent small draughts of iced carbonic acid or Vichy water. Such measures are also useful to allay nausea and vomiting, but if these symptoms are at all obstinate, a mustard sinapism, just strong enough to redden the skin, should be applied to the epigastrium, and the following prescription ordered:—

> ℞. Liquor. Calcis,
> Aquæ Cinnamomi, ... āā f℥ij.
> M.
> S.—One to two teaspoonfuls, according to the age, at intervals of 15 to 30 minutes, as necessary.

Frequently repeated small doses of the effervescing citrate of potassium, or of the effervescing draught already mentioned (page 189), are efficient. A good plan, too, is to divide the contents of both packages of a Seidlitz powder into a number of equal parts, about twelve for a child of three years; dissolve a portion from each in a small tablespoonful of water, pour them together, and administer in a state of effervescence. This may be repeated, at first, every half-hour, later at longer intervals; rarely more than six or eight doses are required to check the vomiting. This method has the additional advantage of acting gently on the bowels.

In those exceptional cases in which, after an unsuitable meal, there is headache, fever, epigastric discomfort, and nausea without vomiting, it is necessary, as a preliminary measure, to induce emesis by draughts of warm water or a sufficient dose of syrup or wine of ipecacuanha.

When vomiting has ceased and nausea disappeared, the patient must begin to take food. At first one ounce of sound milk diluted with half an ounce, or even an ounce, of lime-water or barley-water may be given every two hours; and the quantity increased and the dilution lessened as the stomach regains its functional powers. Weak mutton, veal, or chicken broth, free from grease, and diluted with one-half or an equal quantity of barley-water, sometimes suits when milk cannot be retained.

While attention is paid to the diet, care must be taken to secure a free evacuation of the bowels by a mercurial followed by a saline laxative. Beyond this, all that is required is to administer properly-proportioned doses of bicarbonate of sodium and pepsin before each meal, for three or four days, and to gradually increase the diet to its normal standard as healthy digestion is restored.

2. CHRONIC GASTRIC CATARRH.

This affection presents so many points of dissimilarity, according to the age of the patient, that it is desirable to study it under two heads, namely, chronic gastric catarrh in infants, and chronic gastric catarrh in children who have passed the period of first dentition. Further, since chronic catarrh of the stomach is always attended by imperfect gastric digestion, and since food imperfectly digested in the stomach is unfitted for intestinal digestion, and must act as an irritant and lead to intestinal catarrh, it is impossible to absolutely isolate the two conditions in a clinical description. This is so markedly the case in older children that it seems best to defer the study of the second division of the subject to a later section, headed "chronic gastro-intestinal catarrh," and at present to consider only—

CHRONIC GASTRIC CATARRH IN INFANTS.

This dangerous affection, sometimes termed "chronic vomiting," is of common occurrence.

MORBID ANATOMY.—In the earlier stages there is a simple hyperæmia of the gastric mucosa, but a long continuance of this condition thickens and loosens the membrane, changes its color to an ashen-grey, and leads to an excessive formation of tenacious mucus or muco-pus, while greatly lessening the secretion of efficient gastric juice. Coincident enlargement of the gastric glands also gives the appearance of roughness to the surface of the mucous membrane.

ETIOLOGY.—The period of life between the third and seventh months furnishes by far the greatest number of cases. Sex and season are not influential. Infants fed entirely at the healthy breast are very rarely affected.

The predisposing causes belong to the class of influences that lower the readily depressed vitality of early infancy; for instance, over-crowding, filth, want of sun-light and fresh air in dwelling-rooms, insufficient clothing, and too early weaning.

The one great exciting cause is the administration of unsuitable food. Sometimes the breast-milk departs so much from its normal quality that it acts as an irritant upon the delicate mucous membrance and produces catarrh; or it may flow so freely that the child swallows more than he can digest, and the surplus, having undergone chemical change in the stomach, produces a like result. But the harm commonly arises from the use, in artificial feeding, of food that is either, by its nature, unsuited to the feeble digestive ability of infancy, or which, though good in itself, is rendered hurtful by being kept in unclean vessels, and given from foul or badly constructed bottles.

Of the first or essentially bad articles of diet, the farinaceous foods are the most harmful, because, for the digestion of starch, both saliva and pancreatic juice are required, and these secretions are absent until the fourth month and not fully established for some time later. Further, when subjected to the action of a ferment, as mucus, in the presence of heat and moisture—conditions existing in the stomach—these substances readily undergo fermentation resulting in the formation of acid which acts as an irritant to the susceptible mucous membrane. Consequently, such a diet used, as it too often is, to the exclusion of milk, must be a very active cause of gastric catarrh. The habit of allowing or encouraging infants to bolt bits of table-food and drink tea is quite as injurious; perhaps, though, this indiscretion is more apt to produce chronic diarrhœa than chronic vomiting.

Perfectly pure milk will be quickly changed and rendered irritating and unfit for use by being poured, when delivered by the milk-man, into pitchers or cans not properly cleansed from the remains of the supply of the day before. The smallest quantity of sour milk is sufficient to rapidly produce a like change in several pints of the fresh fluid when mixed with it. The same is true of unclean bottles and tips to which the dregs of former meals adhere in the form of small white curds. In these the change begins as soon as the fresh milk is added, and advances far before the child finishes the meal.

A knowledge of the etiological factors explains why by far the greatest sufferers are foundlings, foster-children, children born to poverty, and those belonging to women who engage themselves as wet-nurses, or are obliged to earn their living by working away from home.

SYMPTOMS.—The first symptom is vomiting, occurring at irregular intervals, and resulting in the expulsion of curdled, sour-smelling milk, or whatever food is in the stomach, stained yellow or green by bile. The characters of the vomit however, soon change, the bile disappearing and only a clear, watery fluid, containing fragments of food, being ejected. In addition, there are eructations of sour or even fetid gas. The surface of the body is normal in temperature or cool, the skin is harsh and sallow, and an eruption of strophulus may cover the trunk and arms. The lips are red and dry, the tongue is coated by a thick, dry, yellow fur, with dull red fungiform papillæ protruding at intervals; the mouth is parched, thirst is increased, and milk or water is taken greedily only to be quickly vomited again. The bowels are constipated, and when an evacuation does occur, it is attended by great straining, and the fæces appear in small, round, hard, light-colored lumps, often enveloped in mucus; sometimes moderate diarrhœa alternates. The abdomen is distended and tympanitic, and there is great tenderness over the epigastrium. Flesh is rapidly lost, the anterior fontanelle becomes sunken, the child is very fretful, has an aged and anxious expression of face, and a deep furrow may be noticed passing downward from the alæ of the nose to encircle the mouth, giving to the lips the appearance of projecting.

This condition continues, with occasional brief periods of improvement, for several months. Then the vomiting becomes more constant, occurring both after food and in the intervals of feeding. It is excited by any disturbance, such a trifling act as wiping the mouth, for example, being sufficient, to bring on an attack. The stomach seems now to have lost its power to even begin the digestion of the blandest food, for if milk be given, it is vomited uncurdled and in the same state as when swallowed. Emaciation progresses very rapidly. The skin, dry and inelastic, hangs in loose folds from the limbs, and is apparently too large for the wasted body. It has a muddy color, and exhales an offensive, sour odor. The face is pinched, the eyes are sunken, though bright, with pearly sclerotics; the nose is sharp, and the cheeks hollow. The infant lies with the knees drawn up against the abdomen, and to this position they are at once returned when straightened out; often the legs are moved about uneasily, indicating abdominal pain. There is little sleep either by day or night. Fretfulness is constant, with an occasional breaking out into loud, painful cries, or as weakness increases, into low wailings. The tongue is dry and heavily coated, the bowels continue constipated, and, toward the end, the abdomen becomes retracted. The pulse grows weak and frequent in proportion to the failure in general strength, and the temperature falls below normal; the thermometer, placed in the rectum, often registering but 97° F. The breath is sour, and the scantily secreted saliva, perspiration and urine are all very acid. As death draws nearer, the surface is perceptibly cool to the touch, the hands and feet become blue, patches of thrush appear upon the inside of the lips and cheeks, the little patient lies utterly exhausted, dozing or half unconscious, and for several days before the fatal termination, the only evidences of life are the gentle rise and fall of the chest in breathing and the occasional expression of pain that flits across the face.

Sometimes, in the last few weeks of the attack, certain of the symptoms become exaggerated, constituting what is termed "spurious hydrocephalus." In this condition there is deep depression of the fontanelle, dilated pupils, transient flushing of the face, great languor, heaviness of the head, drowsines, semi-stupor, and even coma with stertorous breathing. Indications of pain and fever are, however, absent. The sunken fontanelle shows a deficiency in the amount of blood in the brain, but, as suggested by Parrot, there may be, in addition to this source of the symptoms, some toxic element analogous to that of uræmia. Thrombosis of the cerebral sinuses and intracranial hemorrhages are also occasionally found after death, but their connection with the ante-mortem phenomena is by no means uniform.

When the disease terminates favorably the vomiting occurs at longer and longer intervals, and finally stops entirely, though there is great liability to a return on the slightest indiscretion. Afterward all the other symptoms disappear except the constipation, which is apt to be obstinate. An excessive development of fat is a frequent sequel.

DIAGNOSIS.—The protracted course, the frequent and obstinate vomiting of sour liquid, and the excessive emaciation, mark the disease with sufficient distinctness. The association of vomiting and constipation, and the development of the features of spurious hydrocephalus, are suggestive of tubercular meningitis. This disease is to be excluded by the depressed condition of the fontanelle, the regularity of the pulse, the tympanitic abdomen and the apyrexia.

PROGNOSIS.—Chronic vomiting is a dangerous affection, even

under the best circumstances and an unfavorable result may be expected when the attack begins during the first three months, or occurs in a child who has been hand-fed from birth. The course is prolonged, extending from two to four or even six months.

TREATMENT.—The first and most essential step in the successful management is a careful regulation of the diet. There are two ends to be attained; first, to give the stomach as much rest as possible, and second, if a sour odor of the breath and body indicates that fermentation is going on in the viscus, to stop this process by withholding fermentable materials.

In cases of moderate severity, where the vomiting has followed premature weaning, with a substitution of farinaceous food for the natural, a return to the breast is indicated. Or, if this be impracticable, the food must consist exclusively of sterilized milk guarded with lime water or diluted with barley water. For a child of three months a good proportion is two parts of milk to one of lime-water, or equal parts of milk and barley-water. Of either of these mixtures two fluidounces may be given every two hours, though the only guide to the proper quantity is the power of retention ; and if one measure be rejected, less must be given at the next feeding, until the proper amount is ascertained. Subsequently, it may be increased as the stomach becomes retentive.

In more severe and long-standing cases, attended by symptoms of acid fermentation, it is still advisable, with young infants, to try a return to the breast. In doing this, the fact that the mere act of sucking is sometimes sufficient to excite vomiting, must be remembered. So, before discarding the mother's milk as a food, an effort should be made to administer it with a spoon, after pumping it from the breast. It may then be retained and digested. However, the majority of patients in this stage of the disease can digest neither breast-milk nor any of the ordinary preparations of cows' milk, and time and even life may be saved by adopting at once an unfermentable diet, as a mixture of—

Fresh Cream,	f ℥ ss.
Whey,	f ℥ j.
Barley-water,	f ℥ j.

Or—

> Weak veal broth (half a pound to the pint).
> Thin barley-water; in equal quantities.

Either food is best given cold and in small quantities at short intervals. One teaspoonful at a time is enough in bad cases; but when the amount is so small, the dose must be repeated every ten or fifteen minutes. As improvement occurs, the amount of food and the length of the intervals should both be increased. It is important always to forbid the use of a bottle and feed with a spoon. A careful observance of these details is frequently rewarded by a rapid cessation of the vomiting. After the stomach has been retentive for forty-eight hours an effort may be made to return to a milk diet and the bottle. The change may be begun with what is known among dairymen as "strippings." This is the milk obtained by re-milking the cow after the udder has been once unloaded. It contains much cream and little casein. A combination of this sort:—

> Strippings, ... f ℥ j.
> Water, ... f ℥ ij.;

administered every two hours agrees well. This mixture should be used for several days in gradually increasing quantities, until as much as six fluidounces every three hours can be borne with ease, then food of which sterilized milk is the basis may be safely resumed. For example:—

> Milk, ... f ℥ iv.
> Cream, ... f ʒ ij.
> Milk sugar, ... ʒ j.
> Water, ... f ℥ ij.;

given from a perfectly clean bottle, every three hours. The substitution of lime-water or barley-water for water is advisable in case of slow digestion with colic; so, too, is the addition of a teaspoonful of caraway-water if there be flatulence. Another good combination is—

Milk,	f℥ iv.
Cream,	f℥ ij.
Mellin's food,	℥ j.
Water, hot, to dissolve the Mellin's food,	f℥ ij.

These foods are proportioned for infants of about three months. The importance of preparing each meal separately, and immediately before it is served, must not be overlooked.

The second necessary step is to attend to the clothing and hygiene. A light, long-sleeved, woolen shirt, drawers of the same material, and thick worsted stockings, must be worn; the latter especially should be insisted on, as it is essential to keep the feet warm. In addition, it is well to envelop the abdomen with a flannel binder. The clothing must be changed at reasonable intervals. Should it become soiled by vomit, it must be taken off at once and carried out of the room. The frequency of such accidents can be much lessened by placing a towel under the child's chin and over his chest, to receive the vomited matter. This, too, when soiled, is to be removed immediately and replaced by another, perfectly dry and fresh. The sick-room must be light and well ventilated, and no articles of body or bed clothing moistened with vomited matter should be allowed to remain in it a moment; the proper temperature is 68° F.

If the feet remain cold in spite of stockings, they should be rubbed from time to time with the dry hand, or with some stimulating liniment—oil of turpentine, f℈ij, and olive oil, f℥ij; if this does not warm them, the legs, as far as the knees, may be put in a hot mustard foot-bath for five minutes. Hot flaxseed poultices, made light and dashed with mustard, will, when worn over the belly, relieve pain and fretfulness; the same result follows repeated applications of the stimulating liniment. To promote free action of the skin, the whole body should be sponged with warm water twice a day, and afterward anointed with warm olive oil, which must be gently rubbed into the surface with the pulps of the fingers. If there be great prostration, a full bath of 100° F., with or without mustard, may be resorted to, the body being immersed from one to three minutes.

Under such circumstances, it may also be necessary to envelop the legs in cloths wrung out of hot mustard water, and to keep bottles or rubber bags filled with hot water in close contact with the body in order to encourage reaction and maintain a normal temperature.

Of medicines, wine of ipecacuanha, in a sufficient dose thoroughly to rid the stomach of acid contents, prepares the way admirably for other measures, but should never be used when the strength is exhausted. Marked sinking of the fontanelle is always a contraindication.

The ordinary means of relieving gastric irritability are of little avail in checking the vomiting in chronic catarrh of the stomach. The remedy that seems to possess most power to accomplish this is liquor potassii arsenitis. The proper dose for a child of three months is half a drop, three times daily, administered simply in a teaspoonful of water or combined with an alkali and aromatic, as:—

    ℞. Liquor. Potassii Arsenitis, ... ♏xij.
        Sodii Bicarbonatis, ... gr.xxiv.
        Aquæ Menthæ Pip., ... q. s. ad f℥ iij.
    M.
    S.—One teaspoonful, in a little water, three times daily.

When Fowler's solution fails there are several other drugs that may be tried. These are vinum ipecacuanhæ, in drop doses every three hours; calomel, one-sixth of a grain, every four hours; salicylate of sodium, half a grain every two hours; and tinctura nucis vomicæ, half a drop three times daily, combined with bicarbonate of sodium and an aromatic, as in the prescription just given.

While these medicines are being administered, the bowels should be evacuated by laxative enemata.

Prostration demands stimulants. The best is old whiskey, which may be given in ten-drop doses every two hours; but the guide for the dose, as well as for the proper time to commence administration, is the condition of the fontanelle.

When convalescence begins, half a drop of tincture of nux

vomica, or fifteen drops of the ferrated elixir of cinchona, may be prescribed, and the tonic effects of fresh air and sun-light must be utilized by taking the child out of doors when the weather permits.

3. ULCER OF THE STOMACH.

This disease is not very uncommon in new-born infants, but is decidedly rare afterward. It may occur as a single, minute, round ulcer, with a perforating tendency as in adults, or as numerous small scattered erosions which stud the surface of the mucous membrane and assume the appearance of ulcerated follicles. The perforating ulcer has been ascribed to all the various causes which are held to be potent in producing the gastric ulcer of adult life, and it is probable that for children after they are weaned the pathology of the two may be the same; but for newborn infants, circulatory disturbances which ensue somewhat suddenly at birth, the sudden arrest of the placental stream, the gradual development of the pulmonary circulation, associated as it often is with partial atelectasis, so potently predispose to venous stagnation in the abdominal viscera as to give much ground for the belief that congestion, and even ecchymosis, are at the root of the ulceration.

The scattered ulceration has been found under such varied clinical conditions that it is impossible to attach any definite meaning to it, although one may suppose with reason that it is the result of some chronic catarrh.

SYMPTOMS.—Vomiting of blood and melæna are the only indications which point to the existence of an ulcer of the stomach in the infant. A healthy child within a few hours of its birth who begins to vomit blood and to pass pitchy matter per anum, may have a gastric ulcer. More than this we cannot say, for the same symptoms may certainly be present without any ulcer. In the few cases in which a gastric ulcer is present in older children, the symptoms, if definite, should be as in adults—epigastric pain and vomiting. The follicular ulcer cannot be diagnosed, and has always been found accidentally upon the post-mortem table.

TREATMENT.—The bleeding in many cases is so quickly fatal that nothing is available; cold alum whey may be given, and some castor oil, which, by acting upon the bowels, may do something to relieve any local plethora which might exist.

Tubercular ulceration of the stomach is occasionally met with, but it has no symptoms apart from those of tabes mesenterica.

4. SOFTENING OF THE STOMACH (GASTROMALACIA).

This condition has received a great deal of attention, and some of the most distinguished writers upon the diseases of children have credited it with being a distinct disease, but, to my mind, with insufficient reason. Of symptoms it has none which are in any way characteristic, and the appearances found after death are identical with those of post-mortem solution. Whether this, as well as other changes which are cadaveric in their nature, may not at times commence during the last hours of life may perhaps be an open question, but that the change is, in all cases, essentially what has been described as post-mortem solution there is no doubt.

Goodhart has twice found evidence of a gastric solution of the lung, which had gone on during the life of the patient. Into the appearances of the parts it is needless to enter further than to say that they showed a distinctly peculiar broncho-pneumonia, and that in each case there had been a moribund condition associated with vomiting for some days before death. Now it is obvious that such a condition has no right to the position of a disease; it would never have occurred had the circulation of the patient been at its proper tension. It was the result of an ebbing life, not a disease, which caused death. So it is with the gastromalacia of children. It is the result of exhausting disease of any kind, and is virtually, if not literally, a post-mortem change.

5. CHRONIC GASTRO-INTESTINAL CATARRH.

This disease is common in children who have passed the first dentition, and bears to them somewhat the same relation that chronic vomiting does to infants. Among the latter it is very uncommon, perhaps because the anatomical position and greater irritability of the stomach in the early months of life favor the rapid expulsion of improper or partially digested food, and the irritating products of gastric fermentation, which would otherwise, as in older children, pass through the pylorus and induce catarrh of the intestinal mucous membrane. The disease is met with in two forms, differing merely in the degree of catarrh. For convenience, they may be considered separately; as, habitual indigestion, in which the catarrh is moderate in degree; and mucous disease, in which it is intense.

HABITUAL INDIGESTION.

In the rare cases of this disease, where death has resulted from an intercurrent affection, post-mortem examination has revealed the gastro-intestinal mucous membrane, finely injected, reddened in patches, flabby, swollen and covered with a layer of tenacious mucus of variable thickness. In the majority of cases, though, it is probable that the catarrh does not extend beyond the grade that would leave no gross change after death.

ETIOLOGY.—The predisposing agencies are deficient functional activity of the stomach, either existing simply as a factor of a weak constitution, or resulting from previous disease or ill-directed hand-feeding. Residence in large cities, and dark, close and damp houses; too little out-door exercise, and too much confinement and pushing at school; and finally, the eruption of permanent teeth, belong to this class of causes. They all act by lowering the capacity to digest, and the best food imperfectly digested undergoes chemical changes rendering it irritant and capable of transforming the normal hyperæmia of digestion into the congestion of catarrh. Fewer cases are met with in summer than in winter, because during the former season children live more out of doors, and the functions of the skin are more active, keeping a larger quantity of blood at the surface—a great safeguard against catarrh. Season, then, may be added to the predisposing influences.

The prime exciting cause is unsuitable food. As a rule, especially with children of the poorer classes, among whom the disease is very rife, the fault lies in the food being too strong. These children are allowed to sit at table and partake of whatever the elders eat, such as meat two or three times a day, with potatoes, bread and butter and tea, none too well prepared or of too good quality. This coarse food, of itself irritating to the delicate lining of the stomach, is also very difficult to digest. The child may have force enough to maintain a fair degree of health against this odds for a while, and some even win in the race, but for most, the time of trouble surely and soon comes. Some portions of the food begin to escape, more or less completely, the solvent action of the gastric juice. The starches and fats, influenced by the heat of the parts and the organic matter present, undergo fermentation, and are converted into acids with the liberation of carbonic acid gas; the albuminoids become partially decomposed and acrid. These not only irritate the mucous lining of the stomach, but passing into the intestine, act upon its mucous membrane, and cause the same catarrhal lesions there.

At first an attack of vomiting and purging, by cleaning out the alimentary canal, puts an end to the catarrh, and the patient is free from symptoms so long as the resulting anorexia restricts his appetite. But a return to the old diet is quickly followed by a relapse, culminating in another natural effort at relief; and so the attacks recur, growing more and more frequent and easily induced, until what was originally an acute and passing indigestion becomes chronic.

As soon as the catarrh is established and the interior of the canal is covered with tenacious mucus, the disease begins to react upon and increase itself. For, whatever food is taken is soon enveloped by mucus, and this coating prevents the free access of the gastric and intestinal juices, which are solvents and antiferments. Mucus, too, is in itself a powerful ferment and increases the formation of irritating substances; further, by covering the interior of the alimentary canal, it prevents the absorption of what little food is digested, leading to malnutrition, with a deterioration in the quality of the gastric juice and succus entericus, and leaving more material for chemical change. Thus there is a direct and an indirect reaction.

Well-to-do children are spared a coarse diet and, in consequence, do not suffer so severely. In them bad food takes the form of rich dishes, pastry, sweets and so forth.

Exposure to wet and cold has some excitant influence, though, without the aid of bad diet, it is scarcely sufficient to induce an attack.

SYMPTOMS.—When the disease is fully developed, the patient has a spare, delicate appearance, the face wears a languid exprespion and is pale; the pallor at intervals increasing very much, or again giving place to flushing of one or both cheeks. The hair is crisp and lustreless. The conjunctivæ are sometimes natural, but more often slightly yellow. The skin is cool, dry and rough to the touch, and somewhat sallow in hue. The pulse is weak, but otherwise unaltered. The mucous membrane of the mouth is less pink than normal; the breath has a heavy, disagreeable odor; the tongue is pale, broad and flabby, frequently indented by the teeth, and covered with a thin, white frosting, which grows thicker, and more yellowish toward the posterior part of the dorsum. Through this coating the enlarged fungiform papillæ project, and are redder than the rest of the mucous membrane, but not so highly colored as in acute gastric catarrh. Moderate hypertrophy of the tonsils can frequently be observed, and, as a rule, the cervical lymphatic glands are slightly enlarged.

The appetite is variable and perverted, the desire being for highly-seasoned food. After eating, eructations of flatus occur, and small quantities of partially digested food, mixed with thin mucus and intensely sour, are from time to time regurgitated from the stomach. Tympanites is a constant symptom, and when the child is stripped the distended abdomen contrasts markedly with the spare trunk and limbs. Pain is uniformly present. It may be constant or paroxysmal, severe and colicky, or only amounting to discomfort, and either general or confined to certain parts of the abdomen. Usually it is paroxysmal, beginning from two to three hours after meals; if constant, it is subject to exacerbations at these periods. Generally, too, it is only moderately severe, and is confined to the left or right hypochondriac region. The reason for this limitation being, that in both positions, but especially in the first, the colon makes a sharp turn where the gases, liberated by fermentation, become lodged. On account of the mucus covering the fæcal masses as well as the interior of the bowel, bringing two slippery surfaces together, the peristaltic contractions are less efficacious, and constipation results. Intervals of two, three, or even nine days elapse between the movements, which are attended by considerable straining, and result in the expulsion of a small number of dark, hard lumps enveloped in mucus.

The urine, at times, is scanty and high colored, at others, over abundant and light colored. The diminution is apt to attend exacerbations of abdominal pain.

During the day the child is listless, disinclined to play and easily tired, while at night he tosses about the bed in a dreamy sleep.

To the above symptoms catarrh of the nasal and bronchial mucous membranes is often added.

It is usual for the even course of the disease to be broken by vomiting and diarrhœa. In such attacks there may be slight fever, the tongue becomes more heavily coated, the appetite fails and thirst is increased. The vomited matter at first is composed of acid, partially digested food, mixed with stringy mucus; afterward, if there be much retching, of more or less bile-stained mucus alone. The purging, primarily, unloads the bowel of a large quantity of lumpy fæces, apparently the collection of several days; afterward, the stools are made up of mucus and liquid fæces. Such attacks last one or two days, and are followed by a brief period of improvement.

The DIAGNOSIS is easy.

The PROGNOSIS is favorable, though, when left to itself, the disease runs a protracted course, improving in summer to return in winter. By the general debility that it produces, it opens the way to intercurrent affections, or the development of hereditary tendencies, and renders both more fatal.

MUCOUS DISEASE.

This form of chronic gastro-intestinal catarrh occurs much less frequently than the other. It consists of a mucous flux from the whole internal surface of the alimentary canal, which interferes mechanically with the digestion and absorption of food, and so impedes nutrition as to suggest the presence of tubercles. The lesions are identical in kind with those of habitual indigestion, but are much greater in degree.

ETIOLOGY.—The affection usually arises between the fourth and twelfth years, and has the same predisposing and exciting causes as the milder form. There are two conditions, however, under which the disease is especially apt to arise, namely : the eruption of the permanent teeth, and attacks of whooping cough.

The influence of the former is explained by the intimate sympathy existing between the different portions of the digestive tract, on account of which the irritation of the mouth during dentition is reflected throughout the intestinal tract, producing increased secretory activity and greater susceptibility to irritants.

During the course of whooping cough, the gastro-intestinal mucous membrane is always in a catarrhal state. Much of the tenacious mucus expelled at the end of each paroxysm comes from the stomach. When vomiting occurs, most of the matter ejected is mucus, and the stools contain a quantity of the same substance. As the cough subsides, the secondary catarrh usually disappears, but after severe attacks, and in feeble children, it may continue, and pass into mucous disease.

SYMPTOMS.—As might be expected from what has already been said in regard to lesions and causation, the symptoms, in the main, are those of habitual indigestion greatly magnified.

The child is emaciated and muscularly weak. His face is uniformly pale, though subject to great changes in color, and at times a circumscribed crimson flush appears on one or both cheeks; at others, there is so much pallor, especially about the lips, that fainting seems imminent, and, indeed, it does sometimes occur. The eyes are surrounded by bluish circles, which deepen when the face pales. The conjunctivæ are muddy, and there is occasional squinting. The skin is markedly sallow, dry and rough to the touch and, by light friction, numerous fine scales of dead epidermis can be removed, and the hair has a lustreless, faded appearance. The cervical lymphatics are noticeably swollen, though painless.

The oral mucous membrane is pale. The tongue, besides being flabby and indented by the teeth, presents an appearance characteristic of the disease. The dorsum, with the exception of an oval space in the centre, is covered with a light gray coating, scarcely thick enough to obscure the natural pale-pink color, and shows clearly the slightly redder fungiform papillæ. The oval bare spot, about as large as a cent, is still deeper red, and shines as though varnished. This glossy look, in very severe cases, extends over the whole dorsum, and is due to an excessive secretion from the mucous glands of the mouth. Such a tongue does not lose the natural velvety appearance arising from the fungiform papillæ. (See *a*, Plate 2.)

Chronic hypertrophy of the tonsils, with plugging of the follicles by retained secretion, is common, and, in part, accounts for the disagreeable odor of the breath.

The appetite in the beginning fails, then becomes capricious, and, finally, almost insatiable. The increased desire for food is due partly to a morbid craving, excited by the irritation of the fermenting contents of the stomach and intestines, and partly to the demand of the tissues generally for more nutriment than is supplied by the imperfect digestion and impeded absorption.

Eating is followed by a sensation of drowsiness, and by eructations of flatus and acid liquid.

Tympanitic distention of the belly is always marked, and the child complains of pain in this portion of the body. The pain may be general, when it amounts to little more than a sensation of soreness, but more frequently it is limited to the left hypochondrium, and is stitch-like in character. Either variety may be constant, or present only after meals; in the former case there is a temporary increase of discomfort after eating. In some instances, paroxysms of severe pain in the neighborhood of the umbilicus occur early in the morning, and occasionally after meals. These are unattended by nausea, purging, or doubling of the body to secure relief, as in colic, but while they last, the pallor of the face is extreme.

Constipation is the usual condition of the bowels. Evacuations take place at intervals of several days, with much straining, and at times rectal prolapse; they are scanty, and composed of small, hard, dark-colored lumps, with a large proportion of mucus, and often contain intestinal parasites or their ova. Sometimes the constipation lasts for a week or more at a time, to be followed by a number of free evacuations in quick succession, relieving the bowel of the accumulated fæces; then comes another period of confinement, another relief, and so on.

By day, the patient suffers from headache; is languid, ill-tempered, and disinclined for study or play. At night, he is restless; grinds his teeth; starts from sleep in terror caused by frightful dreams, and often screams or talks incoherently, and for a time is seemingly unconscious of his surroundings. Somnambulism and nocturnal incontinence of urine are quite common. Stammering is another nervous symptom occasionally encountered.

There is no alteration in the temperature; the pulse is feeble, and there is frequently a slight, dry, hacking cough, entirely independent of pulmonary disease. The urine is diminished during the continuance of severe pain, but is voided in excessive quantities at the termination of the paroxysms.

At intervals of two or three weeks violent vomiting and purging occur. During these attacks, which last from one to three days, a large quantity of mucus is rejected; there is slight fever, and the tongue is changed in appearance, and for the second time assumes a characteristic aspect. It becomes less flabby, more pointed, and covered with a thick, white, feathery fur, except along the sides, where there are several smooth, bright-red, glazed patches of variable size and shape, with irregular, indented edges. A few red fungiform papillæ show through the coating. Sometimes the whole dorsum is clean, red, and glazed, as if denuded of epithelium. (See *b*, Plate 2.) Temporary improvement follows the clearing-out process, but soon the symptoms return, and slowly grow worse to culminate in another attack.

The course of the disease is very chronic, extending over months. There is no regular progression, though the tendency is for the symptoms to grow more and more severe as time elapses.

DIAGNOSIS.—Tuberculosis is the condition most likely to be confounded with the disease in question, and the mistake is especially apt to be made when a dry, hacking cough is present. The appearance of the symptoms after whooping cough or during second dentition; the state of the tongue; the mucous stools; the condition and color of the skin; the absence of pyrexia except during the attacks of vomiting and purging; the periodicity of these attacks; the diurnal drowsiness and nocturnal terrors, and the irregularity in the course are the distinguishing features.

PROGNOSIS.—Mucous disease is not in itself mortal, and is perfectly amenable to treatment. It is, nevertheles, dangerous from its power to reduce the general nutrition, thus opening the way for more serious intercurrent affections.

As the plan of managing both forms of chronic gastro-intestinal catarrh is the same, it is unnecessary to divide the subject of—

TREATMENT.—Since the exciting cause is perfectly well known and removable, relief may be confidently promised, provided it be possible to regulate the diet. There are two rules to be insisted upon: first, to stop the supply of all those articles of food that readily undergo fermentation; and, second, to allow only a moderate quantity of food at a time, so as not to overdistend the stomach, while the meals are increased to four a day, to insure the ingestion of a proper amount of nourishment.

All farinaceous substances must be excluded from the dietary save stale or toasted bread, and this, even, must be restricted in amount. Potatoes, peas, beans, turnips, carrots, parsnips, fruit, cakes, pastry, sweetmeats and butter are all in the proscribed list.

Of permissible articles, milk, eggs, and lean meat are the chief, though fresh fish, raw oysters, cauliflower tops, spinach, asparagus, lettuce and celery can be used without ill effect. With such food to select from, it is easy to write out a suitable diet list and make changes sufficiently often to avoid cloying the appetite by monotony. In writing such lists, it is best to fix the hour, as well as the ingredients, of each meal. For example:—

Breakfast, at 7 A.M.—One or two tumblerfuls of milk guarded by lime-water * (f℥ij to f℥vj), the yolk of a soft-boiled egg, and a single thin slice of stale, unbuttered bread.

Luncheon, at 11 A.M.—A cup (f℥iv) of beef-tea, or mutton broth, entirely free from fat,† and a thin slice of dry toast.

Dinner, at 2.30 P.M.—Broiled mutton chop, entirely free from fat, one or two, according to the size; a large spoonful of well-boiled spinach, and a slice of stale, dry bread.

Supper, at 7 P.M.—One or two tumblerfuls of milk guarded by lime-water, and a slice of dry toast.

Filtered water must constitute the drink, though, if the child will take it, half a tumblerful of Vichy at luncheon and dinner can be recommended.

---

* The lime-water is added both for the purpose of retarding coagulation and for its effect upon the mucus in the alimentary canal.

† The fat can be completely removed by allowing the broth to stand for a few minutes after it is made, and picking off the globules of oil as they rise to the surface with a fragment of blotting-paper.

Should failing appetite demand a change, another menu must be made, as:—

Breakfast.—Milk, a bit of boiled fresh fish, and a thin slice of unbuttered toast.

Luncheon.—The soft parts of six or eight small oysters, seasoned with salt alone, and a Boston cracker.

Dinner.—A bit of the breast of a roasted or boiled fowl, a moderate portion of well-boiled cauliflower tops, and a slice of stale, dry bread.

Supper.—Milk and dry bread.

Further variety can be had by substituting a thin slice of cold roast mutton or beef for the egg or fish at breakfast; at dinner, by running the changes on roast mutton, broiled beef-steak, roast beef, plainly cooked game, and such vegetables as stewed celery, boiled asparagus tops, spinach and cauliflower; by using different sorts of meat broths, and by changing the manner of cooking the eggs.

When, in mucous disease, there is great debility, stimulants are indicated. They should be given well diluted and with the meals. Whiskey and old dry sherry are the best. Of the first, one or two teaspoonfuls in a fourth of a tumbler of ice-water may be given with lunch and dinner; of the second, one or two tablespoonfuls with twice as much water at the same meals.

Next to regulating the diet it is important to maintain the activity of the skin. This is to be accomplished by baths, inunctions and proper clothing. Each morning the patient, being in a warm room, should be sponged with water at a temperature of 60° F., then thoroughly rubbed down with a coarse towel, and the whole body anointed with warm olive oil, which ought to be gently rubbed into the skin with the finger pulps. At bedtime a full bath of 100°, of five minutes' duration, must be given, and the inunction repeated, after careful drying with friction. In severe cases, where the skin is very dry and rough, the first warm bath should contain a heaped tablespoonful of soda, and with this and soap the whole surface must be thoroughly scrubbed.

Woolen underclothing, to cover completely the trunk and

limbs, and woolen stockings are to be insisted upon. The weight may be changed with the weather, but not the material. This not only keeps the skin warm, full of blood and functionally active, but it also maintains the heat of the whole body and saves force. Children dressed for beauty with four or five inches of bare leg, nine times out of ten suffer from chronic indigestion or bronchitis. First, because chilling of the surface drives the blood toward the interior and puts the mucous membranes in the most favorable conditions for catarrh ; secondly, because so much force is consumed in maintaining the normal temperature in the face of constant chilling that other functions, notably the digestive, must suffer. Parents would appreciate this better if they could be persuaded to try the experiment of sitting, for an hour or so, even in a warm room, in the same degree of nakedness that they inflict on their children, who are less robust and less able to resist cold.

Exercise in the open air on suitable days in winter, and an almost complete out-door life in summer, hastens recovery. The sleeping and living rooms should be large, light, dry, well ventilated and properly warmed.

Medicinal treatment is of minor importance, but by no means to be neglected. The indications to be fulfilled are to check the secretion of mucus ; to neutralize the acids formed by fermentation of the food ; to restore the mucous membrane to a normal condition, thereby improving secretion, digestion and appetite, and to secure regular action of the bowels and the expulsion of collected mucus and fæces. These accomplished, strength and health return, though it may be necessary to call in the aid of tonics.

Alkalies are the best remedies to check the secretion of mucus, and to liquefy it so that it may more readily be removed. They are also most efficient in neutralizing the acid products of fermentation. Simple bitters, too, have some power in lessening the formation of mucus, and considerable influence in arresting fermentation ; at the same time they give tone to the mucous membrane and stimulate digestion. Laxatives keep the bowels clear. Of the first class, bicarbonate of sodium; of the second, gentian or calumba; and of the third, senna or aloës, are to be preferred in treating this disease.

In habitual indigestion, a combination like the following will be all that is required:—

 ℞. Sodii Bicarbonatis, ... ℨj.
   Ext. Sennæ Fluid., ... f℥iij.
   Inf. Gentianæ Comp., ... q. s. ad f℥iij.
  M.
  S.—Two teaspoonfuls three times daily before eating, at the age of seven years.*

Should there be yellowness of the conjunctivæ and marked sallowness of the skin, indicating a slight degree of catarrhal jaundice, it is well, at first, to substitute equal doses of chloride of ammonium for the bicarbonate of sodium in this prescription.

In mucous disease a similar prescription, with minute doses of iodide of potassium to increase the salivary secretion, may be ordered before meals, as:—

 ℞. Potassii Iodidi, ... gr. vj.
   Sodii Bicarbonatis, ... ℨj.
   Ext. Sennæ Fld., ... f℥iij.
   Inf. Calumbæ, ... q. s. ad f℥iij.
  M.
  S.—Two teaspoonfuls three times daily before eating.

After food, it is well to order from ten to twenty drops of tincture of myrrh in a little water, for its powerful tonic action on the intestinal mucous membrane.

Aloës is valuable not alone as a laxative, but in arresting the mucous flux and bracing the mucous membrane. It can be administered in the form of tincture of aloës and myrrh, in doses

---

* All of the subjoined prescriptions are proportioned for children of this age.

of twenty drops, three times daily after eating. Or, if the child be able to swallow a pill, it may be combined thus:—

℞. Pulv. Ipecacuanhæ, ... gr. iv.
Pil. Aloës et Myrrhæ, ... gr. xij.
Ext. Gentianæ, ... gr. vj.
Ext. Taraxaci, ... gr. xij.
M. et ft. pil. No. xij.
S.—One pill three times daily an hour after eating.

When there is much debility iron is demanded, and if the proper form be selected, it may be given in spite of a coated tongue, the usual contraindication. A good formula is:—

℞. Ferri Sulphatis Exsiccati, ... gr. xij.
Tr. Aloës et Myrrhæ, ... f ʒ iv.
Syr. Rhei Aromat., ... q. s. ad f ℥ iij.
M.
S.—One teaspoonful three times daily after meals.

From this prescription there is an astringent action, by the iron and rhubarb, which tends to check the formation of mucus; a laxative action, by the aloës and rhubarb, keeping the bowels clear of mucus and fæces; while the myrrh is a direct tonic to the relaxed mucous membrane.

If, as the tongue cleans, the improvement under this plan comes to a stand, it is advisable to change to an acid treatment. There are several useful prescriptions, for instance:—

℞. Pepsin. Saccharat., ... ʒ ij.
Acidi Muriatici dil., ... f ʒ ij.
Aquæ Cinnamomi, ... q. s. ad f ℥ iij.
M.
S.—One teaspoonful three times daily after eating.

Or the acid may be combined with a bitter:—

℞. Acidi Muriatici dil., ... f ʒ ij.
Inf. Gentianæ Comp., ... q. s. ad f ℥ iij.
M.
S.—One teaspoonful three times daily after meals.

℞. Quiniæ Sulphatis, ... gr. xij.
　Acidi Muriatici dil., ... f℥ij.
　Aquæ Cinnamomi, ... q. s. ad f℥iij.
M.
S.—One teaspoonful three times daily after meals.

All of these prescriptions must be well diluted and taken through a glass tube.

During the periodical attacks of vomiting and diarrhœa, so apt to occur in both forms of the disease, the child must be put to bed, restricted to a diet of milk and meat broths, and ordered the following prescription:—

℞. Pepsinæ Saccharat.,
　Sodii Bicarbonatis, ... āā ʒj.
　Pulv. Aromatici, ... gr. xij.
M. et ft. chart. No. xij.
S.—One powder four times daily.

The diarrhœa must not be interfered with unless it become excessive, when it may be held under control by adding five grains of subcarbonate of bismuth to each of the alkaline powders.

After the tongue becomes normal and the active symptoms have disappeared, the general strength must be built up by a course of tonics. The best, are tincture of nux vomica, ferrated elixir of cinchona, and bitter wine of iron. In order to prevent a relapse, mixed diet must be avoided for at least two months after convalescence is fully established, and to confirm the cure, change of air, by a trip to the sea-shore or mountains, is advisable.

Both habitual indigestion and mucous disease are occasionally attended by a troublesome symptom that demands brief consideration. This is a peculiar cough, which is dry, paroxysmal, and unattended by lesions of the throat or lungs. The paroxysms are due to reflex causes; they commence in the early evening, and may, by their repetition, prevent sleep for half the night. On the following day the patient is as well as usual, or coughs only at long intervals, but about bedtime the trouble begins again. So the symptom continues for weeks at a time, unless its true nature as a "stomach cough" be recognized and it is properly treated. The paroxysms suggest those of pertussis, though they may be distinguished by the absence of whooping, and of the characteristic expulsion of tenacious mucus at the end of the kinks. Questioning often reveals the fact that the cough is worse after a rich and heavy supper.

If proper clothing be worn, the diet carefully regulated, and alkalies prescribed, as for an ordinary case of chronic gastro-intestinal catarrh, improvement is rapid, for in this way the cause is removed. Ordinary cough mixtures do more harm than good, from their tendency to derange digestion; still, the fatiguing cough must be relieved. This can be done by letting the child wear a small bean-shaped belladonna plaster over the larynx, and administering a dose of one of the following mixtures every two hours, beginning at four o'clock in the afternoon:—

    ℞. Pulv. Aluminis, ... gr. xlviij.
        Potassii Bromidi, ... ʒij.
        Syrupi Zingiberis,
        Aquæ, ... āā f ℥iss.
    M.
    S.—Dose, one teaspoonful.

Or—

    ℞. Ext. Belladonnæ, ... gr. ss–j.
        Pulv. Aluminis, ... gr. xlviij.
        Syrupi Zingiberis,
        Aquæ, ... āā f ℥iss.
    M.
    S.—Dose, one teaspoonful.

6. ACUTE INTESTINAL CATARRH.

The condition intended to be indicated by this title is usually called simple or non-inflammatory diarrhœa, and classed as a functional disease. But from its etiology, and from the fact that in certain patients and under certain circumstances it so readily lapses into entero-colitis, it is more than probable that it depends upon a distinct, though passing lesion—a hyperæmia or catarrh of the intestinal mucous membrane. This is difficult to demonstrate, partly because the opportunity for post-mortem inspection is rare in simple diarrhœa, and also on account of the well-recognized rapidity with which the appreciable manifestations of mild forms of catarrh disappear after death. Nevertheless, even those authors who advocate the functional character of the affection, state that in some instances of death, in feeble children or from intercurrent disease, autopsy shows injection, swelling, and relaxation of the mucous membrane, and tumefaction of the intestinal glands.

ETIOLOGY.—Constitutional feebleness and unfavorable hygienic surroundings, especially residence in crowded, damp, and filthy houses and quarters of cities, increase the liability to attacks of diarrhœa. Many more cases occur in summer than at other seasons of the year. Children of either sex, or of any age, may be affected, though the younger the patient the more serious the disease.

In infancy there are numerous exciting causes. Over-feeding, even with healthy breast-milk or well-prepared cows' milk, is one. Ordinarily, in such cases vomiting is so easy that the child gets rid of the surplus and no harm is done; but if this does not happen, the excess remains undigested; undergoes change; acts as an irritant to the intestinal mucous membrane, and causes diarrhœa. Another cause is food of bad quality; either poor and cholesterin-laden breast-milk, or unsound cows' milk and farinaceous preparations. Here the action is the same as in over-feeding, though more rapid and violent; this is especially true of the farinacea, on account of their readiness to undergo acid fermentation. Again, exposure to cold and wet, by chilling the surface and determining the blood to the interior of the body and mucous membranes, may lead to an intestinal catarrh in the same way that it does, more frequently, to a bronchial catarrh. Hyperæmia, too, of the mucous membrane of the alimentary tract is attended by a diminution in the secretion of digestive solvents and an increased production of mucus; two conditions most favorable to incomplete digestion and fermentation of the food with the formation of irritant products. These, as already seen, are quite capable, in themselves, to cause looseness of the bowels, and must greatly add to the ill effects of exposure. High atmospheric temperature is much more influential than low, particularly when associated with excessive moisture. Such conditions are powerful depressants to the vital forces; the digestion shares in the general weakness, and much of the food is left to ferment and become irritant. Finally, dentition is a frequent cause. During this process, the whole digestive tract sympathizes with the condition of the mouth, and becomes less able to perform its functions and more susceptible to irritants.

After the eruption of the milk teeth the use of unsuitable food and the disturbing influences of second dentition are the chief causes.

It is almost unnecessary to call attention to the lesson taught by this study of the etiology. There is, on the one hand, the presence of an irritant as a constant factor; on the other, a mucous membrane naturally delicate and functionally very active. The conclusion is inevitable, that the ordinary effect must follow, and hyperæmia or catarrh be produced.

SYMPTOMS.—In infants, the attack may begin suddenly, or be preceded for twenty-four hours or more by peevishness, languor, faded cheeks, slight abdominal pain, indicated by moaning or fits of crying, and restless, disturbed sleep.

Next, the bowels become disturbed. The movements number from four to eight in the twenty-four hours, and usually occur only while food is being taken—from six in the morning to ten o'clock at night. At first they differ from the normal, merely in being more liquid and copious, and having a more offensive odor. As the disease progresses they undergo various changes. Sometimes they are composed of a yellowish liquid containing white or yellowish flakes resembling curdled milk. At others, distinct white lumps of undigested curd are mixed with the liquid. Still again, green flakes may appear in a stool having the characters of the first; and finally, the whole may be of a deep green color, and contain small masses of mucus. In exceptional cases a small amount of bright-colored blood may be seen in the evacuations. Often the movements are preceded, for a short time, by pain, but this disappears as soon as the act is accomplished. Occasionally, if the stools be acid, considerable tenesmus attends their expulsion, and it is under such circumstances that blood is most likely to be voided.

The tongue is lightly coated; there is anorexia; increased thirst, and occasionally nausea and vomiting. The abdomen is natural in shape, and is soft and painless on palpation. The urine is somewhat lessened in quantity, and high colored. There is no pyrexia, and the pulse is but slightly increased in frequency.

The evil effect of several days' continuance of diarrhœa upon the general condition of the child is shown by the pallor of the face, the sunken eyes, the loss of weight and the flabbiness of the muscles. Under proper management the attack terminates in from four to seven days, and strength is soon restored.

Simple diarrhœa is more uncommon in older children and much milder in its manifestation. There is slight furring of the tongue, loss of appetite, and abdominal pain of a colicky nature, with more or less frequent evacuations of light yellow, offensive, semi-solid or liquid fæcal matter, at times containing masses of partially digested food. The patient is weak and disinclined to exert himself. These attacks last for three or four days, and are followed by little constitutional depression.

DIAGNOSIS.—There is no difficulty in distinguishing the disease. The only conditions for which it could possibly be mistaken are tubercular diarrhœa and entero-colitis. The former is excluded by the history and course of the case and by lack of evidence of tuberculosis of other portions of the body; the latter, by the apyrexia and the non-existence of symptoms indicating intestinal inflammation.

PROGNOSIS.—The result of even the more serious attacks in infants is, in the great majority of cases, favorable; nevertheless, it must not be forgotten that an acute catarrhal diarrhœa, when it occurs in a weak, ill-fed and badly cared-for child during hot weather, has a tendency to run into entero-colitis, and thus prove fatal. An infant, too, may be so debilitated by previous illness as to be carried off by an attack of ordinary severity.

TREATMENT.—Before entering into the details of the management of this disease, it is necessary to draw attention to the conservative nature of the diarrhœa. The frequent, loose and copious stools clear the intestines of irritant matter, and remove the cause of further trouble. Consequently, it is never advisable, early in the course, to completely arrest the evacuations, although at the same time they must be kept well in hand, lest the attack pass into entero-colitis. During dentition, particularly, this caution must be observed, for when there are three or four loose passages daily, cerebral symptoms are much less apt to arise.

As in other digestive disorders, the most essential step is to attend to the feeding. With infants nursed at a healthy breast it is enough to see that they are not fed too frequently, and to lessen the quantity taken by shortening each act of sucking. If, from any cause, the breast-milk be unsuitable, the babe must be weaned and carefully fed by bottle. In hand-fed babies it is necessary, first, to insist upon the use of the old-fashioned bottle and tip, and to see that they are kept absolutely clean. Next, to banish all farinaceous preparations, used purely as foods, from the diet. This does not preclude the employment of small quantities of arrow-root or barley-water for the purpose of breaking up the milk curd. Thirdly, to direct that the daily supply of milk —the only food to be allowed—must come from one dairy; be received fresh in the morning, and kept in a separate, perfectly clean vessel, and, if possible, in an especial refrigerator. And finally, to give careful, written orders as to the manner of preparing the milk food, and to make a rule that each bottle shall be mixed separately and only immediately before it is required. In hot weather it is advisable to sterilize the whole supply of milk, but this does not affect the principle of the separate preparation of each portion.

As guides to the manner of preparing the food, two formulæ

may be given; they are proportioned for children of four to six months.

> Unskimmed milk, ... f℥ iijss.
> Cream, ... f℥ ss.
> Lime-water, ... f℥ lj.

Mix these in a clean bottle, and warm by standing in hot water. Five to six bottles to be taken during the day.

Or—

> Unskimmed milk, ... f℥ iijss.
> Cream, ... f℥ ss.
> Arrowroot water,* . . . . f℥ ij.
> Sugar of milk, ... ℨj.

Mix and treat as before.

The quantity is to be reduced and the dilution increased in proportion to the youth of the infant, and the reverse as age increases. Sometimes in children of one or two months a cream and whey mixture suits better, as:—

> Fresh cream, ... f℥j.
> Whey, ... f℥ ij.
> Hot water, ... f℥ ij.

When there is thirst, cool water and bits of ice ought to be given with moderate freedom.

The sleeping room should be airy, well ventilated, and, in hot weather, the coolest the house affords. Soiled diapers, or the vessel containing a stool, must not be left about. In summer the patient should pass the mornings and evenings in the open air, and the hot mid-day in a cool room. A day's excursion on a steamboat, or to the country, if the journey be short, is very beneficial, while a trip to the sea-shore works wonders; a single day passed in salt air often removing every trace of the disease. Even in winter, if an attack occurs, the child, well wrapped up,

---

* Take one and a half teaspoonfuls of arrowroot, rub it down with a tablespoonful of cold water until smooth, and add, with stirring, a pint of boiling water.

should be taken out for an hour at noon on warm, sunny, still days.

The daily bath must be continued, and in hot weather a bath morning and evening is none too much. Woolen drawers and shirts of the lightest texture must be worn in summer, and if the diarrhœa prove at all obstinate, the abdomen must be enveloped in a light flannel bandage.

The condition of the mouth must always be investigated, and if the gums be hot and swollen, from approaching teeth, lancing is indicated. If there be much pain, with hardness of the gums, relief can be obtained by rubbing them gently at intervals with paregoric and water, ten drops to the teaspoonful, or with a solution of chloride of zinc, one grain to the fluidounce.

When these measures are carefully carried out in mild cases, medicines are often unnecessary. In those more severe, it is well to assist nature and begin the treatment with a laxative. Pain, green stools and the presence of blood always indicate this course. The best laxative is castor oil. This not only efficiently clears away the irritating contents of the intestines, but has a secondary, soothing action upon the mucous membrane. For a child of six months, the dose is a teaspoonful, with five drops of camphorated tincture of opium to prevent griping.

After this has operated, a teaspoonful of chalk mixture every two hours will complete the cure in some instances. A more efficient prescription, however, is:—

> ℞. Sodii Bicarbonatis, ... ℨss.
> Syrupi Rhei Aromat., ... f ℥ss.
> Aquæ Menthæ Pip., ... q. s. ad f ℥ iij.
> M.
> S.—Teaspoonful every two hours.

The great value of rhubarb depends upon its combined laxative and astringent action, precisely what is required in simple diarrhœa.

Should the stools still fail to become less frequent and more natural in color and consistence, resort must be made to opium and astringents. A very good formula is:—

 ℞. Tr. Opii Deod., ... ♏vj.
   Bismuth. Sub-nitrat. (Squibb) , ... ʒj.
   Syrupi, ... f℥ss.
   Misturæ Cretæ, ... q. s. ad f℥ iij.
 M.
 S.—Teaspoonful every two hours.

The value of calomel in certain cases where the evacuations obstinately remain green and acrid must not be overlooked, though the necessity for its use is rare. It must be employed cautiously and in small doses, and combined with an alkali, thus:—

 ℞. Hydrargyri Chloridi Mit., ... gr. j.
   Cretæ Præparatæ, ... ʒj.
 M. et ft. chart. No. xij.
 S.—One powder every two hours.

Its good effect should be noted in twenty-four hours, then it must be discontinued, and one of the other prescriptions given.

When the stools become normal, wine of pepsin must be ordered for a week or more until the digestion is put upon a sound footing.

In older children the treatment is very simple. All that is required is a bland diet, perhaps a dose of castor oil, and some mild astringent mixture. For example, let the patient take for breakfast—a soft-boiled egg; milk guarded with lime-water, and stale, dry bread; for dinner—some meat broth, free from fat; stale, dry bread, and rice and milk pudding; and for supper—milk, and stale, dry bread.

The opium and bismuth mixture already given, increased in dose proportionately to the age, is very serviceable, or a combination of aromatic syrup of rhubarb and chalk mixture may be used. As with infants, a course of wine of pepsin, or pepsin with muriatic acid, should terminate the treatment.

7. CHRONIC INTESTINAL CATARRH—CHRONIC ENTERO-COLITIS.

Chronic diarrhœa, as this condition is frequently termed, is a common and fatal disease in infants. When it occurs after the completion of the first dentition it is less dangerous to life, though it runs a protracted course and interferes greatly with nutrition.

MORBID ANATOMY.—As with other catarrhs, the absence of appreciable lesions is quite possible; but usually the mucous membrane of the colon is studded with minute, dark spots—the shaven-beard appearance—which the microscope shows to depend upon rings of vascular injection around the orifices of the follicles. In some instances there is deep congestion, limited principally to the summits of the longitudinal plicæ, while in others, ulcers are also found. These ulcers are shallow, and either elongated and narrow, when they occupy the summits of the plicæ, or small and circular, when they are seated between the folds. They are best seen by looking obliquely at the surface of the gut. Together with the ulcers there are numerous pearl-like projections, surrounded by narrow rings of congestion. These are enlarged solitary glands, and it is to their suppuration that the round ulcers are due. The whole mucous membrane is softened and thickened, unless the disease has been of very long duration, when it becomes extremely thin. The mesenteric glands are swollen and may even be caseous. In exceptional cases, the lower portion of the ileum presents the same changes as the large intestine.

ETIOLOGY.—Entero-colitis, or a series of attacks of simple diarrhœa, may establish chronic diarrhœa; but the disease frequently arises insidiously from the constant action of the great exciting cause—improper food. This cause is most operative in hand-fed infants, and at the time of weaning, but it affects nurslings who are supplied with poor breast-milk or allowed to eat bits of table food, and also older children.

Exposure to wet and cold is another excitant; so, too, are

various acute diseases, notably measles, croupous pneumonia, typhoid fever, variola and scarlet fever.

The predisposing agencies are bad hygienic surroundings, particularly over-crowding. In regard to age, the period of greatest liability as well as greatest fatality, is from birth to the end of the second year; afterward it grows less common as age advances. In our climate the greater number of cases originate in early spring and autumn, when the weather is most changeable; and late winter, when it is cold and damp.

SYMPTOMS.—The first indication of the disease is an alteration in the character of the stools. These assume the color and consistence of putty, and are composed of curd and farinaceous matter, with semi-solid fæces, and, at times, mucus and streaks of blood. They are voided with much pain and straining, but are little, if at all, increased in frequency. Their odor is offensive and sour. The face is pale and listless in expression, though the child is sufficiently lively, takes his food well and has no fever.

These symptoms continue with trifling change for two or more months, the patient gradually becoming thinner, paler and more languid. Then for the first time diarrhœa, sufficiently marked to arrest the nurse's attention, sets in. The evacuations now have a putrid odor, but vary considerably in other characters from day to day. They may be thin, liquid and brownish like dirty water; or clay-colored, of the consistence of thin mud; or watery, with particles of grass-green matter (from altered blood); and finally they may be slimy and contain whitish masses of undigested curd or particles of other food. The number of movements varies from ten to thirty in twenty-four hours; their frequency depending upon the amount of food taken and, to some degree, upon the weather; being greater on moist, cold days, than on warm, dry ones. They are preceded by pain, indicated by crying or uneasy movements of the legs, and are attended with straining, sometimes sufficient to cause prolapse of the rectum.

The tongue is, usually, natural, though at times the tip and edges are too vividly red and the fungiform papillæ too prominent. The appetite is normal, or even increased; nevertheless wasting is continuous. The skin grows pale, dry and harsh, and assumes a peculiar earthy tinge, which is deepest over the abdomen. The eyes are sunken and surrounded by dark circles; the lips are bloodless and thin; the nasal lines of Jadelot are marked, and the fontanelle is depressed. The abdomen may be soft and flaccid, but oftener is distended with flatus, and then is the seat of pain, manifested by moaning and twitching of the corners of the mouth. Palpation is painless unless there be ulceration; in the latter case there is tenderness, and the contact of the hand causes borborygmus. The skin on the internal aspect of the thighs and the nates is reddened by intertrigo, due to the irritant action of the fæces and urine. Prostration is so great that the child lies perfectly passive; the pulse is feeble and frequent; the temperature is not elevated, but, on the contrary, the hands and feet often feel cold, and have a bluish color.

The urine is diminished in quantity and retained for long periods.

With occasional brief intervals of improvement the condition gradually grows worse. The stools become more watery; look like chopped spinach floating in brown, putrid water, and may contain mucus and pus with blood, in brownish-yellow masses. Abdominal distention, tenderness and gurgling, the signs of intestinal ulceration, are present. The appetite is capricious or lost. The face becomes thin and pinched; the forehead is wrinkled; the hair dry and lustreless, and the whole expression that of a puny, weak, old man. General wasting progresses until the body seems to consist of little more than the bones, which stand out prominently with the muddy, harsh, flaccid skin hanging from them in folds. To this emaciation the distended belly stands in marked contrast. The fontanelle, at this stage, is deeply depressed; the pulse feeble; the breathing superficial, and the temperature sub-normal, being sometimes as low as 97.5° F. in the rectum.

As the end approaches, the nasal lines increase in depth; the lips are red, fissured and encrusted with scales; the tongue dry, red and rasp-like from enlarged fungiform papillæ, and the whole oral mucous membrane is covered with aphthæ or thrush patches. A fetid odor hangs about the body. The feet and hands are cold, purple and œdematous. The little sufferer lies quiet, with half-shut, lustreless eyes; from time to time an expression of pain flits over his face, but he is too weak to cry. Finally, there is no evidence of living, save the slow rise and fall of the chest as the breath comes and goes, and gradually this ceases, so gently that it is difficult to decide upon the exact moment at which life passes away. It is not uncommon for the discharges from the bowels to stop entirely for several days before the fatal termination. This circumstance alone has no favorable significance.

Death may result from exhaustion, or several complications may arise and hasten this event. These are serous effusions, hypostatic pneumonia, exanthemata, convulsions, and thrombosis of the cerebral sinuses.

Serous effusion may take place into the pleuræ, peritoneum and pericardium, but usually occurs in the form of œdema of the feet, hands, and, at times, the face. It is due to the impoverished condition of the blood and want of tonicity in the vascular walls.

Hypostatic pneumonia, due to the constant dorsal decubitus, is a common cause of death.

The exanthemata are very prone to attack the subjects of chronic diarrhœa, probably on account of the attendant prostration reducing the power of resisting contagion.

Convulsions are only dangerous in the early stages of the attack; later, the nervous irritability is so blunted that this complication is rare.

Thrombosis of the sinuses of the brain depends upon the withdrawal of the liquid elements from the blood by the diarrhœa. Water is then absorbed from the brain, lessening its bulk. The resulting vacuum, together with atmospheric pressure from without, leads to depression of the fontanelle, and even overlapping of the cranial bones in young subjects. If this be insufficient to compensate, the cerebral sinuses and blood vessels become engorged with blood, and as the naturally sluggish current in the sinuses is rendered more slow by inspissation of the blood and feebleness of the heart, the conditions for clotting are most favorable. At the autopsy, the clot is usually found in the longitudinal sinus, completely obliterating the channel ; it is laminated, whitish, and adherent to the walls of the sinus, which are free from signs of inflammation. The veins that enter the sinus are distended with blood. The symptons preceding death from this complication are difficult respiration ; stupor; dilatation of the pupils and strabismus ; spasms of the posterior cervical muscles ; fulness of the jugular veins, and unilateral facial paralysis.

When the case tends to recovery, the evacuations become more solid and natural in odor and color ; the latter change being caused by the reappearance of bile. The semi-stupor disappears, and the child grows very irritable, often crying out and shedding tears—a most favorable omen. The flesh, also, begins to return, the buttocks being the first part of the body to show the improvement. Diarrhœa is, after a time, succeeded by a constipated condition of the bowels. Convalescence is protracted.

Children over two years of age, when affected with chronic diarrhœa, are pale, thin, languid and readily fatigued. Irritability of temper, night terrors, and nocturnal incontinence of urine are common. The tongue is red at the tip and edges, with prominent papillæ, and perhaps light frosting. The appetite may be normal, craving or capricious. The stools vary in number from three to twelve in twenty-four hours ; in the former case they are semi-solid, light colored, and mixed with minute masses of green or colorless mucus ; in the latter, they consist of dark liquid, containing lumps of clay-colored fæces ; this variation bears some relation to the state of the weather. The evacuations are always fetid in odor, and the act of defecation is attended by pain and straining. The abdomen is distended by flatus. Feebleness of the pulse is proportionate to the general weakness ; respiration is unaltered, and there is no pyrexia.

In some instances the stools are limited to four or five a day, and are composed almost completely of undigested food and mucus. One evacuation occurs in the morning, soon after rising ;

the others during or immediately after meals. They are preceded by griping pain and by so urgent a desire, that the patient has difficulty in waiting for the chamber or reaching the closet. The condition undoubtedly depends upon great irritability of the intestine and exaggerated peristalsis.

DIAGNOSIS.—The diarrhœa of chronic catarrh is to be distinguished from that of tuberculosis of the intestines; the only condition with which it is likely to be confounded. Should it begin soon after birth or at weaning; if there be a history of bad feeding or exposure, and if there be no constant elevation of temperature, the affection is probably catarrhal. A temporary rise in temperature may be caused by the eruption of teeth or other passing irritation, and is of no diagnostic importance.

Tuberculous diarrhœa, on the contrary, occurs after the third year, and is attended by pyrexia and enlargement of the mesenteric glands. On pressure there is tenderness and gurgling in the right iliac fossa, and tension of the abdominal wall over this region. There is also evidence of tuberculosis of the lungs. The evacuations, too, are distinctive; they are intensely fetid, brown and liquid, when passed, but, on standing, deposit a dark sediment, composed of flocculent matter, with small, black clots of blood, and little masses of mucus, and pus. The presence, therefore, of these features or their absence, while the symptoms of catarrhal diarrhœa are observed, will determine the nature of the affection in children who have passed the age of infancy.

PROGNOSIS.—Chronic intestinal catarrh is fraught with great danger when it attacks children under the age of two years. It is particularly fatal when it follows an acute disease; when it occurs in syphilitic, rachitic or feeble subjects, and when it is complicated by measles or other exanthem. Unfavorable symptoms are dryness and roughness of the tongue; thrush; anasarca; features indicating intestinal ulceration, great depression of the fontanelle, and extreme emaciation. Favorable symptoms are normal progression of dentition; the reappearance of tears; intermissions in the diarrhœa, and improvement in the character of the stools and general symptoms.

TREATMENT.—As the disease is produced by over-crowding, neglect, exposure, and unsuitable food, the initial measures of treatment must be the regulation of the hygiene, clothing and diet.

The sleeping-room must be kept at a uniform temperature—between 64° and 68° F.—it must be dry, well-ventilated, and, if possible, heated, in cold weather, by an open wood-fire, and occupied by no one but the patient and nurse. During the day the patient must be moved to another room, being wrapped in a blanket if cold halls have to be passed. This room should be large, well-ventilated, dry, and kept at the same temperature as the first. After the removal, the windows of the sleeping-room should be opened, and the bed and its linen thoroughly aired and freshened. Soiled diapers or chambers containing stools are to be removed at once, and no cooking is to be done in either room. The child's person must be kept clean, and it is especially important to sponge the perineum and nates with warm water after each movement of the bowels; and, if there be any redness of the skin, to anoint the parts with oxide of zinc ointment, or powder them. It may be impossible to carry out this plan among poor patients, but it can be approximated by keeping the baby clean, out of the kitchen and away from the door-step.

As to clothing, the body must be clad in woolen from the neck to the toes, and, as an additional protection, a broad flannel, abdominal belt must be worn. So clothed, the patient may be taken into the open air on dry days, during the early stages of the attack. Soiled garments are to be replaced at once by fresh ones, and diapers must be washed when soiled; not simply dried and used over again.

The diet should vary with the age of the patient. The great principle being to maintain the general nutrition with the least amount of irritation of the intestinal mucous membrane.

Infants partly nursed and partly bottle-fed do best when restricted to the breast, provided the latter be healthy. If the diarrhœa does not improve under the change, both the intervals and the time of nursing must be shortened.

If the infant be hand-fed, every precaution must be taken to insure purity of food and perfect cleanliness of the feeding apparatus. The latter must consist of a simple bottle and tip, unless the amount to be given be very small, when a teaspoon can be used. The quantity of food and intervals of feeding always depend upon the degree of diarrhœa; thus, in very severe cases, not more than a teaspoonful every fifteen minutes can be allowed. The quality depends upon the age.

For an infant under six months, cows' milk and lime-water, in the proportion of one part to two, or in equal quantities, may be tested. If this undergoes acid fermentation, fresh whey and veal or chicken broth, with equal quantities of barley-water, may be substituted. A teaspoonful of Mellin's food, dissolved in whey, barley-water, or diluted broth, makes an admirable food. At the age of six months, a good scale of diet when milk cannot be taken is:—

First meal, 7 A.M.—One teaspoonful of Mellin's food dissolved in six ounces of veal broth and barley water, in equal parts.

Second meal, 10 A.M.—One tablespoonful of cream in six ounces of freshly prepared whey.

Third meal, 1 P.M.—Same as first, with chicken broth in place of veal broth.

Fourth meal, 5 P.M.—Same as second.

Fifth meal, 10 P.M.—Same as first.

After a week or more of improvement, milk may be resumed gradually, in the beginning at the first meal only; then at the first and last, and so on.

Should these foods disagree, they must be discontinued and the child fed upon meat juice. This is prepared by chopping a piece of sirloin steak, free from fat or tendon, into small bits, and, after slightly warming, pressing out the juice with a lemon-squeezer. A teaspoonful, with a little salt, is to be given four times a day, and the quantity gradually increased as the peculiar fetid odor which it imparts to the stools disappears.

After the age of six months, the yolk of a raw egg, well beaten with ten drops of brandy, a teaspoonful of cinnamon water, and a little white sugar, may be administered once or twice a day, together with the milk, whey, or broth-food.

After twelve months, if milk can be taken, the following diet is suitable:—

First meal, 7 A.M.—Two teaspoonfuls of Mellin's food dissolved in six ounces of milk and barley-water, equal parts.

Second meal, 10 A.M.—Four ounces of veal broth with two ounces of barley-water.

Third meal, 2 P.M.—The yolk of a raw egg, beaten up well with twenty drops of brandy, a teaspoonful of cinnamon water and a little white sugar.

Fourth meal, 6 P.M.—Same as second, or four ounces of fresh whey with a tablespoonful of cream.

Fifth meal, 10 P.M.—Same as first.

It is most important to remember that if the evacuations be very frequent and watery, there can be no set meals, but the food must be given by the teaspoonful at intervals of ten or fifteen minutes. Also, that between set meals and these minimum quantities, there is a wide range in the amounts and intervals, according to the grade of the symptoms.

From older children it is necessary to withhold potatoes and farinaceous vegetables generally; fruits, sugar, sweetmeats, pastry, hot bread or cakes, butter and all made and highly-seasoned dishes; at the same time the bulk of each meal must be somewhat restricted. A good diet is:—

For breakfast, at 7.30 A.M.—One or two tumblerfuls of milk warmed and diluted by the addition of a fourth part of hot water; the yolk of a soft-boiled egg, salted, and a slice of thin, dry toast.

For luncheon, at 12 M.—The soft parts of eight raw oysters, flavored by lemon juice, and a Boston cracker. Or in summer a small teacupful of junket, with a cracker.

For dinner, at 3 P.M.—A bit of the breast of chicken cut up very fine, or a tender piece of roast beef or beef-steak treated in the same way; with a tablespoonful of well-boiled spinach, asparagus tops, cauliflower tops, or stewed celery, and a thin slice of dry, stale bread.

For supper, at 7 P.M.—A glass of milk, warmed as at breakfast, and a slice of well-made cream toast.

An important rule in all cases is to watch the diet carefully until all danger of a relapse has passed.

Baths and external applications are useful. Infants who are not much prostrated should be placed in a hot bath (95°–100° F.) every evening for three minutes, then quickly dried, anointed over the whole body with warm olive oil, wrapped in a blanket and put to bed. If there be much prostration, the bath must contain mustard, one teaspoonful to the gallon, and the child kept in until the supporting arms of the nurse begin to tingle.

When intestinal ulceration is suspected, the belly should be enveloped in a light flax-seed poultice, or, what answers as well, a layer of carded cotton covered with oiled silk.

Medicines are to be selected according to the stage of the attack. Early, while the stools are little increased in number, but putty-like and of sour odor, the bowels must be gently acted on by:—

   ℞. Pulv. Rhei, ... gr. vj.
     Sodii Bicarbonatis, ... gr. xij.
   M. et ft. chart. No. vj.

   S.—One powder three times daily, for an infant of three to six months.

Afterwards—on the succeeding day, usually—the following powder may be administered:—

   ℞. Pulv. Ipecacuanhæ Comp., ... gr. iv.
     Cretæ Præparatæ, ... gr. xxxvj.
   M. et ft. chart. No. xij.

   S.—One powder every four hours.

AFFECTIONS OF THE STOMACH AND INTESTINES. 245

Or often better:—

 ℞. Tr. Opii Deod., ... ♏vj.
  Sodii Bicarb., ... gr. xlviij.
  Syrupi, ... f ℥ ss.
  Aquæ Menthæ Pip., ... q. s. ad f ℥ iij.
 M.
 S.—One teaspoonful every four hours.

When the stools become frequent and green, the mixture of opium, bismuth and chalk, already given (page 234), is very useful; and if tenesmus be very severe, a sedative enema must be used, as:—

 ℞. Tr. Opii, ... gtt. iij.
  Potass. Bicarb., ... gr. iij.
  Mucilag. Amyli., ... f ℥ ss.
 M.
 S.—To be injected into rectum.

This may be repeated every six or twelve hours, according to the necessity, taking care that the child—and all children are very susceptible—does not get too much opium.

Should the diarrhœa still continue, and the stools become watery and very fetid, astringents are required; for example:—

 ℞. Acid. Sulphurici Aromat., ... ♏xxiv.
  Liquor. Morphiæ Sulph., ... f ʒ ij.
  Elix. Curaçoæ, ... f ℥ ss.
  Aquæ, ... q. s. ad f ℥ iij.
 M.
 S.—One teaspoonful every three hours.

Or—

 ℞. Argenti Nitratis, ... gr.ss.
  Syr. Acaciæ, ... f ℥ ss.
  Aquæ, ... q. s. ad f ℥ iij.
 M.
 S.—One teaspoonful every three hours, midway between meals, if possible.

Nitrate of silver is most valuable when the stools contain mucus and blood, and aphthæ or thrush are present. It may be also used as an injection, if there be evidences of ulceration, thus:—

  ℞. Argenti Nitratis, ... gr.j.
   Aquæ, ... f ℥j.
  M.
  S.—Inject once or twice daily after cleaning out the rectum with an injection of warm water.

These injections must be suspended for twenty-four hours after being continued for three days.

Prostration and depression of the fontanelle demand stimulants. Ten drops of whiskey in water every two hours is about the average dose, but it may be given oftener and in larger quantities as circumstances require.

As soon as the stools become normal in character and frequency, the child must be ordered tonics, as:—

  ℞. Liquor. Ferri Nitratis, ... ♏xxiv.
   Glycerinæ, ... f ℥ss.
   Aq. Menthæ Pip., ... q. s. ad f ℥ iij.
  M.
  S.—One teaspoonful three times daily.

Or—

  ℞. Ferri et Ammonii Citratis, ... gr.xij.
   Tr. Gentianæ Comp., ... f ʒj.
   Spt. Lavandulæ Comp., ... f ʒij.
   Syrupi Limonis, ... q. s. ad f ℥ iij.
  M.
  S.—One teaspoonful three times daily.

For the constipation of convalescence very small doses of castor oil—twenty drops—may be ordered once or twice daily, but it is best not to interfere unless the bowels have been indolent for twenty-four or forty-eight hours.

With older children the medical treatment is more simple. Ordinarily, either of the following prescriptions will suffice:—

R. Syr. Rhei Aromat., ... f ℨ vj.
Sodii Bicarbonatis, ... f ℨ ij.
Tr. Opii Deod., ... ♏ xxxvj.
Aq. Menthæ Pip., ... q. s. ad f ℥ iij.
M.
S.—One teaspoonful every three hours, for a child from four to six years.

Or if the diarrhœa resist:—

R. Tr. Krameriæ, ... f ℨ iij.
Tr. Opii Camphoratæ, ... f ℨ ij.
Spt. Lavandulæ Comp., ... f ℨ ij.
Misturæ Cretæ, ... q. s. ad f ℥ iij.
M.
S.—Two teaspoonfuls every three hours.

The lienteric form of diarrhœa should not be treated by astringents but by nux vomica followed by arsenic. For instance, until the stools become less frequent and urgent and the griping pain diminishes, a good prescription is:—

R. Tr. Opii Deod.,
Tr. Nucis Vomicæ, ... āā ♏ xlviij.
Aq. Menthæ Pip., ... q. s. ad f ℥ iij.
M.
S.—One teaspoonful before each meal, at the age of six years.

Afterward:—

R. Liq. Potassii Arsenitis, ... f ℨ j.
Inf. Gentianæ Comp., ... q. s. ad f ℥ iij.
M.
S.—One teaspoonful after each meal.

Washing out the intestine is also a useful method of treatment: this will be fully described in the next section (Entero-colitis).

During convalescence from chronic diarrhœa, older children do well upon the same tonics as infants, the doses being proportionately increased.

8. ENTERO-COLITIS.

(Summer Diarrhœa—Febrile Diarrhœa.)

Entero-colitis, or inflammatory diarrhœa, is the scourge of our large cities during the summer months, when it brings death to hundreds of children, especially among the over-crowded, ill-fed poor. To it is due the popular dread of that period of an infant's life termed " the second summer," and justly, for among those unfortunates who are obliged to pass this time in crowded houses, and narrow, filthy streets, the instances of complete escape are very rare.

MORBID ANATOMY.—The anatomical lesions consist in inflammatory hyperæmia of the intestinal mucous membrane. This may be distributed over the whole tract, but commonly it is limited to the ileum and colon, and is most intense in the neighborhood of the ileo-cæcal valve and the sigmoid flexure. The mucous membrane is reddened, swollen and softened. Redness is either general or in the form of arborescent patches about the follicles; while swelling and softening are proportionate to the degree of congestion. The former is sometimes so great at the lower end of the ileum as almost to occlude the valve; to this has been attributed the vomiting which, in the absence of gastric lesions, is otherwise difficult to explain. The isolated glands are enlarged, and more opaque than normal, having the appearance of grains of white sand scattered over the mucous surface, and the Peyer's patches are tumefied and projecting, with punctated surfaces. On the peritoneal aspect, the gut, in positions corresponding to the inflamed glands, presents areas of arborescent injection. There is moderate enlargement of the mesenteric glands.

From this condition it is but a step to the state of ulceration seen in chronic intestinal catarrh— a not infrequent result of entero-colitis.

The stomach, as already hinted, is usually normal in appearance; occasionally its mucous membrane is reddened and thickened, and it is quite possible that this viscus is often the seat of a catarrh so moderate in degree as to leave no evidences after death, though sufficient to give rise to vomiting during life.

ETIOLOGY.—Season, age and locality of residence are important factors in the causation. Only isolated cases occur in the winter months, and these are met with among the poor, with whom it is a habit, for convenience in watching, to keep infants in the living room, which is also the kitchen; this is heated by the cooking-stove, and is either intensely hot when the room-door is closed, or too cold when it is left open, in the frequent excursions of the older members of the family to the yard or street. There is, therefore, a constant exposure to sudden and marked changes in temperature. At the same time the air of such a room is contaminated by cooking, by re-breathing, and by the exhalations from soiled clothing and dirty bodies. These are sufficient causes for an attack of entero-colitis. About the middle of May or June, according to the character of the individual season, cases become more common, and as the summer heats are established, in July, August, and the first half of September, the number is augmented to the proportions of an epidemic. Late in September or in October, according, again, to the season, there is a marked diminution, and this increases as winter approaches. During the summer the number of cases and deaths varies with the range of the thermometer; several successive days with a temperature above 90° F. being attended by a great increase, while a similar period with a temperature below 80° is followed by a decided decrease. Hot, damp weather is the most productive, and of all months August is the most fatal, both on this account and because a high temperature is maintained during the night.

Infants between the ages of six and eighteen months are by far the commonest sufferers. Their liability depends upon the sympathetic irritation of the alimentary tract attending the cutting of the teeth; the increased tendency to inflammation produced by the rapid development that the intestinal glands and follicles are simultaneously undergoing, and the fact that weaning, with its consequent change of diet, usually takes place during this interval. From the eighteenth month to the end of the second year, about one-fourth as many cases occur, and the third period of greatest frequency is from birth to the sixth month. Children over three years are not often attacked.

Residence in large cities is almost an essential etiological condition; the vast majority of cases occur where the streets are narrow and more than ordinarily filthy, and where the houses are overcrowded and dirty, and the people poor, ill-fed and unclean. There must be another factor at work here besides the elevated temperature, since in the open country immediately surrounding affected cities, where the thermometer ranges nearly as high, the disease is of exceptional occurrence. This factor is an atmosphere polluted by poisonous gases and containing countless bacteria, the result of decomposing organic substances.

Another potent agent is bad food. Infants, hand-fed from birth, are the most frequent sufferers; next, those who are weaned early; in both, the chief injurious articles of diet are sour milk, farinaceous preparations in excess, and "tastes" of table-food. Nursing infants are more exempt, but even with them, too frequent and continuous feeding, or breast-milk of abnormal quality, when other conditions are unfavorable, often produce entero-colitis.

SYMPTOMS.—For one or two days prior to the actual attack, the infant is restless and fretful; his sleep is disturbed by moaning or fits of crying; he is paler than usual, and his head and, perhaps, the palms of his hands, feel hot. He also ceases to empty his bottles; after feeding, eructations of very sour-smelling material are apt to occur, and the stools are somewhat more numerous and softer than usual.

Next, vomiting and diarrhœa set in. The former occurs after feeding, and, in bad cases, is so obstinate that nothing is retained. The matter rejected consists of sour, acid and curdled milk, or other food imperfectly digested.

The stools range from six to twenty or more in twenty-four hours, and vary in character from day to day, and even from hour to hour. At first, they are semi-solid, homogeneous, yellow in color and neutral in reaction; then they become more liquid and green, though still homogeneous and neutral, and then the reaction becomes acid without change in the other characters. Often they are semi-fluid, heterogeneous, green with little masses of yellow fæces, and neutral; or semi-fluid, heterogeneous and green, with fragments of yellowish-white caseine and acid; or watery, with floating flakes of white, yellow or green matter, and acid. Mucus and blood may be mixed with any of these stools; the first in stringy masses; the second, in bright red streaks or merely tingeing the mucus. In severe cases the passages become watery and so colorless as hardly to stain the diapers. The odor at first is fæcal, then sour, and finally offensive. The act of defecation is preceded by pain, manifested by the expression of the face, by crying, and by twisting of the trunk and drawing up of the legs. Sometimes there is tenesmus and slight prolapse of the rectum; it is under these circumstances that blood appears in the stools.

The tongue is dry, red at the tip and edges and covered in the centre with a light white coating; the appetite is diminished and the thirst increased. The abdomen is distended by flatus, and, at times, there is tenderness on pressure.

With these features there is pyrexia, moderately high and continuous for the first three or four days, afterward remittent; the head is especially hot, and the palms of the hands are dry and burning to the touch. The pulse is weak and frequent, beating 120, or even 140 times per minute. The urine is reduced in quantity and passed at long intervals, sometimes only two or three times a day.

As the diarrhœa continues the face becomes pale; the eyes are surrounded by dark circles; the nasal lines appear; the fontanelle, if still membranous, is depressed; the fat disappears from the body; the muscles grow soft and flabby; the buttocks and inner surfaces of the thighs are reddened by the acid stools and concentrated urine, and there is great feebleness and languor. In grave attacks these changes take place in an incredibly short space of time, twenty-four hours being ample to reduce an active, robust infant to a mere shadow of himself.

If death approach, the patient, in some cases, grows fretful; has a dry, burning skin; rolls the head from side to side; vomits incessantly; has strabismus and indolent pupils, and may have convulsions, which are more frequently unilateral than general. In others, there is drowsiness, an apathetic refusal of food, cessation of vomiting and diarrhœa, and coolness of the extremities. This difference depends upon the acuteness of the attack, for upon this rests the preservation or loss of nervous irritability.

The great diminution of the urinary excretion suggests the possibility of the fatal termination being, in some instances, due to uræmic poisoning.

When the attack tends to recovery the vomiting stops; the motions are less numerous and more fæcal; the skin becomes cooler and more moist; the urine is excreted freely; the eyes grow bright; the child again shows interest in his surroundings; takes his food better, and rapidly regains flesh and strength.

DIAGNOSIS.—The pyrexia, the vomiting, and the frequency and character of the stools, taken in conjunction with the early age of the patient; the season and locality of occurrence; and the almost epidemic prevalence of the disease, make its distinction an easy matter. The portion of the intestinal canal chiefly involved is not so readily determined, though the presence of mucus and blood in the evacuations points to the colon as the seat of inflammation; their absence, by inference, to the small intestine. It is important to differentiate this disease from cholera infantum, which is an infinitely more serious disease. Cholera infantum is sudden in its onset, characterized by a high temperature, from 105° F. to 108° F., uncontrollable vomiting; frequent and profuse serous evacuations; embarrassed respiration; frequent and irregular pulse; marked involvement of the nervous system, and rapid collapse. Often a case will pass in the course of twenty-four hours from blooming health into a condition of almost ante-mortem decomposition. We do not see these changes in entero-colitis.

PROGNOSIS.—Inflammatory diarrhœa ranks among the most dangerous of the affections of infancy, both from its inherent nature and its tendency to run into chronic entero-colitis. Nevertheless, under appropriate management, a large proportion of cases recover. The outlook is most discouraging when the infant's lot has been cast in poverty; when it has been hand-fed from, or soon after birth; and when it has had the bad fortune to be born in the late winter or spring, so that weaning and the time of cutting the more troublesome teeth come together in the second summer.

The unfavorable features are high fever, very frequent and watery evacuations, rapid collapse, cerebral symptoms and convulsions.

An attack may prove fatal in four or five days, or it may be protracted for two weeks. The latter is about the duration of severe cases that terminate in recovery. One attack predisposes to another, an important point to remember in the treatment by change of climate.

TREATMENT.—People with means avoid the dangers of summer diarrhœa by taking their children to the country, sea-shore or mountains, where the air is uncontaminated, the heat less intense and the milk pure. Such escape is not open to the children of the poor; nevertheless, much may be done to preserve their health by keeping them during the day in the fresh air of public parks; by bathing in cool water; by proper, cleanly clothing; good food—for good milk is as cheap as bad—and by attention to the cleanliness of beds and sleeping rooms. This the parents can, and in many cases will do, and if they would only secure well-paved and decently clean streets—for it is impossible to clean ill-paved ones—entero-colitis would become a far less common disease.

When an attack occurs during the hot months, the patient, if possible, must be sent at once from the city to the sea-side or country. The locality selected should be near at hand, or the journey will be too fatiguing; still, it is important to fix upon a place affording a decided change of air and a lower temperature.

From Philadelphia the infant may be taken to Atlantic City, Cape May, Point Pleasant, Avon or any of the many resorts on the Jersey coast, kept there for two or three weeks and then removed to the New Hampshire hills for the remainder of the summer. A long stay is essential, since a return to town in hot weather is almost certain to be followed by a relapse.

If circumstances render it impossible to carry out this most potent of all prescriptions, fresh air must be secured by taking the child to the public squares in the cool of the morning and evening, or by spending the day in the Park, or, better still, by a morning and evening trip on one of the river steamboats. The heat of the day must be spent in as cool a room as can be had. It is of great moment to let the little sufferer rest in bed and not on the hot lap or shoulder, and when out, to wheel him in a coach rather than carry him. Many a stout mother has hastened her infant's death by too fond and constant nursing.

The clothing must be as thin as possible, provided, always, that woolen be worn next the skin.

Twice or three times a day, in very hot weather, the whole surface of the body should be sponged with water at a temperature of 80° F., and dried with gentle rubbing. The bracing effect of these baths is greatly increased by the addition of rock salt, or concentrated sea-water if the purse can afford it. These cool spongings must be supplanted by full warm baths when there is much prostration.

In regulating the diet, it must be remembered that the presence of fever, with increased thirst, leads the child to take more liquid food than is needed or can be digested; consequently, it is necessary to specify the quantity as well as the quality of the food. Infants at the breast are to be suckled only at intervals of two or three hours, according to their age, and taken away before they have completely satisfied themselves.

Hand-fed babies are to be similarly restricted. As cows' milk must constitute the bulk of their food, it is important to see that it is obtained fresh every day from a reliable dealer, promptly sterilized, and administered from an absolutely clean bottle fitted with a simple tip. For example:—

Milk,	f℥iij.
Cream,	f℥ss.
Lime-water,	f℥ijss.
Sugar of milk,	ℨj.

Mix in a clean tin-cup, pour into bottle, adjust tip, and warm by plunging into hot water.

Milk,	f℥iij.
Cream,	f℥ss.
Mellin's Food,	ℨij.
Hot water,	f℥ijss.

Dissolve the Mellin's food in the hot water, add the milk and cream, and, if necessary, warm as before.

Milk,	f℥iij.
Cream,	f℥ss.
Flour-ball,	ℨj.
Water,	f℥ijss.

Either one of these foods may be given every three hours to a child of ten or twelve months old. The quantity is less and the dilution greater than for a healthy infant of the same age, because enfeebled digestion demands a proportionate reduction in the amount and strength of the food.

When preparations of milk are vomited or passed undigested from the bowels, a whey mixture or strippings can be resorted to, and if these fail, beef-juice, or—

Flour-ball,	ℨij.
Water,	f℥vj.

Mix, and add—
  Half the white of a fresh egg.

Bits of ice and cool filtered water can be allowed, in moderation, to relieve the thirst.

If vomiting be persistent, all food must be stopped for from twelve to twenty-four hours, and the thirst quenched by thin barley-gruel or Vichy water,—cold, and in small quantities. If the child be at the breast, as soon as vomiting is checked, it can gradually be brought back to its accustomed diet, care being taken that too much food be not taken. In bottle-fed children under two years, it is better to *withhold milk entirely;* wine-whey, chicken and mutton broth, Mellin's food with barley-gruel, the juice expressed from raw beefsteak or roast beef, and sometimes raw-scraped beef, should constitute the "no-milk diet."

The indications for medical treatment may be grouped under four heads: 1. To clear out the bowels; 2. To stop decomposition; 3. To restore healthy action in the alimentary tract; 4. To treat the consecutive lesions.

1. The bowels should be emptied as completely as possible, as the first step in the treatment, and for precisely the same reasons that the surgeon cleanses a wound thoroughly before applying antiseptic dressing. This rule holds good not only where there is a history of antecedent constipation, or the evidence of the presence of indigestible food in the alimentary tract, but in every case in which there are altered secretions undergoing putrefactive changes. The only instances in which the process of cleansing should not be undertaken, because unnecessary, are those where, after two or three fecal or semi-fecal evacuations, the discharges consist of almost pure serum, large in amount, alkaline in reaction, and odorless.

To sweep out the intestinal canal nothing compares in efficacy with castor-oil. Should the stomach be very irritable, however, it will be necessary to substitute enemata. These should consist of pure water at a temperature of 65° Fah., and to be efficient must be copious enough to reach the cæcal valve,—about one pint in a child of six months, and two pints in one of two years. The injection must be given slowly, with a fountain syringe, the abdomen meanwhile being gently manipulated.

Many mild cases can be cured, if taken at the start, by castor-oil and a strict diet alone.

2 and 3. To stop decomposition and restore a healthy action

in the intestines, the administration of antiseptics and attention to diet are necessary.

Antiseptics must be given in small doses lest the stomach reject them, and frequently to maintain a continuous action. The best are calomel, salicylate of sodium and naphthalin.

Calomel may be prescribed in the following combination :—

R. Hydrargyri Chloridi Mit., ... gr. ½.
Bismuthi Subcarbonatis, ... gr. xxxvj.
Pulv. Aromatici, ... gr. vj.
M.
Et ft. chart. No. xij.
S.—One powder every two hours.

Salicylate of sodium is prescribed in doses of from one to three grains every two hours, according to the age, from three months to three years. An aqueous solution is tasteless, and can readily be given in the food or drink; it has a tendency to check rather than occasion vomiting. It may also be substituted for the calomel in the above prescription.

Naphthalin, although possessing a strong odor, is not disagreeable to the taste. On account of its insolubility, it is best administered rubbed up with some moist powder, like sugar of milk. The doses should be larger than those of the salicylate of sodium,—one to five grains, according to the age.

Resorcin and bichloride of mercury are also useful antiseptics. Resorcin is bitter, and though freely soluble in water, not easily administered; the dose is one-half a grain to two grains. The bichloride is given in doses of $\frac{1}{120}$ to $\frac{1}{100}$ of a grain, but even in these minute quantities frequently causes vomiting.

Counter-irritation by mustard plasters to the belly is useful. Stimulants are required when prostration sets in, and must be given in doses and at intervals adapted to the demands of the case.

Applications of oxide of zinc ointment, with cleanliness, cure the intertrigo of the buttocks and thighs most quickly, or, at least, keep it in check until the cause is removed.

4. The essential consecutive lesions are in the colon, and

consist practically of a follicular colitis. When the condition of ulceration is reached, astringents by the mouth are useless, with the possible exception of bismuth.

Three things are valuable :—

First. As careful attention to the diet as during the acute stages, and in recent cases. Deviation from dietetic rules is the most frequent cause of relapse.

Second. The continuance of antiseptics to check intestinal decomposition, and hence stop irritation.

Third. The whole large intestine should be washed out once every day, either with pure water at 65° F., or with weak antiseptic or astringent solutions. Of the former the best are benzoate or salicylate of sodium; of the latter, nitrate of silver or tannic acid.

Attention to diet and hygiene is not to be relaxed when convalescence is established, and after the measures calculated to check diarrhœa are unnecessary, digestants, as wine of pepsin, and tonics, as the ferrated elixir of cinchona, are still required, to restore health.

The exceptional cases that occur in cold weather should, of course, be treated at home in a well-ventilated and warm room; otherwise, the only alteration to be made in the general plan of management is to envelop the abdomen with light linseed poultices, or with cotton covered by oiled silk.

9. CHOLERA INFANTUM.

This affection occurs in teething children during hot weather, and is characterized by a sudden onset, high fever, irritability of the stomach, frequent serous evacuations, changes in the respiration and pulse, marked symptoms of nerve involvement, and rapid collapse. It is a far less common disease than enterocolitis, and is the analogue of cholera morbus in the adult.

MORBID ANATOMY.—In cases that run the ordinary course and die early, the gastro-intestinal mucous membrane is congested, thickened and softened, and the follicles and Peyer's patches are enlarged. In other words, the appearances indicate the early stage of inflammation, which passes into lesions identical with those of entero-colitis, when the patient, as sometimes happens, survives the choleraic stage and dies, subsequently, from a more protracted diarrhœa. But, in addition to inflammation, there is probably—and this is the important point—some involvement of the sympathetic nerves, leading to dilatation of the capillaries and transudation of serum into the intestine, and to alterations in the pulse, temperature, respiration and urinary excretion. The nature of this is paralytic, so far as the intestine is concerned, and resembles in its results experimental section of the sympathetic nerves. It is due, in part, to direct over-stimulation by the irritant contents of the canal, and in part to the nerve exhaustion produced by high atmospheric temperature, one of the essential causes of cholera infantum. The changes in calorification and in the functions of the heart, lungs, and kidneys, depend upon reflected irritation, and also, perhaps, upon the depressing effects of heat on the governing nerve centres.

ETIOLOGY.—Like entero-colitis, this is a disease of cities, finding its victims chiefly among those who live in poverty and squalor. Almost exclusively confined to hot weather, it may occur at any time between the middle of May and the end of September, though the greater proportion of cases originate during the latter half of July, August, and the first half of September. Infants from six to twelve months are the most susceptible subjects; it may, however, occur at any age up to two years. The susceptibility of the former age is due to the great functional activity of the intestinal mucous membrane and the rapid development of the follicles that accompany dentition, rather than the mere act of cutting teeth. The direct causes are high temperature (85°–95° F. or more) sustained for several days, and especially if associated with a moist atmosphere, exposure to an atmosphere rendered impure by noxious gases and bacilli generated from filth by heat, impure water and bad food.

SYMPTOMS.—An attack may arise in the midst of health, or it

may be preceded by diarrhœa. In either case the onset is sudden. The infant begins to void copious stools. These at first, if there has been no premonitory diarrhœa, contain more or less fæcal matter, but they soon become watery. Sometimes they are so serous as to soak away into the diaper without leaving any stain; at others, they contain a few yellow or green flocculi or little masses of mucus, and, in both instances, are odorless. Again, they are composed of yellow or brown liquid, containing a small proportion of thin, fæcal matter, and have a peculiar musty and offensive odor, which clings to the napkins and clothing, and even to the body of the child, in spite of the utmost efforts at cleanliness. The number varies from eight to thirty in twenty-four hours, and they are evacuated with considerable force.

At the same time, or soon after, the stomach becomes so irritable that everything, even to a mouthful of ice-water, is rejected as soon as swallowed, and there is violent retching. Appetite is lost, but there is intense thirst, the patient eagerly drinking when the opportunity offers, and following the glass, as it is removed, with greedy eyes. The tongue, originally moist and lightly frosted, soon becomes dry and pasty, and protrudes from the parched lips. The abdomen is flaccid and indolent.

There is great restlessness; the temperature is elevated to 105° or even 108° F.; the pulse is small and very frequent, counting from 130 to 150 beats per minute; the breathing becomes irregular and anxious, and the urine is greatly diminished in quantity.

With these symptoms there is a marked and appalling change in appearance. Within a few hours, the infant, perhaps plump and rosy before, can scarcely be recognized; the face becomes pale and pinched; the eyes and cheeks sunken, and the eyelids and lips permanently parted from loss of muscular contractility; the fat melts from the body; the muscles grow flabby; the bones appear prominent, and the skin, often greenish or cadaverous in hue, hangs in loose folds.

Soon the features of collapse appear. The hands, feet, nose, and even the breath, become cool, the pulse is thready and so frequent as to be uncountable; the respiratory movements are more unequal, and there is drowsiness, apathy, and suppression of urine. As life ebbs away, the vomiting stops; the surface becomes cold and clammy; the face is set with the lines of death; the respiration is quickened and shallow; the pulse scarcely perceptible, and the patient sinks into a state of semi-coma, with bleared eyes and contracted pupils. In this condition the end may come quietly or be preceded by slight convulsions.

The course of the disease, whatever the result, is always very short. It may prove fatal in from one to four days, or the character of the attack may change and death result later from a secondary inflammatory diarrhœa.

In case of recovery, the stools, after four or five days, gradually become less copious, frequent and watery, more fæcal and of better odor; vomiting stops; thirst diminishes; appetite returns; the urinary excretion is reëstablished; the temperature and pulse fall; the respiratory movements become rhythmical; emaciation ceases, and the child, though very feeble, again notices his surroundings, and after a week or more of simple diarrhœa, regains a moderate degree of health.

DIAGNOSIS.—The character of the stools, the extreme irritability of the stomach; disturbed respiratory rhythm; high temperature; intense thirst and rapid emaciation and collapse, distinguish this from entero-colitis, and from other forms of diarrhœa.

There is a certain resemblance between cholera infantum and sunstroke, and, by some, the two conditions have been considered as identical. The forms of similarity, as well as those of dissimilarity, may be seen in the following table:—

CHOLERA INFANTUM.	SUNSTROKE.
Temperature 105° to 108° F.	Temperature often 108° F.
Pupils contracted.	Pupils contracted.
Evacuations watery.	Evacuations watery.
Respiration embarrassed.	Respiration embarrassed.
Urine scanty.	Urine scanty.

CHOLERA INFANTUM.	SUNSTROKE.
Cerebral symptoms marked, but secondary	Cerebral symptoms marked, but primary.
Gastro-intestinal symptoms, precedent and prominent.	Gastro-intestinal symptoms, secondary
Onset rapid.	Onset sudden, by a *stroke*.
Preceded by diarrhœa or uncomfortable sensations.	No such previous history.
Restlessness at onset.	Stupor from beginning.
Occurs at any time of day or night.	Occurs only during excessive heat of day.
Inflammatory lesions of intestines.	No such lesions.

Between epidemic cholera and cholera infantum it is impossible to make a diagnosis.

PROGNOSIS.—The prospect is most discouraging, and even in seemingly favorable instances the opinion as to the result must be guarded, for though the choleriform symptoms be survived, there is danger from the succeeding diarrhœa. The disease is most fatal in children of the poor, who are badly fed and subjected to the worst hygienic influences; conversely, it is more apt to terminate in recovery in the rich, who can be treated in large, airy rooms, fed on good food, and removed to healthy localities.

TREATMENT.—The large and frequent watery evacuations are such a strain upon the system, that it is of the first consequence to replace the waste by food and drink, and at the same time check it by appropriate treatment.

The irritability of the stomach is a formidable barrier to alimentation; nevertheless, every effort must be made to give food in small quantities and at short intervals. Should the infant be at the breast, it may be allowed to nurse for a few minutes, every half-hour or hour. If hand-fed, it may be given the same foods recommended for entero-colitis, or chronic vomiting, in such quantities as can be retained, and at intervals corresponding in frequency to the smallness of the amount. Bits of ice and water should be allowed freely, even though they be rejected as soon as swallowed.

To check the diarrhœa opium and astringents are necessary.
A very serviceable formula is :—

R. Liquor. Morphiæ Sulph., ... f ʒj.
Acid. Sulphurici Aromat., ... ℔xxiv.
Elix. Curacoæ, ... f ℥ ss.
Aquæ, ... q. s. ad f ℥ iij.
M.
S.—One teaspoonful every two hours, for a child of six months old.

With this, two drops of laudanum, suspended in two teaspoonfuls of starch-water, should be administered, by the rectum, every three hours. Two or three times daily a mustard plaster (one part of mustard to five of flour) must be applied over the whole surface of the abdomen, long enough to redden the skin, and the whole body should be sponged several times a day with water at a temperature of 95° F.

The clothing, diapers and person must be kept perfectly clean; the sick-room must be as large and airy as can be commanded, and the infant must lie upon a bed, and not be constantly nursed upon the lap. If it be possible, the patient should be sent early to the sea-shore or country, as this affords by far the best chance for recovery. Failing in this, morning and evening airings in a coach, or daily steamboat excursions, must be resorted to.

Stimulants are needed from the first to ward off prostration. From five to ten drops of whiskey in a teaspoonful of lime-water may be given every two or three hours at the age of six months.

When collapse sets in, the quantity of alcohol must be increased, and, if the stomach can bear it, a combination of stimulants is useful, as :—

R. Spt. Frumenti, ... f ℥ ss.
Ammonii Carbonatis, ... gr. xxiv.
Syrupi Acaciæ, ... f ʒj.
Aq. Menth. Pip., ... q. s., ad f ℥ iij.
M.
S.—One teaspoonful p. r. n.

The temperature must be maintained by hot flannel wraps and

hot water bottles, and the child kept in a horizontal position and disturbed as little as may be.

In this stage astringents are still indicated; but opium must be used with great caution, or even discontinued entirely, when there are cerebral symptoms and semi-coma.

In the fortunate instances in which this plan is successful, it is still necessary to treat the succeeding diarrhœa, and finally, to build up the general health by good food, tonics and fresh air.

10. INFLAMMATION OF THE COLON AND RECTUM—DYSENTERY.

Dysentery is not a very frequent disease in children, but it may occur in an endemic or epidemic form, and as a sequel of measles, scarlet fever, or variola.

MORBID ANATOMY.—The mucous membrane of the colon and rectum is swollen, red, softened and even loosened by diffuse suppuration, and the solitary glands are enlarged and ulcerated, while the corresponding peritoneal surface is congested. In severe and epidemic cases, the inflamed surface presents more or less adherent pseudo-membranous patches, which, when removed, leave ulcers of irregular outline and variable depth. Perforation and cicatricial contraction of the intestine are occasional results.

ETIOLOGY.—Sporadic and endemic cases are produced by the same causes as entero-colitis—excessive heat, bad food, and exposure to cold and wet. The epidemic form is certainly infectious, and there are grounds for believing that it is also contagious, although the last fact is not yet definitely established. The disease is most common in the second and third years of life, and seems to attack boys more frequently than girls.

SYMPTOMS.—Nausea, vomiting, high fever, and acute abdominal pain usher in the attack. Then the bowels become distended. The evacuations are numerous, ranging from four to forty a day; small in quantity and voided with much straining. At first they contain fæcal matter, but after a short time are composed entirely of mucus and blood, mixed with yellow or green flocculi, fragments of false membrane and pus. The blood may appear in dark red streaks or clots; in black masses; as a substance resembling the washings of meat, or merely diffused through the mucus, giving it a uniform red color. Their odor is most offensive.

The face wears an anxious expression; there is great restlessness, sleeplessness, muscular weakness, and rapid emaciation. The tongue is dry, red at the tip and edges, and covered in the centre with a brownish coating. There is anorexia and urgent thirst. The abdomen is distended, tympanitic, and painful on pressure, particularly over the course of the colon.

As the attack progresses, tenesmus becomes the most permanent symptom; it occurs without the passage of stools, and is often attended with prolapse of the rectum. Fever gives place to coolness of the surface; restlessness, to semi-stupor; the eyes and cheeks sink; the face becomes pinched, and death may take place quietly or be preceded by slight convulsions.

The duration varies from two or three days in grave cases, to about two weeks in those that result favorably.

The DIAGNOSTIC features are high fever, tenderness along the track of the large intestine, tenesmus and the number and character of the stools.

The PROGNOSIS is favorable in the sporadic form and when there is only slight elevation of temperature and moderately frequent stools. Quite the reverse, if there be high fever, great tenesmus, frequent evacuations containing much blood or false membrane and pus; when there is a tendency to collapse, and when the disease is epidemic. Relapses frequently occur.

TREATMENT.—Children affected with dysentery must be kept at rest in the best room—so far as ventilation and coolness are concerned—that the house affords. Their diet should be liquid, and even this form of food must, on account of the irritability of the stomach, be given in moderate quantities. From four to six ounces of whey and cream mixture ; of flour-ball and milk, or arrow-root and milk, may be given every three hours to a child of two years. A good preparation is :—

  Arrow-root, ... ʒj.
  Hot water, ... f ℥ ss.
   Mix thoroughly, and add to—
  Milk, ... f ℥ ij.
  Cream, ... f ℥ ss.
  Water, ... f ℥ j.

Small pieces of ice and moderate quantities of iced filtered water can be allowed to relieve thirst.

Two or three times daily the body should be thoroughly sponged with water at a temperature of 95° F., and the abdomen must be kept covered with a light flaxseed poultice, over the surface of which a little mustard has been sprinkled ; this must be covered with oiled silk and changed as often as it becomes cold.

If the patient be seen early in the attack, the medicinal treatment may be begun with a laxative, as :—

  ℞. Ol. Ricini, ... f ʒ ijss.
   Pulv. Acaciæ, ... ʒ ij.
   Tr. Opii, ... ♏ viij.
   Aq. Menth. Pip., ... q. s. ad f ℥ ij.
  M.
  S.—One teaspoonful every three hours, at three years of age.

After this has been continued for twenty-four hours, there should be marked improvement in the evacuations. If this be not the case, it is well to order the following :—

  ℞. Pulv. Ipecac. Comp., ... gr. vj.
   Bismuthi Sub-carb., ... ʒj.
   Pulv. Aromat, ... gr. vj.
   M. et ft. chart. No. xij.
  S.—One powder every three hours.

With an enema of laudanum—gtt. iij to f ʒ ss of warm water—

every four hours; or a suppository of opium and acetate of lead :—

> ℞. Pulv. Opii, ... gr. ss.
> Plumbi Acetat., ... gr. j.
> Ol. Theobromæ, ... ʒj.
> M. et ft. supposit., No. vj.
> S.—One to be used every four or six hours.

Should these fail, nitrate of silver may be administered by the mouth or rectum. If there be great rectal irritability and quick expulsion of the caustic injections, it is best to follow them with enemas of laudanum. A good formula for administration by the mouth is :—

> ℞. Argenti Nitrat., ... gr. j.
> Tr. Opii Deod., ... ♏xxiv.
> Syr. Acaciæ, ... f℥j.
> Aquæ, ... q. s. ad f℥iij.
> M.
> S.—One teaspoonful every three hours.

For an enema :—

> ℞. Argenti Nitrat., ... gr. j.
> Aquæ Dest., ... f℥ij.
> M.
> S.—Inject twice daily, and allow an interval of twenty-four hours after three days' successive use.

To ward off prostration, it is necessary to employ stimulants, in doses and at intervals proportionate to the demands of the case. Should collapse occur, alcohol and artificial heat to maintain the body temperature are the main resources.

When convalescence is established, it is still necessary to guard the diet carefully and to build up the general health with tonics. Of these, the best are quinine with dilute nitro-muriatic acid, or tincture of nux vomica with compound tincture of gentian, followed by ferrated elixir of cinchona, or citrate of iron and quinia.

11. TUBERCULAR ULCERATION OF THE INTESTINES.

This form of ulceration commonly occurs as a complication of pulmonary and abdominal tuberculosis. Gray granulations may or may not be present at the site of lesion, but the special degenerative process in the intestinal mucous membrane is so intimately associated with tubercle that the term "tubercular ulceration," seems to be warranted.

MORBID ANATOMY.—The lesions are chiefly confined to the ileum, and primarily affect the solitary follicles and Peyer's patches, particularly those about the ileo-cæcal valve. The follicles become enlarged from multiplication of their cell elements, then undergo caseous degeneration and softening, with the formation of isolated ulcers in the case of the solitary glands, and clusters of coalescing ulcers in that of the patches of Peyer. From having, at first, the shape of the follicles and patches, they gradually extend by a similar process of corpuscular infiltration, caseation and softening in the surrounding tissues. The fully-formed ulcers are irregularly oval in shape, with their greatest diameter directed transversely to the axis of the gut; their edges are indented, thick and somewhat undermined; their floors are red or gray, and formed by one or the other tissue of the intestine, as far down as the peritoneum, according to the depth of destruction. Perforation is rare on account of localized adhesive peritonitis. Gray granulations—a secondary product—may be found in the tunica adventitia of the small arteries and lymphatics, or on the reddened and cloudy peritoneal surface corresponding to the ulcers. Cicatrization takes place rarely, but may be the cause of stricture.

The uninvolved mucous membrane is congested, thickened and softened. The mesenteric glands are enlarged and cheesy, and miliary tubercles are usually found in the lungs or elsewhere.

ETIOLOGY.—The disease is met with in children who have passed the fourth year, and in whom the tubercular or strumous diathesis exists. Bad hygiene, bad food, and exposure, act as predisposing causes, by interfering with general nutrition and paving the way for the development of the diathetic tendency. An unsuitable diet, too, may indirectly lead to this form of ulceration, by bringing about an abnormal condition of the lining membrane of the bowel.

SYMPTOMS.—In addition to the features indicating a scrofulous or tubercular tendency, the child, after suffering for a variable time from the symptoms of simple intestinal catarrh, begins to have fever and to pass excessively offensive stools, composed of dirty-brown liquid that, on standing, deposits flocculi, mucus, pus and small, black clots of blood. There is colic preceding the evacuations; moderate distention of the belly, with tension of the parietes over the right iliac region, and tenderness on pressure there. Abdominal palpation also reveals enlargement of the mesenteric glands, and physical examination of the chest the evidences of pulmonary phthisis. Such cases usually result fatally, after a more or less protracted course, the direct causes of death being tuberculosis of the lungs or of the meninges of the brain.

TREATMENT.—Pure air, warm clothing, good food and tonics comprise the measures of treatment. The best of the tonics is cod-liver oil, which, in these cases, often seems to lessen the tendency to diarrhœa. Half a teaspoonful three times daily is quite enough for a child of five years. It may be given combined with maltine, or in an emulsion with lacto-phosphate of lime, or the compound syrup of the hypophosphites. The following is an admirable formula:—

R. Olei Morrhuæ, ... f℥ij.
 Ext. Malt (dry), ... ℨiv.
 Calcii Hypophos.,
 Sodii Hypophos., ... āā gr. xvj.
 Potassii Hypophos., ... gr. viij.
 Glycerinæ, ... f℥ij.
 Pulv. Acaciæ, ... ℨij.
 Aquæ, ... q. s. ad f℥iv.
M.

S.—One teaspoonful three times daily.

In addition to this general treatment, attention must be paid to the intestinal condition. A light flax-seed poultice should be placed over the right iliac region or over the whole abdomen, or a dressing of cotton, covered with oiled-silk, may be used. Internally, sub-nitrate of bismuth with compound ipecacuanha powder and nitrate of silver are useful; at the same time it is well to administer clysters of laudanum.

12. COLIC.

Colicky pains frequently attend dysentery, constipation and other intestinal disorders; but colic with flatulence so uniformly occurs as a functional affection in children from birth to the end of the third month, and gives so much discomfort both to the infant and its attendants, by causing fretfulness, crying and wakefulness, that it demands separate consideration.

ETIOLOGY.—In studying the causation of this condition, it must be remembered that after birth the infant, previously nourished through the blood of its mother, begins to take food through a new channel. Hence a new habit has to be formed, in addition to the development of a secreting and absorbing apparatus hitherto inactive. It is during this transition state that food of the best quality may be imperfectly or slowly digested and flatulence and colic result.

Food that is at all difficult to digest almost always occasions colic, and hand-fed babies are especially liable to it. Other causes are fulness of the stomach in over-feeding, or the opposite condition of emptiness after nursing at a breast that affords milk in small quantity, and, finally, inherited feebleness of digestive power, and over-sensitiveness of the mucous membrane to the contact of food.

SYMPTOMS.—Soon after feeding, the infant becomes restless, kicks his legs about uneasily, twists his body, grunts, or utters a series of piercing cries. The face is congested at first, from the effort of crying, but soon becomes pale, with a tinge of blue around the lips. The belly is full and hard, the hands and feet are cold, and, in bad cases, the fontanelle is more or less depressed. After a time, varying from a few minutes to an hour, eructations of flatus or of curdled milk occur, and the symptoms disappear for awhile. Such paroxysms may occur at any hour of the day, but are most frequent and severe in the evening and night.

There is usually, also, a moderate degree of constipation, or the bowels are irregular. At night the rest is broken by uneasy tossing and whimpering, and during sleep a smile or an expression of pain often flits over the face; but, in spite of the fretfulness and discomfort, the infant suffers little in general health, and increases in flesh and strength almost as rapidly as is normal.

TREATMENT.—When the infant is fed at a healthy breast, it is of great importance to insist upon the rule of feeding only at proper intervals, and absolutely to forbid the habit of putting the child to the breast whenever it cries. Food will be taken whenever it is offered, and the warm milk entering the stomach relieves the pain for a time, only, however, to increase it later by giving the viscus more work to do, and filling it with material to undergo fermentation with the production of flatus. Consequently, it is much better to resort to one of the preparations to be hereafter given for the relief of the pain.

Should the child draw but a poor and scanty supply of milk, and the colic be due to emptiness, the breast must be supplemented by hand-feeding. Under these circumstances, and when the whole feeding is by bottle, much may be done to prevent or relieve the attacks of pain by attention to cleanliness of the feeding apparatus; by carefully selecting the ingredients of the food, and by adding an aromatic to the latter. A good food for a child of one month old is:—

Milk,	f ℥ jss.
Cream,	f ℥ ij.
Barley-water,	f ℥ jss.
Caraway-water,	f ℥ j.
Sugar of milk,	ℨ ss.

These ingredients are to be mixed in a clean vessel, poured into a perfectly clean bottle, and heated to a temperature of about 98° F. in a water-bath.*

A little pancreatin and bicarbonate of sodium added to the bottle of food just at the time of its administration produce good results by aiding intestinal digestion.

When the bowels are inclined to constipation, the barley-water may be replaced by a gruel made of ground oatmeal (Bethlehem brand). One or two teaspoonfuls of the meal to the quantity of water necessary for each bottle is the proper proportion. In place of this, a teaspoonful of Mellin's food may be added to the requisite quantity of water.

The belly should be anointed twice a day with warm olive oil, and enveloped in a broad flannel binder. It is even more important to keep the feet warm, and for this purpose thick socks or long woolen stockings should be worn, and in bad cases, artificial heat must be applied by hot water bottles.

Medicines are indicated chiefly during the attacks of pain. A simple and serviceable prescription is ten drops of gin in a teaspoonful of sweetened warm water. Another is:—

 ℞. Sodii Bicarb., ... gr. xvj.
  Syrupi, ... f ℥ ss.
  Aq. Menth. Pip., ... q. s. ad f ℥ ij.
 M.
 S.—One teaspoonful p. r. n. for a child of one month.

This is rendered more efficient by the addition of two drops of aromatic spirit of ammonia to each dose, or, in severe cases, one drop of spirit of chloroform.

Bromide of potassium and chloral are most useful; they may be combined as follows:—

 ℞. Potassii Bromidi, ... gr. xvj.
  Chloral Hydrate, ... gr. viij.
  Syrupi, ... f ℥ ss.
  Aq. Menthæ Pip., ... q. s. ad f ℥ ij.
 M.
 S.—One teaspoonful for a dose.

---

* A tin-cup half filled with water, placed on an ordinary stand over a gas-burner, makes a good water-bath for nursery use.

Of this preparation, it is rarely necessary to give more than two or three doses, at intervals of half an hour. It is well to reserve this mixture for severe attacks, and in ordinary cases, to use the gin or the soda mixture.

Should the paroxysm be so violent as to lead to depression of the fontanelle and threaten collapse, the infant must be placed in a warm bath for five minutes; after being removed and carefully dried, he must be wrapped in a blanket; a flax-seed poultice with a dash of mustard placed over the abdomen; a hot-water bottle applied to the feet; the bowels relieved by an enema of warm water, and ten drops of gin or brandy in warm water administered by the mouth. If the fontanelle still remains depressed, the stimulant must be continued in doses and at intervals proportioned to the urgency of the symptoms; at the same time the soda and ammonia mixture may be given.

As a routine treatment to improve digestion, it is well to order fifteen drops of essence of pepsin (Fairchild's) three times daily.

13. HABITUAL CONSTIPATION.

In addition to the locking of the bowels that results from mechanical causes, as intussusception, peritoneal adhesions, and so on, or from paralysis of the muscular coat of the intestine in certain nervous diseases, constipation of a functional character is a frequent and often an obstinate condition during childhood.

ETIOLOGY.—Before the completion of the first dentition, it is more common in hand-fed babies than those nursed at the breast, and is due to the use of milk over-rich in casein; the abuse of starchy food; an insufficient supply of water, and often to the action of popular remedies given to relieve colic. With children who have passed the first dentition, constipation arises from faulty habits, and from the employment of a diet that is either bad in quality or unsuitable from its too great sameness. In all cases, inherited sluggishness of the peristaltic movements must be remembered as a possible cause.

SYMPTOMS.—These vary greatly in degree. Thus, an infant, instead of the normal number, may have but one evacuation a day, or one, two, and even three days may intervene between the movements. The stools are scanty; composed of hard, dry, whitish lumps, and are voided with much pain and straining. Should the last symptom be severe, it is frequently attended by rectal prolapse and hemorrhage. Other features are colic, abdominal distention, diminished appetite, occasional vomiting, feverishness, fretfulness, restless sleep, and, in bad cases, convulsions.

In older children there may be one scanty passage each day, or a week at a time may elapse without relief. The stools, while lumpy and hard, are dark colored and mixed with mucus. The abdomen is the seat of pain, and may or may not be distended with flatus; in the latter event, palpation often reveals the presence of hard masses along the course of the descending colon. The tongue is coated; the appetite capricious; there is nausea and a sensation of discomfort in the rectum, leading to frequent, though unproductive, straining efforts at defecation. There is also languor, irritability of temper, headache and restless sleep; a muddy complexion and general spareness of frame.

DIAGNOSIS.—There is little difficulty in establishing the existence of habitual constipation. One must be cautious, however, not to place too much reliance upon the statement that "the child's bowels are open every day," for in obstinate cases, it is not unusual for daily evacuations of thin, worm-like masses to take place whilst the bulky and hard fæces are retained.

PROGNOSIS.—Proper management rarely fails in regulating the action of the bowels, but constipation may prove serious in two ways: first, by leading to fæcal accumulation; second, by generating a condition of general ill-health, during which the child is more exposed to the attack of acute and dangerous disease.

TREATMENT.—In every case the relief of the actual state of retention of fæces in the rectum and the breaking up of the costive habit are the ends to be accomplished.

For the former purpose, I prefer the use of purgative enemata and suppositories to the administration of the same class of remedies by the mouth, particularly when abdominal palpation or digital examination of the rectum show that the retained mass is large and hard. The author's plan is to inject into the rectum, according to the age of the patient, from one to four teaspoonfuls of warm sweet oil; allow it to remain for six hours, and then use one or more clysters of soap and warm water, or olive oil, soap and warm water.* The preliminary injections of oil soften the fæces, while the clysters—which must vary in bulk from one to six fluidounces, to be adapted to the capacity of the gut—have the additional effect of distending the walls of the rectum, and thus bring about muscular contraction and expulsion of its contents. Should the mass present at the anus but be too bulky to escape, more liquid may be injected, and if this fail, it must be broken up by the finger and its passage assisted by gently supporting the perineum during the straining efforts. In severe cases little result may follow a single application of this method, though a course of one or two oil injections and purgative clysters daily for several successive days will rarely fail to empty the bowel.

When the soapy water and oil fail to produce expulsive efforts, the enemata may be rendered more efficient by the addition of a teaspoonful or more of castor oil or oil of turpentine. To make such an enema for a child of two years:—

> Take—One teaspoonful of oil of turpentine,
> Two teaspoonfuls of olive oil,
> The yolk of one egg.
> Mix thoroughly, and add, with constant stirring, to
> Four fluidounces of warm water.

Another enema which rarely fails to act quickly and efficiently is from one to two fluidrachms of pure glycerine with half a fluidounce of water.

---

*An enema composed of one teaspoonful of common salt to four fluidounces of warm water is very efficient.

All injections must be thrown in gently, and the action of the syringe stopped as soon as pain is produced.

In infants, unless the rectum be very full, clysters give no better results, and are far less convenient than suppositories. At the age of two months the following prescription may be ordered:—

℞. Saponis, ... gr. vj.
　　Olei Theobromæ . . . . ʒj.
　M. et. ft. supposit. No. vj.
　S.—One to be inserted every morning or morning and evening.

Or a small glycerine suppository may be used.

Careful regulation of the diet is often all that is required to remove the tendency to constipation, and is a most important element of the treatment even in those cases where it is necessary to call in the aid of medicines.

Bottle-fed babies must be fed upon cows' milk, so modified by the addition of cream, sugar of milk and water as to be as nearly like human milk as possible; and, should the bowels still remain confined, some laxative article, as Mellin's food or oatmeal, can be added. An admirable mixture for a child of three months is:—

Milk, ... f ℥ ii·s.
Cream, ... f ℥ ss.
Sugar of milk, ... ʒj.
Bethlehem oat-meal (fine powder), ... ʒij.
Water, ... f ℥ iss.

In preparing this, the water must be heated—just short of boiling—in a tin vessel, and the oat-meal added slowly, with stirring, until a smooth, white mixture is obtained; the other ingredients are then to be added, and the whole administered from a perfectly clean feeding-bottle. It is usually unnecessary to add the oatmeal to every bottle; one or two meals of it, each day, being sufficient.

During childhood the food selected must be of good quality, thoroughly digestible and varied. Starches and meat are to be allowed in moderation; pastry, salt meat and sweets forbidden, and a judicious use made of such articles as oat-meal or cracked wheat in the form of mush, well-cooked spinach, celery, cabbage and peas, baked apples, stewed prunes, thoroughly ripe peaches and pears, or the juice of oranges.

To encourage peristalsis, warm sweet oil may be gently rubbed into the skin of the infant's abdomen twice daily, the natural course of the colon being followed; and with children more advanced in age, cold spongings of the belly, followed by frictions with a coarse towel until the surface is red, are very beneficial.

The ordinary cathartics, castor oil and rhubarb, are not adapted to the treatment of habitual constipation, because their primary laxative action is followed by a secondary binding effect, and they consequently increase the original trouble. There are, however, other medicines of the same class that are free from this disadvantage, and one of them, or, better, a combination of several of them, may be employed.

For infants a very serviceable prescription is:—

℞. Mannæ Opt.,
Magnesii Carb., ... āā ʒij.
Ext. Sennæ Fld., ... f ℥ ss.
Syrupi, ... f ℥ j.
Aq. Menth. Pip., ... q. s. ad f ℥ iij.
M.
S.—A teaspoonful once, twice or three times daily for a child of six months.

Or should a sallow skin, yellowish conjunctivæ and loaded tongue indicate torpor of the liver:—

℞. Resinæ Podophylli, ... gr. ss.
Alcohol, ... ℔xlviij.
Syrupi, ... q. s. ad f ℥ iij.
M.
S.—A teaspoonful two or three times daily for a child of one year.

If it be difficult to make the infant take medicine, manna— which imparts only a sweet taste—may be dissolved in the food, and given from the bottle as often as required. Phosphate of sodium—an admirable laxative—can also be administered in the same way, in doses of two to five grains three times each day, at the age of six months.

Children of three or four years and upward do best upon aloës and belladonna. Tincture of aloës and myrrh in doses of five drops thrice daily, or in a single dose of ten drops at bedtime, acts well; but if the patient be old enough to swallow a pill, the following prescription is to be preferred:—

 ℞. Ext. Belladonnæ, ... gr. ss.
  Pil. Aloës et Myrrh., ... gr. vj.
  Ol. Cari, ... gtt. iij.
  Ext. Taraxaci, ... gr. xij.
 M. et ft. pil. No. xij.
 S.—One pill at bedtime for a child of six years.

Or the aloës and belladonna may be combined in a mixture,* thus:—

 ℞. Tr. Belladonnæ, ... f ʒ j.
  Tr. Aloës et Myrrh., ... f ʒ ss.
  Mucilag. Acaciæ, ... q. s.
  Aquæ Menth. Pip., ... q. s. ad f ℥ iij.
 M.
 S.—One teaspoonful for a dose.

In using aloës and myrrh, it is usually necessary to reduce the dose after a time, as its purgative action increases rather than diminishes with repetition.

---

*A clearer mixture may be made by using a solution of aloës and myrrh instead of the officinal tincture. The following is the formula:—

 ℞. Aloës,
  Myrrh, ... āā gr. ijss.
  Alcohol, ... f ℥ ss.
  Glycerinæ, ... f ℥ j.
  Aquæ, ... q. s. ad f ℥ iij.
 M.

This solution was compounded, at the author's request, by Mr. J. J. Ottinger of Philadelphia. The dose is the same as the tincture.

Another useful laxative is cascara, in the form of a fluid extract or an elixir; of the first preparation ten drops, of the second, twenty drops may be given, once or several times daily to a child of six. It does not quickly lose its effects by repetition.

I have lately used with much satisfaction a laxative confection, composed of tamarind pulp (gr. xxxvj) and senna in powder (gr. iv), aromatized with aniseed and lemon, and acidulated with tartaric acid. One of these may be eaten every evening, or as often as necessary, by a child of three years of age. They are regarded as sweets rather than medicine, and the little patients eat them readily. Glycerine suppositories may also be used once or even twice a day if occasion require.

14. SIMPLE ATROPHY.

Simple atrophy, or, as it is often termed, marasmus, is a condition in which there is extreme wasting of the soft tissues of the body, either without special organic lesions or with catarrhal inflammation of the mucous membrane of the gastro-intestinal canal.

MORBID ANATOMY.—After death, the muscular and other tissues are found in a state of atrophy, and there is a total disappearance of normal fat from the body. Fatty degeneration of the kidneys, lungs and brain may be discovered; the stomach is sometimes ulcerated, and hemorrhagic effusions into the cranium are not uncommon.

ETIOLOGY.—Wasting usually occurs during the first twelve months of life, though it may begin in the second year, and is most frequently encountered among children of the poor. It arises both in breast-fed babies and in those brought up by hand, being, in either case, due to insufficient nourishment.

Food can be insufficient in two ways: first, when it is supplied in amounts too limited to meet the demands of the system; and second, when it contains a minimum of the elements essential to nutrition, or presents them in a form ill adapted to the feeble digestive powers of infancy. For example, nursing infants waste in consequence of feeding either from a breast that yields too little good milk, or from one that secretes abundantly a poor, watery fluid entirely unfit for nourishment. With artificially fed children, on the other hand, it rarely happens that the quantity of food is too small; the fault lies, rather, in the direction of quality. Undiluted cows' milk; milk thickened with starchy materials; farinaceous foods, and even table food—meat, vegetables, and bread—are given to babies a few weeks or months old. Now, all of these are highly nutritious, but the digestive apparatus is not sufficiently developed to prepare them for absorption. They are strong foods, adapted to nourish and strengthen much older children and adults, but as the infant cannot appropriate them, he starves as surely, if more slowly, than when taking no food at all. Such aliment, also, while remaining undigested in the stomach and intestines, undergoes fermentation with the formation of irritant products, causing vomiting or diarrhœa; conditions that still further lower the vital powers and hasten atrophy.

It is often possible to trace the disease directly to want of cleanliness in the feeding apparatus, and especially to the use of a form of bottle that has lately been very popular. This bottle has, in place of a plain gum tip, an arrangement of glass and rubber tubing of small calibre. One extremity of the rubber tubing, which is eight or nine inches long, terminates in a small nipple-shaped tip and bone shield; the other, after penetrating an ornamental rubber cork, is fitted to a bit of glass tubing long enough to extend quite to the bottom of the bottle. By this plan, the trouble of holding the bottle and keeping it at a proper angle during feeding is avoided. The seeming advantage, though, is counterbalanced both by the minor drawback that the child, left to itself, is apt to continue suction long after the bottle is exhausted, thus swallowing a quantity of air, and by the greater disadvantage that the tubing can never be kept clean.

For a number of years the author has made it a rule to ask for the bottle of every hand-fed infant presented for treatment, and few days have passed without his seeing several of the complicated contrivances referred to. In almost every instance, notwithstanding the most careful and frequent cleansing, a sour odor could be detected, and if milk were present, it contained numerous small curds; while in cases of carelessness the odor was intolerable, and the interior of the tubing was encrusted with a layer of altered curd. With ordinary bottles, on the contrary, alterations in the character of the milk and coating of the interior of the tips were very infrequent. As there is little difficulty in keeping the bottles themselves clean, there can be only one reason for this difference, namely, in the old-fashioned instrument the nipple is readily removed and as easily inverted and cleaned, but in the other there is no way of cleaning thoroughly the twelve or more inches of fine tubing. The latter cannot be inverted, and the passage of a stream of water, or of a small, stiff brush, only imperfectly removes the milk clinging to the interior. This, of course, soon undergoes decomposition, and in this state quickly inaugurates change in the next supply of milk placed in the bottle. It is evident that a constant supply of food, no matter how good originally, thus rendered acid and partially curdled, must, like an excess of farinaceous or other unsuitable food, produce-irritation of the alimentary canal, interfere with the processes of nutrition, and lead to a state in which the features of wasting and disordered digestion are combined.

The custom of preparing in the morning a supply of food sufficient for the whole day is another fruitful cause of atrophy. If this be done, no matter how carefully the mixture be proportioned, or how well adapted to the age and digestion of the child, it becomes unfit for consumption after standing eight or ten hours. The change may or may not be appreciable to the senses, but test-paper will always show acidity and the microscope demonstrate the existence of actively moving bacteria.

Finally, food upon which a child has thrived for three or four months, perhaps, can become unsuitable and, consequently, lead to wasting, if the digestive powers be suddenly reduced by an intercurrent disease.

SYMPTOMS.—The clinical features differ materially, according to whether the element of insufficiency be one of quantity or quality. They may, therefore, be divided into two classes, viz: those developed by food that is suitable but not sufficient; and those resulting from unsuitable food.

The first group of symptoms is most frequently encountered in children who have been nursed at the breasts of feeble or over-worked mothers, in whom the milk is often both scanty and of poor quality. There is a gradual loss of plumpness; the muscles grow flaccid, and there seems to be an arrest of growth. The face is white; the lips pale and thin, the skin harsh and dry or too moist, and the anterior fontanelle level or slightly depressed. The temper is irritable and sleep restless and disturbed, or the child is abnormally quiet, dozing constantly, and sucking his fingers until they become raw. When nursed, the child seizes the nipple ravenously; then, if there be little milk, he quickly drops it to cry passionately, as if disappointed at not being able to satisfy his hunger; but if the milk be abundant, though thin, he will lie a long time quietly at the breast and often fall to sleep with the nipple in his mouth. The bowels are inclined to constipation, the stools being scanty, hard and dry. Physical signs connected with the chest and abdomen are negative, and no indication of disease of any special organ of the body can be detected.

In the second class, features of wasting are associated with those of irritation of the alimentary canal, and the symptoms altogether are much more grave than in cases of the preceding group. The subjects are almost invariably hand-fed infants.

Emaciation progresses with a rapidity and to an extent depending upon the original strength of the child's constitution; the age at which artificial feeding is begun, and the sort of food employed. It is often so extreme that an infant several months old weighs less and appears smaller than at birth, and this, even after a large quantity of food, such as it is, has been consumed.

The combination of great wasting with a voracious appetite, is very striking, and is only apparently contradictory, since hunger —the demand of the tissues for reparative material—cannot be appeased by food which, from its bad quality, is incapable of digestion or proper preparation for absorption and assimilation. Unsuitable food, too, by irritating the mucous membrane of the stomach, creates a fictitious appetite.

Sooner or later the face becomes pinched, the eyes are sunken, the lips pale, and when moved display a deep furrow about the angles of the mouth; the facial expression is uneasy or languid, and the anterior fontanelle is deeply depressed. The skin, generally, is dry, harsh and yellowish; hangs in loose folds over the bones, and may be mottled by an eruption of strophulus or urticaria, or present red patches of intertrigo in the neighborhood of the genitalia and over the buttocks and inner surface of the thighs. The extremities are cold and the hands claw-like. The tongue is heavily furred or red and dry. With the mucous membrane of the mouth, it may be the seat of aphthous ulceration or thrush deposit. As already stated, the appetite is often ravenous, and the cries of hunger are violent, oft repeated; and only temporarily silenced by food; thirst is increased; colic is common; the bowels are constipated, and the stools, which are voided with difficulty and straining, are composed of a few light-colored, cheesy lumps, partly covered with greenish mucus.

Attacks of acute vomiting and diarrhœa often interrupt the regular course of the disease. At such times there is moderate fever during the night, though, ordinarily, the temperature is subnormal. Again, chronic vomiting and chronic diarrhœa* are apt to arise as complications, and greatly increase the danger of a fatal termination.

Sleep is restless and disturbed, and many hours, particularly during the night, are spent in fretful crying. A common group of symptoms connected with the nervous system is "inward spasms." When these occur, the upper lip becomes livid, some-

---

* See chapters on Chronic Vomiting and Diarrhœa.

what everted, and tremulous; the eye-balls rotate or there is a slight squint, and the fingers and toes are strongly flexed. They frequently usher in true convulsions.

Sometimes the nervous manifestations are much more complex. Thus, I have seen cases where there was retraction of the head; boring of the head into the pillow; an approximation to the "gun-hammer" decubitus; general hyperæsthesia and the tache cérébrale; all suggestive of tubercular meningitis. Such symptoms disappear under an appropriate diet, with proper medicinal treatment, and are to be referred to an intensely excitable nervous system; a condition depending upon insufficient nourishment, and differing merely in degree from that leading to "inward spasms."

There is, of course, extreme prostration, the cardiac action is weak and the respiration shallow. The urine is citron-colored or very dark yellow; has a specific gravity of 1009 to 1012.5; a strong, characteristic odor, and is diminished in quantity. It is always cloudy or milky, only becoming clear on the approach of recovery. The sediment deposited on standing, contains variously shaped cylinders; fatty elements with tinted nuclei; mucus; colored uric acid; urates in a crystallized or amorphous condition; pigment, etc. The reaction is sometimes highly acid. The proportion of urates is decidedly, that of uric acid notably, and of coloring matter and extractives somewhat increased. Albumen is always present in variable quantity and sugar may be also frequently detected.*

Death may be preceded by convulsions or the symptoms of spurious hydrocephalus, or may result from prostration.

DIAGNOSIS.—Great emaciation may result from inherited syphilis or acute tuberculosis, but both of these conditions are attended by characteristic symptoms, rendering their diagnosis a matter of little difficulty.

When symptoms resembling those of tubercular meningitis are present, it is often necessary to delay a definite opinion. In

---

* "Parrot and Robin."

AFFECTIONS OF THE STOMACH AND INTESTINES. 285

simple atrophy, however, the open fontanelle is level or depressed; the belly is never scaphoid; the bowels, though frequently constipated, are never locked; vomiting is apt to be associated with diarrhœa; the respiration and pulse are regular in rhythm; the temperature, as a rule, is sub-normal; there is no hydrencephalic cry, and the antecedent history and the course are different from the tubercular disease.

PROGNOSIS.—A vast number of cases die annually in our large cities, yet the results of appropriate management are often rapidly and surprisingly successful. Patients should never be given up unless there be extreme wasting and prostration, or unless the symptoms of spurious hydrocephalus arise; convulsions occur; or obstinate chronic vomiting or diarrhœa be developed.

TREATMENT.—For the arrest of wasting from insufficient nourishment, the first thing to be attended to is the diet. Without entering at length into this subject,* it may be stated, as a uniform rule, that in selecting a diet the object should be to fix upon one suited to the age and digestive powers of the child, so that he may be able to digest, and, therefore, be nourished by, all the food consumed.

Generally, infants under twelve months who have to be either partially or entirely "brought up by hand," do well upon sterilized cows' milk, with lime-water or with barley-water. The food should be administered from a bottle capable of holding half a pint, made of colorless glass, so that the least particle of dirt can be seen, and provided with a soft India-rubber tip. The whole quantity of food intended to be given in a day should never be prepared at once, but each portion must be made separately at the time of administration. Thus, a bottle of the sort described, absolutely clean, may be filled with a mixture of one part of lime-water to two or three of sound milk, or with one part of barley-water to two or three of milk, to either of which may be added from one to two tablespoonfuls of cream and a teaspoonful of pure sugar of milk. The bottle must next be placed in hot

---

* For the details of diet and general management, see Part II.

water until the contents become warm, when it is ready for the child.

The degree of dilution of the milk and the proportion of cream added vary with the age and the feebleness of digestion. Lime-water is the preferable diluent when there is frequent vomiting or acid eructation. Both it and barley-water are of service in preventing the formation of large, compact curds.

After digestion has been brought into good condition by such a diet, the food may be cautiously increased to a standard suitable for a healthy child of the same age. At eight or ten months, from two to four fluidounces of thin mutton or chicken broth, free from grease, may be allowed each day in addition to the milk; at twelve months, the yolk of a soft-boiled egg, rice and milk, and carefully mashed potatoes moistened with gravy; and at the end of the second year, a small quantity of finely minced meat.

Once daily the patient should be bathed in warm water, or, at least, sponged over with warm water, and every morning and evening a teaspoonful of warm olive oil or of cod-liver oil should be rubbed into the skin over the abdomen and chest. At the same time, the belly must be completely covered with a soft flannel binder, and the feet and surface generally kept warm by woolen clothing. In this way attacks of colic, if not entirely prevented, are rendered much less frequent and severe.

If there be intertrigo, cleanliness and the free use of oxide of zinc ointment usually suffice to effect a cure.

Of medicines, bicarbonate of sodium, pepsin and cod-liver oil are, perhaps, the most useful. Cod-liver oil should not be given until the digestive powers have been brought into a comparatively normal state by proper food, antacids and digestants. The oil is most easily borne when given in emulsion, and may be advantageously combined with lacto-phosphate of lime or with the hypophosphites.

Such symptoms as constipation, diarrhœa and vomiting demand, of course, appropriate treatment.

15. TYPHLITIS AND PERITYPHLITIS.

Destructive inflammation of the coats of the cæcum or vermiform appendix—typhlitis—and of the post-cæcal areolar tissue—perityphlitis—are so closely associated that it is best to study them together. In the cæcum and appendix the inflammatory process may stop short of ulceration; may proceed to ulceration with perforation and the production of perityphlitis, or, infrequently, may lead to the latter by simple extension of the morbid process without perforation. The independent and primary origin of perityphlitis is possible, perhaps, but must be extremely rare. Neither condition is, strictly speaking, an affection of childhood; nevertheless, children between four and twelve years of age are liable, particularly to that form in which perforation of the vermiform appendix occurs.

MORBID ANATOMY.—In typhlitis without ulceration, a large extent of the mucous lining of the cæcum or appendix is the seat of catarrhal inflammation, while the investing peritoneum is opaque and injected, and may form adhesions to the neighboring intestinal loops. When the inflammation advances to ulceration, though a more limited area be affected, there is a tendency to involvement, with destruction, of all the coats of the bowel; and during this process the peritoneal adhesions may become firmer and more extensive.

The result of the perforating ulcer depends upon its position. Should it be situated on the anterior wall of the cæcum, the intestinal contents escape into the peritoneal sack, in spite of the adhesions—which are rarely firm enough to offer an efficient obstacle—and produce a rapidly fatal general peritonitis. When, on the contrary, it occupies the posterior aspect, where the wall of the cæcum is devoid of peritoneum, the escaping fæcal matter enters the post-cæcal connective tissue, and causes inflammation, with suppuration, and forms a "fæcal abscess." Such abscesses may reopen into the intestine; may extend down the sheath of the psoas muscle, reaching the surface below Poupart's ligament, or may point in the lumbar region or in any situation along the iliac crest. Sometimes the ulceration stops before perforation occurs, and the inflammation assumes a chronic form. In such cases, dense peri-cæcal adhesions form; the cæcum is contracted in calibre; has its walls thickened, and its mucous membrane either "almost entirely destroyed or converted into a retiform and trabecular fibroid tissue." *

In the appendix the ulcer may be situated at the free extremity, or, more frequently, at some point in the lower third of the canal. As to extent, the loss of substance may be small or involve the whole circumference. A collection of pus may be present in the cavity of the appendix, and it is usual to find a foreign body or intestinal concretion near the position of perforation. While the ulceration is going on, firm adhesions are occasionally formed with the cæcum; the anterior wall of the abdomen, or the tissues of the right iliac fossa. If the first event occurs, the resulting circumscribed abscess opens into the intestine; if the second, it points in the abdominal wall; and if the third, perityphlitis is set up, with the results already described. Usually, however, there are no adhesions or weak ones, and fatal general peritonitis follows the exit of the concretion and pus.

The concretions resemble in shape and size cherry or date stones. They are hard; often laminated in structure; have a smooth, waxy-looking surface; are grayish or brown in color, and are composed of earthy phosphates combined with inspissated mucus and fæcal matter. Pins, shot and splinters of bone, strawberry seeds, hairs and little masses of hardened mucus, may form the nidus of these calculi. These articles, too, illustrate the class of foreign bodies which cause perforation when arrested in the vermiform appendix.

Sir William Jenner, quoted by Eustace Smith, attributes the arrest of calculi in the appendix to malposition. "This process, owing to its length and the attachment of its mesentery, may be bent at an angle (instead of being directed upward and inward) so that hardened particles can slip readily into it, but are prevented from returning."†

---

* Meigs and Pepper. † "Diseases of Children," 2d Ed., p. 723.

ETIOLOGY.—In addition to peculiar anatomical relations and physiological attributes, which render the segment of intestine in question very prone to disease, a constipated habit is the chief predisposing cause. The existence of the strumous diathesis, too, while it has little influence in increasing the susceptibility to typhlitis, does augment the tendency to ulceration and perforation after inflammation is established.

Retention of hardened fæcal matter in the cæcum, the so-called "typhlitis stercoralis"; accumulation of the seeds of certain fruits, as strawberries or raspberries, in one of the pouches of the cæcum; the passage of these, or of intestinal concretions, or foreign bodies—shot, pins and bone spiculæ—into the appendix, and the habitual use of coarse, undigestible food, are the most common excitants. Cold and exposure, blows upon the abdomen, and violent exertion with strain of the abdominal muscles, are also sometimes determining causes.

In perityphlitis the inflammation is generally produced by the escape of fæcal matter into the peri-cæcal connective tissue. The perforation occasionally results from the ulceration of typhoid fever or of intestinal tuberculosis.

SYMPTOMS.—Simple typhlitis begins suddenly, with pain in the right iliac region, and vomiting. The pain is constant and severe, and is increased by coughing, sneezing, vomiting and by efforts to stand or walk. The vomiting is attended by distressing retching; is often repeated, and the ejections consist, first, of food, and afterwards, of bile-stained fluid. The patient has an anxious face; lies on his back slightly inclined to the right side, with the right thigh drawn up, and complains if an attempt be made to straighten it. Abdominal respiratory movements are partially suppressed; the right iliac region is full and even prominent; very tender to the touch and dull on percussion. Palpation, when it can be practiced, reveals a resistant mass occupying the site of the cæcum. There is fever, indicated by a coated tongue, extreme thirst, a frequent and somewhat wiry pulse, and a temperature ranging about 101° or 102° F. The bowels are confined.

When properly managed, these symptoms disappear in from four to twelve days; the bowels yield and move freely with the expulsion of masses of hardened fæces; the vomiting ceases; the pain abates, and the tenderness and swelling slowly subside.

Inflammation of the appendix alone is attended by the same symptoms. The pain, however, is more intense, and evacuation of the bowel is not followed by the same rapid relief.

Perityphlitis may be ushered in by the marked symptoms just described. On the contrary, the causal ulceration, as it involves a limited area of the cæcum or appendix, may be very latent. In the former instances the vomiting may stop, the bowels may be moved, and the acute pain be superseded by aching, or all discomfort disappear. The tenderness and swelling of the right iliac region, however, remain, although they are materially lessened; the patient looks ill; has a distressed face, and is listless. If, as is ordinarily the case, the onset be latent, complaints are made only of dull aching or discomfort in the cæcal region. These sensations are subject to exacerbations of a few hours' duration, when the suffering becomes acute and there is vomiting and fever. In the intervals, the general health is somewhat below par; the child, while up and about, takes little interest in play; is peevish; has a poor appetite; is, perhaps, restless, thirsty and feverish at night, and has irregular movements of the bowels—attacks of diarrhœa alternating with constipation.

After an indefinite time, in either case, perforation occurs. The event is followed by little change in the symptoms at first, but soon there is more constant and severe pain in the affected region, which is increased by movement or pressure; there is greater fulness, too; the bowels are confined; sleep is more restless and disturbed, and there is more pyrexia. The child takes to his bed, where he lies on his back with the right thigh drawn up. If assisted to stand, he rests his whole weight on the left leg, keeping the right bent at the hip and knee and rotated outward, and limps when he walks. There may be pain in the knee, and any rough attempt to move the leg increases the abdominal pain. As suppuration progresses in the peri-cæcal tissue, hectic fever develops, with rigors or chills, followed by profuse sweating; a dry, brown tongue; diarrhœa; a running, feeble pulse; prostration and rapid loss of flesh. The abdomen becomes distended and very painful; the cæcal tumor increases in size, but becomes softer; there is pain in the right knee and ankle, and sometimes œdema of the leg. Should the abscess point toward the surface, the skin becomes swollen; doughy; dark-red or purple in hue; palpation yields emphysematous crepitation, and an incision gives vent to brownish offensive pus and fetid gas. When the abscess reopens into the intestine, the local pain, swelling and tenderness diminish, and the general symptoms improve.

The uncommon cases in which perityphlitis occurs without previous typhlitis, are marked by pain; deep tenderness and moderate fulness; but there is no tumor, in the right iliac region. The bowels are irregular, and there is colic and moderate fever.

Perforation of the anterior wall of the cæcum gives rise to the symptoms of local or general peritonitis, according to the presence or absence of firm adhesions. Ulceration of the appendix is more frequently followed by symptoms of general peritonitis than by those of perityphlitis, and the former may be the first indication of lesion of this portion of the gut.

DIAGNOSIS.—A sudden attack of pain referred to the right side of the abdomen; vomiting; constipation; a pinched, anxious face; fever; a dorso-lateral decubitus; flexion of the right thigh, and the presence of an intensely tender tumor in the cæcal region, are the characteristic symptoms of typhlitis.

Perforative ulceration may be suspected if these symptoms disappear and reappear several times, or if, after a free evacuation of the bowels, local pain, tenderness and swelling continue.

Intussusception resembles typhlitis in some of its features, but in this condition tenderness is a late symptom; the tumor is situated more to the left of the abdomen; sometimes the lower end of the invagination can be felt in rectal examination, and there is severe tenesmus with the expulsion of blood-stained mucus.

The limping gait, with the pain and tenderness in the right groin which are incident to perityphlitis, may suggest hip-joint disease, particularly if other symptoms be but poorly developed. The distinction, though, can be made without much difficulty. In perityphlitis, although the right thigh is semi-flexed, and cannot be extended without great pain, it is possible, if care be taken, to rotate the head of the bone without causing suffering, and to make pressure on, or behind, the trochanter without giving discomfort. The patient, too, while avoiding extension, will often freely flex, abduct or adduct the thigh as requested. Again, there is no atrophy of the thigh muscles, no flattening of the buttock on the affected side, and no lowering of the buttock fold or obliteration of the fold of the groin. Finally, the history shows an acute course, and there are usually other symptoms which directly indicate that the disease is situated in the right iliac region.

PROGNOSIS.—Simple typhlitis should almost uniformly terminate in recovery. The duration of active illness is, as already stated, from four to twelve days, though several weeks often pass before the local tenderness entirely disappears, and the functions of the intestine are restored.

Ulcerative destruction of the anterior wall of the cæcum generally results in death from general peritonitis in two or three days. Perforation of the appendix is always rapidly fatal.

The termination of perityphlitis resulting from perforation depends upon the direction taken by the pus. When the abscess opens upon the surface of the body the mortality is about fifty per cent.; death resulting from exhaustion or extension of inflammation to the peritoneum. A reopening into the intestine is more favorable, and many cases get well. Under any of these circumstances the course is apt to be prolonged.

TREATMENT.—For prevention, it is necessary to guard against habitual constipation, by a properly selected diet, by regular exercise, and by enforcing the rule of making daily attempts to evacuate the bowels at a fixed hour. Nature may be assisted by a teaspoonful of compound licorice powder at bedtime, or one of the following pills:—

℞. Resin. Podophylli, ... gr. jss.
Ext. Belladonnæ, ... gr. j.
Ext. Taraxaci, ... gr. xij.
M. et ft. pil. No. xij.
S.—One pill every night for a child of six years.

Or—

℞. Ext. Belladonnæ,
Ext. Nucis Vomicæ, ... āā gr. j.
Ext. Colocynth. Comp., ... gr. vj.
Ol. Cari, ... gtt. iij.
Confec. Rosæ, ... gr. vj.
M. et ft. pil No. xij.
S.—One pill every night.

Should there be a tendency to fæcal accumulation laxatives are not to be administered, but the mass is to be removed by purgative enemata. Two teaspoonfuls of table salt to a half pint of warm water, will be efficient for this purpose, or one of the enemata mentioned in treatment of constipation, (p. 275) may be used.

A child attacked by typhlitis must be put to bed, and a small pillow placed under the right knee to support the thigh. The iliac region is to be covered with hot flax-seed poultices, or, if the child be robust and the tenderness and pain excessive, two or three leeches may be applied before poulticing. No food but milk, or milk with a little mutton or chicken broth, is allowable, and these are to be given in small quantities at short intervals. A patient six years old may take every two hours:—

Milk, ... f ℥ ij.
Barley-water, ... f ℥ ij.
Saccharated Solution of Lime, ... gtt. xv.

Saccharated solution of lime is used as an alkali instead of lime-water, on account of its adding no bulk to the food: it is prepared in this way:—

Take of—

Slaked Lime, ... ℥j.
Refined Sugar, in powder, ... ℥ij.
Distilled Water, ... f℥xvj.

Mix the lime and sugar by trituration in a mortar; transfer to a bottle containing the distilled water, cork and shake occasionally for a few hours. Finally, separate the clear solution with a siphon and keep in a stoppered bottle.

When broth is used, it may take the place of the milk at three or four feedings during the twenty-four hours. Should the milk or broth not be retained, whey mixtures, peptonized milk, and meat juice can be tried.

Thirst is best relieved by small quantities of cold carbonic-acid water and bits of ice.

The therapeutic indication is to relieve irritation and arrest excessive peristaltic action of the bowels. This is best accomplished by a combination of opium and belladonna, as:—

℞. Tr. Opii., ... ♏xxiv.
Tr. Belladonnæ, ... ♏xlviij.
Aq. Cinnamomi, ... q. s. ad f℥iij.
M.
S.—One teaspoonful every two hours.

The action of these drugs must be constantly pushed, until pain be relieved or the limit of systemic toleration be reached. The second or third dose usually checks vomiting; but should this be not the case, morphia must be administered hypodermically in doses of one-sixteenth of a grain at intervals of four or six hours, according to its influence on the pain and its narcotic action. As the pain subsides the bowels act spontaneously.

There is one rule in the treatment of typhlitis that must never be forgotten, namely, no purge, no matter how gentle in action, is to be used, either by the mouth or rectum, while the acute symptoms are present. After they disappear, if the bowels be not relieved, enemata of warm water can be given safely, but no purge by the mouth. Furthermore, it is well to withhold the latter for several weeks after convalescence is established, for there may be some latent ulceration in progress that can only result favorably through the formation of firm peritoneal adhesions, and nothing so surely destroys them, while in the process of development, as a purgative.

As soon as the bowels are moved and convalescence begins, the diet may be cautiously increased; a belladonna plaster substituted for the poultices; tonics administered, and the patient allowed to sit up in bed, and after a time, as health returns, to be up and about. Very active exertion should be avoided for several months.

Perityphlitis demands the same rest in bed, careful dieting, local applications, and avoidance of purgatives. As hectic fever appears, the food must be more nutritious; eggs, finely minced meat and beef-tea may be added to the milk. Alcoholic stimulants are also required with quinine, and, when there is great prostration, carbonate of ammonium.

So soon as the abscess points it must be evacuated and a free discharge encouraged. In the after treatment every effort should be made to support the strength by good food, tonics and stimulants.

It would not be well to leave the question of treatment without some more distinct allusion to the intervention of surgery. Peritonitis of every form is passing more and more into the hands of the surgeon, and remarkable successes have of late been recorded in cases which might well have been deemed desperate. Regarded from the point of view of the physician, the subject stands thus: A number—it is difficult to say how many, but probably a large majority—of these cases, if treated judiciously, get perfectly well, and an operation, however successful, might well be called meddlesome. In others the inflammatory process localizes itself; then, if the symptoms indicate no progress toward recovery, or are in any degree urgent, an exploratory incision is not only justifiable, but demanded. Next come those other cases already described, where the peritonitis is general, and in which the life of the child is in the balance. Then it is that the experience of other cases, that have struggled through; the fear that a serious undertaking, such as opening the abdomen, may extinguish the last hope; the doubt that must exist whether, if an operation be begun, any relief can be afforded, and similar considerations, make confusion when we most need calm judgment. We can be wise after the event, and talk glibly of the advantages of early operation, but this is small help to us when the point aimed at is so to time our measures as to be neither too soon nor yet too late. No precise rules can be established; these cases must remain full of anxiety, of doubt and difficulty, and the man of courage and judgment will occasionally save a life by a timely and carefully-conducted operation. So far as advice can be given, it may be said that for a dry peritonitis probably no good will come of surgery. If any evidence can be obtained favoring the existence of pus or of serum—for the serum in these cases is irritant and noxious, and often as urgently calls for removal as pus—here, if the right moment can be seized, an incision, and such steps as may be necessary for cleansing the peritoneum, will sometimes prove successful.

16. INTUSSUSCEPTION.

In intussusception or invagination one portion of the intestine is forced, from above downward, into another portion immediately continuous with it.

Apart from fæcal accumulation, this is practically the sole cause of intestinal obstruction in infancy. For, although instances are on record in which the bowel, in children, has been closed by peritoneal adhesions; by a twisted vermiform appendix, and by morbid growths, these are but pathological curiosities.

Two forms are met with, namely:—intussusception without symptoms; and intussusception with symptoms.

INTUSSUSCEPTION WITHOUT SYMPTOMS.

This condition, which must be regarded rather as an accident than a disease, is frequently encountered in autopsies upon young children who have met death from very diverse affections.

Such intussusceptions occur shortly before, or during, the death agony, and are probably produced by irregular and violent contractions of the muscular fibres of the gut. They consist simply of an involution of the bowel, without evidence of inflammatory action at the site of lesion, and can be readily reduced by traction. Sometimes there is but one inversion, though usually there are several; as many as ten or twelve distinct invaginations, at distances of a few inches from each other, having been found in the same subject. The length of gut displaced is rarely more than three or four inches. The small bowel is the uniform seat; and of this division of the intestines, the lower part of the jejunum and the upper part of the ileum, are most frequently involved.

Without a post-morten examination, it is impossible to recognize the existence of this form of intussusception, on account of the entire absence of symptoms. Nevertheless, its discovery may be anticipated when death has resulted from cerebral or spasmodic diseases, or from acute or chronic entero-colitis.

INTUSSUSCEPTION WITH SYMPTOMS.

True intussusception is, fortunately, not very frequently met with in children, though it is more common in early infancy than in later childhood, youth or adolescence.

MORBID ANATOMY.—The probable mechanism of an intussusception is that a limited portion of the intestine contracts forcibly, and, by elongating and moving forward, enters a non-contracted segment immediately below, drawing in more or less of the latter, together with its mesentery or meso-colon. Next, new peristaltic movements force the invaginated bowel further and further along, until extension is arrested by resistance from the mesentery, or by secondary inflammatory adhesions. The intussusception must, therefore, be made up of three layers of intestine, one above the other. The outer layer is called the sheath, or intussuscipiens; the middle and inner ones, the intussusceptum. Of these, the external and middle have mucous surfaces in contact; the middle and internal, serous surfaces. The involuted mesentery or meso-colon lies between the two last-named layers, and, on account of the firm attachment at its roots, exerts a one-sided traction upon the intussusceptum, curving it upon its axis and drawing the lower opening—which is elongated to a narrow fissure—from the centre toward the side of the sheath. The sheath itself is much folded or puckered, and on this account, with the curving of the intussusceptum, the apparent length of gut involved is always much less than the actual length. This varies from a few inches to several feet; in extreme cases an intussusception beginning at the ileo-cæcal valve, may become apparent to the touch or sight at the anus. Increase in length is accomplished by peristaltic action from behind; it takes place, always, at the expense of the external layer, and depends, for its degree, upon the force of peristalsis, the width and laxity of the mesentery or meso-colon, and the amount and character of the contents of the intestine behind the seat of involution.

The results of an intussusception are, first, occlusion of the lumen of the canal with partial, or generally complete, arrest in the passage of the intestinal contents; and second, obstruction of the blood current in the middle and inner tubes, due to the pressure upon the mesenteric vessels.

The obstruction of the circulation leads to deep congestion of the tissues of the intussusceptum; the mass becomes purple and swollen; the mucous surfaces exude a bloody material, and soon the opposed serous surfaces are glued together by inflammatory adhesions.

Should there be complete strangulation the intussusceptum becomes gangrenous, and, under favorable circumstances, may be detached *en masse* or in pieces, and discharged through the anus. When this occurs, provided firm adhesions have formed, the sheath, being united at its upper extremity to the intestine directly above the point of inversion, forms, with the latter, a continuous tube, notwithstanding the separation of the intervening portion.

Several accidents may happen during this process. Thus, the inflammation in the opposed serous coats may extend beyond the involution, and give rise to general peritonitis. Or, ulceration and perforation of the sheath may be produced by the pressure and irritation of the free end of the intussusceptum. Again, when adhesions are imperfect, the contents of the intestine may escape into the peritoneal cavity through a rent, resulting from the separation of the sloughing intussusceptum ; and, finally, even after the gangrenous mass is expelled, the adhesions may give way and permit extravasation.

Generally, in those fortunate cases in which sloughing is followed by recovery, no permanent injury results from the cicatrization at the point of junction of the sheath and uninvolved intestine. The cicatrix, though at first contracted, gradually stretches and a free passage-way is established.

Sometimes intussusception is attended with so little constriction of the involuted gut that the passage for the involved intestinal contents is quite free enough to allow of the maintenance of life for months, the patient finally dying of exhaustion.

In infants the invagination is almost invariably ileo-cæcal. The end of the ileum with the ileo-cæcal valve is forced into the cæcum, and, as the intussusception increases, penetrates further and further into the colon, drawing along more of the ileum, and doubling in, first, the cæcum, and then the ascending, or even the transverse and descending portions, of the colon. In some cases a few inches of the gut pass through the ileo-cæcal valve before the cæcum is inverted. Occasionally an intussusception involves the colon alone, and, very exceptionally, the small intestine.

Upon opening the abdomen, in an ordinary case, much of the colon appears to be wanting, and a tumor is found occupying the left side or the left iliac fossa. This mass—the intussusception—is slate-gray in color; elongated or sausage-shaped, and doughy to the touch. By more or less forcible traction, the involution may be reduced, though the gut is usually softened and apt to be torn in the effort. If an incision be made through the sheath, exposing the intussusceptum, two orifices will be observed at the lower end of the latter, one leading through the valve, the other into the cavity of the appendix vermiformis. The invaginated intestine is either of a uniform deep red color, resembling a long, firm clot of blood, or presents the appearances common to gangrenous and sloughing tissues. If death has occurred early, there are few evidences of inflammation between the serous surfaces; if later, these are adherent, the adhesions extending a few lines beyond and above the neck of the intussusceptum on to the sound intestine. The gut situated above the point of obstruction is usually greatly distended with accumulated fæcal matter and flatus; whilst that below is collapsed and empty, or at most, contains a small quantity of mucus, stained with blood, pressed out from the capillaries of the strangulated mass.

As the age of the child advances the more likely is the intussusception to be confined to the small intestine.

ETIOLOGY.—As already indicated, early age seems to act as a powerful predisposing cause. Of fifty-two cases in children, recorded by J. Lewis Smith, twenty-three occurred between the ages of three and six months; eight between the sixth and twelfth months; and eighteen between the first and twelfth years. Of Leichtenstern's four hundred cases, one-fourth occurred in the first year, after the third month. The greater liability in infancy is due partly to anatomical peculiarities, and partly to the want of regularity and the energy of the intestinal movements. Thus, in infants, the large intestine holds to the abdominal space that it is forced to occupy the relation of about three to one, necessitating doubling of the gut upon itself. At this time of life, too, the meso-colon is much wider than in later years, except where it passes over the kidneys, in which position it is very narrow, or even almost absent. These two conditions, combined with unrhythmical and violent peristalsis, cannot but favor involution.

Many more males are affected than females. Rilliet and Barthez record twenty-five cases, all but three in boys, and the statistics of other authorities bear out their figures.

The existing causes are imperfectly understood. Attacks have been attributed both to obstinate diarrhœa and prolonged constipation; to the presence of intestinal worms; to the use of irritating and indigestible food; and to external violence.

SYMPTOMS.—These vary considerably, according to the age of the sufferer and the completeness of intestinal obstruction.

In patients under one year the onset is abrupt, whether it occur in the midst of health, or during the course of some derangement of the digestive tract. The child is seized with intense pain in the abdomen, turns excessively pale, screams, and then cries violently, writhing and drawing up his legs. The contents of the stomach are vomited, and usually, unless the bowels have been evacuated just before the attack, there is a single discharge of somewhat liquid feculent matter. After a time the pain passes away, leaving the little sufferer pallid and exhausted. There is now a rest from pain, but not from vomiting; all food or medicines taken into the stomach are returned at once, either by the easy process of regurgitation, or by violent retching; and if the viscus be empty, the ejections consist of a little bile-stained mucus, or even of blood. Sooner or later—the interval varying from a few minutes to several hours—there is another paroxysm of pain, accompanied by violent tenesmus, resulting in the evacuation of blood and mucus.

At this time the abdomen differs little from its normal condition. There is no fulness nor tenderness, nor any tumor appreciable to the touch; on the contrary, gentle friction often relieves the colicky pains, and the child prefers to lie upon its belly. The hands and feet may feel cool, though, otherwise, the temperature of the surface is unaltered. The mind is clear, but the expression of face is anxious, and denotes severe illness. The tongue may be lightly furred, and there is increased thirst, leading to a greedy consumption of the contents of the feeding-bottle, or a ravenous sucking at the breast. There is also restlessness, constant whining or moaning, and an inability to sleep. After a period of twelve or twenty-four hours, in which the paroxysms of pain and tenesmus have grown more frequent and severe, the abdomen becomes full; there is tenderness in the left iliac fossa, and if deep pressure be made in this region, during the absence of pain, a distinct swelling may be detected. The tumor gradually becomes more defined; it is elongated or sausage-shaped; of doughy consistence; and ranges in size from that of a hen's egg to that of a clinched fist. Later, it may change its position, moving toward the left side. When easily detected by external palpation, the tumor may be touched by a finger inserted into the rectum, and under these circumstances feels much like the cervix uteri in a vaginal examination. Occasionally the lower end of the involuted intestine protrudes from the anus, looking not unlike a prolapsed rectum.

While these features are developing, vomiting continues, and bloody mucus is expelled, with great straining, from the rectum, but there is no passage of either fæces or flatus. The amount of blood varies considerably; in some cases there is no more than sufficient to stain the diapers, in others, three or four ounces are voided several times daily. The tongue is red and glazed, or covered with a dry, brown coating; the pulse becomes frequent and feeble; the temperature rises to 102° or 103° F.; the abdominal respiratory play is restricted, and the flesh wastes. The urine may be greatly diminished in quantity; this, however, is by no means a constant condition, and seems to bear no relation to the seat or extent of the intussusception.

By the third day symptoms of collapse set in. The face becomes pinched; the eyes sunken and surrounded by dark circles; the skin feels cool and clammy, and the thermometer indicates a sub-normal temperature. The attacks of vomiting are less frequent; the pain less intense; the bloody evacuations lessen or disappear, and the child lies upon his back, in an apathetic condition, with half-closed eyelids, until the end comes at some time between the fourth and sixth day. Occasionally death is preceded by convulsions.

If, by any means, the invagination be reduced, the vomiting stops, pain disappears, flatus and thin, copious, semi-liquid and offensive stools are voided, and the patient finds rest in sleep. Afterwards there is pallor, languor and weakness, though the appetite quickly returns and flesh and strength are soon regained.

When older children are attacked the picture differs in some of its details.

Thus, abdominal distention appears earlier and is more marked. The gut behind the position of obstruction being filled with fæces and flatus is greatly stretched, and the outline of its coils may be distinctly seen and felt through the tense wall of the belly. This is especially the case during the paroxysms of pain, when, too, waves of peristalsis may be seen, and loud, gurgling sounds heard. The tumor is large and better defined, and the dull percussion sound that it yields contrasts strongly with the general tympany.

Vomiting is more apt to be stercoraceous. This characteristic symptom is absent in many cases, and very naturally so; for, if Brinton's theory be adopted, namely, that fæcal vomiting is due to a reverse axial current in the contents of the intestine, and not to anti-peristalsis, it is apparent that, for the development of this symptom at all, the obstruction must be either in the large intestine or in the lower part of the ileum. The date of its onset must depend upon the distance of the starting-point of the reverse current, or the obstruction, from the stomach; upon the rapidity with which the bowel above is filled by ingestion and secretion; and, should the colon alone be involved, upon the readiness with which the resistance of the ileo-cæcal valve is overcome.

Evacuations of blood are much less uniformly observed. Whether this symptom be present or absent depends solely upon the degree of constriction of the invaginated bowel. When the constriction is just sufficient to obstruct the circulation and over-fill the vessels, hemorrhage is constant; but when there is actual strangulation, with complete arrest of circulation, no blood escapes. In older children strangulation is much more apt to occur than in infants, partly because the invagination more frequently involves the ileum or jejunum—which have a smaller calibre than the colon—and partly from the fact that in them, life being more prolonged after the accident, there is greater opportunity for inflammatory swelling. Hence it is that in this class of cases there is, in many instances, absolute constipation without bloody stools, though in others, where the intussusception is ileo-cæcal or colic, hemorrhage from the anus may be noticed as an early and permanent feature.

The blood appears in a liquid form, mixed with mucus, or in small clots. It always has the venous hue, and is darker in color and smaller in quantity, in proportion to the distance of the involution from the anus.

The older the child the more likely is gangrene, separation and elimination of the invaginated gut, to take place. This result is usually noted during the course of the second week of illness, and can be attributed to the greater power of resistance and tenacity of life displayed by older children than by infants. It is a fortunate ending, but, unfortunately, rarely occurs at any age. After the process of separation is completed, violent straining efforts set in, expelling the black, ill-smelling, gangrenous mass, either in its entirety or in patches and shreds, together with a large quantity of dark, offensive, feculent matter. The child then falls into a deep sleep, and awakes much refreshed. Thirst diminishes; the appetite returns; the paroxysms of pain cease; the face expresses ease and comfort, and the path to health is rapidly traversed.

When death occurs it is usually due to asthenia, and may, as in infants, be preceded by convulsions. General peritonitis is very uncommon.

In addition to the form of intussusception just described, and which may be termed *acute*, a variety having a much more prolonged course, is sometimes encountered.

CHRONIC INTUSSUSCEPTION may occur at any age, and may

exist for weeks, or even months, without producing severe illness. It originates most frequently in ileo-cæcal intussusceptions, and depends upon the inflammatory union of the outer and middle layers, and the restoration of the permeability of the inner tube by the complete disappearance of the swelling. The patient wastes, has periodical attacks of colicky pain, constipation and vomiting, and occasionally passes a little blood. Palpation reveals a tumor that alters its position, shape and density from time to time, and, on account of hypertrophy of the muscular coat above the partial obstruction, the knuckles of the intestine show distinctly through the emaciated abdominal wall.

Such cases may end in recovery, by separation; or in death, by perforative peritonitis; or by steadily increasing marasmus with chronic diarrhœa.

DIAGNOSIS.—Intussusception may be strongly suspected when a child in good health, or previously affected with simple diarrhœa, is suddenly seized with violent, paroxysmal abdominal pain and vomiting, quickly followed by straining efforts, resulting in the evacuation of mucus and blood, and by intense prostration. The suspicion will be reduced to a certainty if, at the same time, it be possible to detect a sausage-shaped tumor on the left side of the belly, or to touch the lower end of the inverted intestine upon rectal exploration. It is necessary to remember, however, that often, at the commencement of the attack, there is a single loose feculent stool, and also that, in older children, bloody, mucous discharges may be entirely absent.

The diseases with which intussusception is most likely to be confounded are simple colic, perforative peritonitis, dysentery and fæcal accumulation within the bowel.

In simple colic, the pain, though often severe, is never distinctly paroxysmal. It is attended by suppression of urine, is relieved by the discharge of flatus per anum, and is followed by copious urination. During the attacks the skin may be hot, and the belly is usually hard and tense. There is no vomiting, tenesmus or discharge of bloody mucus. The misfortune of confounding intussusception with colic can hardly be overestimated; for a laxative, as castor-oil, while relieving the latter by clearing out the intestinal canal, cannot fail to aggravate the former by increasing the force of the peristaltic contractions.

In perforative peritonitis there is pyrexia from the beginning, the abdomen is distended and tense, and pressure in the right iliac region—since the seat of perforation is usually the vermiform appendix—produces pain. On the other hand, tenesmus and the evacuation of blood and mucus from the rectum are never observed, neither is it possible to detect a tumor by abdominal palpation, nor the lower end of an intussusception by rectal exploration.

Dysentery presents, in the character of the dejections and the severe pain and tenesmus, features similar to intussusception, but it lacks the sudden onset, the obstinate vomiting and the abdominal tumor. The two diseases differ, too, in their course.

A fæcal accumulation, as it produces actual occlusion of the intestine, has many symptoms similar to invagination. Thus, there is vomiting, colicky pain, tenesmus, constipation and a tumor. The former accident, however, is preceded, for some time, by the passage of hard and scanty stools; while the attendant vomiting is less obstinate; there are no bloody evacuations; and the tumor is more superficial, more fixed, and of such consistence that it can be indented by moderately firm pressure with the fingers. A purgative enema, also, rapidly leads to the expulsion of the impacted mass and relieves the symptoms.

PROGNOSIS.—Every case of intussusception affords an outlook that is grave in the extreme, though the danger to life depends directly upon two factors, namely: the age of the patient attacked, and the acuteness and consequent severity of the symptoms. In children under one year, death almost invariably results. With those who are older, the constitutional resistance being greater, sloughing of the intussusception is more apt to take place and recovery follow; but the rarity of this fortunate termination at any time of life, has already been alluded to.

TREATMENT.—In no other disease does the prospect of success rest so much upon an early diagnosis and appropriate management.

The indication to be met is the total arrest of all action of the muscular fibres of the intestine. To accomplish this end, the patient must be kept in a state of absolute repose; must be made to take enough opium to relieve pain and check intestinal peristalsis, and must be carefully and properly fed.

Opium may be employed alone or in combination with belladonna, and may be administered by the mouth, by the rectum or, in children over a year old, by hypodermic injection. A combination of the two drugs and their administration by the rectum, is to be preferred in ordinary cases. For a child of one year the following suppository may be ordered:—

    ℞. Ext. Opii, ... gr. jss.
    Ext. Belladonnæ, ... gr. ss.
    Ol. Theobromæ, ... ℥ij.
    M. et ft. supposit. No. xij.

In the beginning, one of these suppositories can be introduced every two hours; but the interval must be lessened or the dose increased to the point necessary to relieve pain and tenesmus.

When the mouth is selected as the channel of exhibition, the opium and belladonna may be prescribed in a mixture, as:—

    ℞. Tr. Opii Deod., ... f℥j.
    Tr. Belladonnæ, ... f℥ss.
    Aquæ Anisi, ... q. s. ad f℥iij.
    M.
    S.—One teaspoonful every two hours (or p. r. n.) for a child
      of one year of age.

When the hypodermic method is selected, it is still best to combine the two drugs, thus:—

    ℞. Morphiæ Sulphatis, ... gr. $\frac{1}{3}$.
    Atropiæ Sulphatis, ... gr. $\frac{1}{20}$.
    Aquæ Dest., ... f℥j.
    M.
    S.—Inject ten minims p. r. n., in a child of six years.

The application of anodynes to the surface of the abdomen

has certainly some power in relieving pain, and should, consequently, not be neglected. For this purpose a piece of soft flannel, large enough when doubled to cover the whole belly, should be dipped in hot water, wrung moderately dry, sprinkled with a teaspoonful of laudanum, laid over the surface and covered with oiled silk.

As to food, only so much as is absolutely required to sustain life should be allowed. For the first twenty-four or forty-eight hours, especially if there be vomiting, food may be altogether withheld, and a teaspoonful of barley-water or a small bit of ice given every fifteen minutes to allay the thirst. Later, milk and barley-water for infants, and milk, beef-juice or strong beef essence for older children, are admissible; but always in minimum quantities and at intervals of two or three hours. Thus, for a child of one year, half an ounce of milk and barley-water, in equal parts, every two hours, will be quite enough; while at the age of six years two ounces of milk, one ounce of beef-essence, or half an ounce of beef-juice, every three hours, will suffice. It is well to peptonize the food, as in this way it is prepared for assimilation, and little residue is left in the intestinal canal. Everything must be taken cold.

When the strength begins to fail, brandy must be administered. The following mixture is a good food and stimulant:—

> Brandy, ... 4 fluidounces.
> Cinnamon Water, ... 4 fluidounces.
> The Yolks of two Eggs.
> White Sugar, ... ½ ounce.
> Rub the yolks and sugar together, then add the cinnamon water and spirit.
> DOSE.—A dessertspoonful to two tablespoonfuls every two hours, according to the age.

Should the patient be seen before the fourth day of the attack, in addition to the above measures of treatment, efforts should be made to reduce the involution mechanically. Mechanical interference can be successful only before inflammatory adhesions have formed, and is contraindicated when there is tenderness over the seat of lesion. There are three possible ways of accomplishing reduction, viz., by taxis, by forced injection of water, and by insufflation of air. To perform either of these successfully the patient, unless an infant under one year, must be previously put thoroughly under the anæsthetic influence of ether.

Taxis consists in kneading and otherwise manipulating the abdomen with the warmed and oiled hand; it is usually employed in conjunction with one of the other methods.

If forced enemata be resorted to, the child is placed upon his back in bed, with the buttocks elevated, so that the trunk is inclined at an angle of 45 degrees. Then, with a Davidson's or Fountain syringe, the physician himself must inject, carefully and slowly, as much tepid water as the capacity of the intestine will permit. While this is being done, the abdomen must be kneaded in such a manner as to force the water upward along the bowel in the direction of the invagination. At times the obstruction can be felt to give way; but the best proofs of this fortunate occurrence are the subsidence of the more urgent symptoms and the onset of sound sleep. Soon, too, there is a discharge of bloody mucus from the rectum, and then a free, offensive, semi-fluid fæcal evacuation.

Insufflation of air is suited to those cases in which the intussusception having descended into the rectum, little or no water can be injected or retained. The apparatus for this purpose consists of a bellows, having its nozzle attached, by means of a piece of flexible rubber tubing, to a caoutchouc tube about a foot in length; the latter is inserted into the rectum, care being taken to secure a perfect fit at the anus by a packing of lint. The air should be pumped in slowly and gently, and during its introduction, taxis is practiced in the same manner as in the forced injection of water.

Should these measures fail, or should the case be seen for the first time after the third or fourth day, the question may arise as to the propriety of laying the abdomen open and reducing the intussusception by direct traction. For the details of the operation of laparotomy, the reader must be referred to works on surgery. Briefly stated, it consists of making an incision in the median line of the abdominal wall; opening the peritoneum; finding the intussusception, and working it back at the neck in the same manner that a hernia is reduced.

The practical results of the operation prove, beyond doubt, that there is very little to be expected from it. This is especially true in the case of infants less than a year old, in whom it has been invariably fatal. By its performance, too, the chance of recovery from separation of the strangulated intestine is lost.

Upon this point, Prof. John Ashhurst, Jr.,* states:—

"Inspection of the table (comprising 13 cases of laparotomy, 5 in children), shows, in the first place, that no encouragement is afforded to repeat the operation in very young infants. The only instances in which it has been resorted to during the first year of life have all terminated fatally (Gerson, Wells, Weinlechner). But when it is remembered that of Pelz's 162 cases (all occurring in children), no less than 91 were in infants less than a year old, it will be seen how large a proportion of cases must at once be put aside as unfitted for operative treatment. It is very true that the fatality of intussusception at this early age is enormous, the mortality being, according to Leichtenstein's elaborate statistics, no less than 86 per cent. But the case is very different from that, for instance, of an operation for imperforate rectum; for in this condition there is necessarily no hope but in an operation; whereas, in the case of intussusception, experience shows on the one hand that, even at this age, a certain number do recover without operation, and that, on the other hand, as might be expected, operative treatment is in such cases of no avail.

"In the second place, the table shows that in what may be called acute cases, those, namely, in which in addition to symptoms of obstruction there are evidences of strangulation, such as peritonitis and intestinal hemorrhage, a resort to operative inter-

---

* American Journal of Medical Sciences, July, 1874.

ference will be productive of no benefit. These cases are, on the other hand, as justly remarked by Mr. Hutchinson, precisely those in which there is most hope of recovery by sloughing of the invaginated portion.

"There remains, then, a limited number of cases, in not very young infants, in which the symptoms are those of obstruction merely, without intestinal hemorrhage or peritonitis, and in which, when other measures fail, the question of operation may properly be considered."

From additional cases collected since the compilation of the table above referred to, Prof. Ashhurst finds no data altering his previous conclusions.

When separation and discharge of the invaginated segment of the intestine takes place, it is necessary to exert the utmost care lest the new-formed adhesions be broken down. The patient must lead a passive life; the food must be readily digestible and restricted in quantity, and all farinaceous and fermentable articles are to be excluded from the dietary.

In chronic intussusception great reliance can be placed upon the free administration of opium and belladonna, with forced enemata ; or, in proper cases, insufflation of air, and as this condition generally occurs in late childhood, laparotomy, when other measures fail, may be undertaken with more hope of success.

17. INTESTINAL WORMS.

The common parasitic worms that find their habitat in the human intestines may be divided into two classes, each including several varieties, namely :—

NEMATODES.
- *Oxyuris vermicularis.*
- *Ascaris lumbricoides.*
- *Tricocephalus dispar.*

CESTODES.
- *Tænia solium.*
- *Tænia saginata.*
- *Bothriocephalus latus* (rare).

312   DISEASES OF DIGESTIVE ORGANS IN CHILDREN.

Of these, the oxyuris vermicularis—small thread-worm—and the ascaris lumbricoides—long, round worm—are most frequently found in children. The tænia, or tape-worm, is uncommon before the age of six or seven years.

DESCRIPTION AND MODE OF ENTERING THE BODY.

OXYURIS VERMICULARIS.—These worms are silvery-white in color and of small size; the males being one-sixth of an inch, the females one-half an inch long. To the unaided eye they present the appearance of small, white threads. They inhabit the cæcum and whole length of the colon, but are most abundant in the sigmoid flexure and rectum, where they derive nourishment from the fæces, and where, alone, they give rise to symptoms or evidences of their presence.

Oxyures enter the body by direct passage of the ova into the mouth. They are introduced clinging to fruit and various articles of food, or are conveyed to the lips by the hands of the child previously used to relieve itching, occasioned by the presence of the parasites in the neighborhood of the anus. Having entered the stomach, the embryos are liberated by the action of the gastric juice and pass into the small intestine, which they descend with the food, developing so rapidly that they become sexually mature by the time they reach the cæcum.

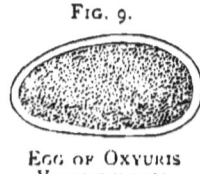

FIG. 9.

EGG OF OXYURIS VERMICULARIS.

The eggs, as seen by the microscope, are $\frac{1}{500}$ of an inch in length, ovoid and unsymmetrical. They are produced in great numbers, each female giving birth to several broods, numbering from ten to twelve thousand, and they are extremely tenacious of vitality.

ASCARIS LUMBRICOIDES.—This parasite has a certain superficial resemblance to the common earth-worm. It is cylindrical, tapering at both extremities, is reddish or brownish in color, has a body marked with fine transverse rings, and possesses a peculiar, disagreeable odor independent of the substance in which it lives.

The head of the worm presents three prominent labial papillæ surrounding the mouth, and the tail is conical. The male is from three to six inches in length and one-eighth of an inch in thickness, with an incurved tail and a penis consisting of a pair of slender, clavate, chitinous spicules, the ends of which protrude from a cloacal aperture at the root of the tail. The female measures from six to fourteen inches in length and one-fourth of an inch in thickness. The genital aperture is situated on the ventral surface, near the anterior third of the body; the ovarian tubes may be observed as long, tortuous canals, and the uterine tubes as short, straight canals; the latter contain many millions of ova. The ripe ova are laid in the intestine, and are expelled, with the stools, in great numbers, sometimes even in large masses. The eggs are $\frac{1}{500}$ of an inch in diameter, and are oval, with a thick, elastic, brownish, double shell and nodulated surface. After expulsion from the rectum, they are very tenacious of life, remaining in a condition capable of development for several years. This is particularly the case when they find their way into water or moist earth. Here the embryos slowly develop.

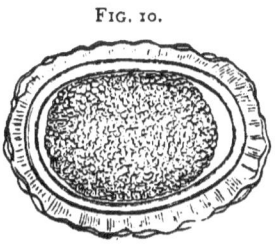

FIG. 10.

EGG OF ASCARIS LUMBRICOIDES.

It is not positively known in what way children become infected with the parasite, but impure water is, without doubt, the medium. Recent experiments on both animals and men have demonstrated that infection cannot be directly produced by taking the recently laid, ripe ova into the stomach. It is probable, therefore, that the ova passed by an infected subject, after entering—through drainage or otherwise—moist soil or running water, and undergoing partial development, are eaten by some common, but unknown, minute aquatic animal. Within the bodies of these they are still further developed, so that, when the animalculi are ingested with impure drinking water, and the embryos liberated by the action of the digestive solvents, the latter are in a position rapidly to assume their mature characters.

Lumbrici inhabit the small intestine principally, though they frequently migrate. Their number in a single individual may vary from two to several hundreds.

TRICOCEPHALUS DISPAR.—The whip-worm, as this parasite is sometimes called, is yellowish-white in color, with the anterior half, or more, of the body attenuated in a hair-like manner. The male is about one inch and a half in length, has the thick portion of the body enrolled, and a blunt tail. The female is two inches long, with a conical tail. The eggs of this worm are laid in the intestine and voided with the fæces; nothing is known of their subsequent history, or of the method in which the human being is infected.

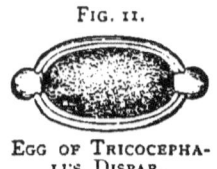

FIG. 11.

EGG OF TRICOCEPHA-
LUS DISPAR.

Whip-worms inhabit the lower end of the ileum, the cæcum and the vermiform appendix, and feed on the intestinal contents. They are met with in small colonies, varying in number from two to twelve. Their occurrence is considered to be exceptional in this country, but this may be explained in two ways: first, as they are rarely voided with the fæces, like other worms, they escape ordinary observation; and second, as their presence gives rise to no symptoms during life, few think of looking for them on the post-mortem table.

TÆNIÆ infest children over the age of six years with almost the same frequency as adults. Of the three varieties, the *tænia saginata*, beef tape-worm, and *tænia solium*, pork tape-worm, are the most common, while of these two the former is met with in by far the greater number of cases.

The *tænia saginata* is a soft, yellowish-white, band-like worm, varying in length from six to twenty or more feet. The head has about the bulk of a yellow mustard-seed, is rounded quadrate, and provided with four hemispherical suckers. Between the head and body is a short, unsegmented, flattened neck, narrowest at its upper extremity and gradually broadening as it merges into the body. The latter is distinctly segmented. The first segments are several times wider than long, but become successively larger until the length exceeds the breadth two or three times; their number may reach twelve hundred, and the largest measure from one-fourth to one-third of an inch in width, and a quarter to a full inch in length. The parasite is usually solitary. It inhabits the small intestine, in which position the segments, as they ripen, are spontaneously detached, some having already expelled their burden of eggs (often numbering 35,000), others still laden. Both eggs and liberated segments then become mixed with the intestinal contents, pass downward into the colon, and are finally expelled with the fæces. The mature ova are brown, oval in shape and of minute size; they have a thick inner shell, and an outer longitudinally striated envelope. Each ova contains an embryo—a spherical or oval body—having at one pole three pairs of divergent, boring spiculæ.

FIG. 12.

EGG OF TÆNIA SAGINATA.

In rural districts, it is a very common habit, with both adults and children, to stool in a fence corner, or other more or less secluded spot, at any time there is a call for evacuating the bowels. Should ova-laden tape-worm segments or liberated eggs be contained in the fæces so carelessly deposited, these are apt to be swallowed by cattle grazing in the neighborhood. Having once entered the stomach, the embryos, liberated from the eggs by the action of the digestive solvents, attach themselves to the mucous membrane of the viscus, perforate it by means of their powerful boring apparatus, and, either directly or through the medium of the blood current, migrate into the tissues of the body, usually the muscles or liver. After attaining their destination, they become fixed and slowly undergo development, dropping the spiculæ and being transferred into the larval form or *scolex*. The scolex consists of a head like that of the mature worm, with a neck terminating in a capacious cyst, within which the head and neck are inverted. In this form each parasite is surrounded by a sack of connective tissue; the new formation depending upon the presence of the larva, acting like an imbedded foreign body. The flesh, liver and other organs of cattle so infected are said to be "measly." Now should measly beef be taken into the stomach insufficiently cooked, or should it be administered raw—a frequent practice in the treatment of certain intestinal diseases—the parasite, during the process of digestion, is liberated from the investing connective tissue envelope, everts its head from the containing sack, attaches itself to the mucous membrane of the small intestine, feeds upon the intestinal contents, and, growing from its caudal extremity rapidly develops into a mature tape-worm. The time required for development has been proved by experiment to be less than two months, and the natural duration of life is very protracted.

TÆNIA SOLIUM.—This species of tape-worm enters the human body through the medium of measly pork. The methods of propagation and development are identical with those of the tænia saginata. In general appearance, also, the two worms are very similar. The pork tape-worm, however, is white in color and broader and shorter, the usual length being between five and ten feet. The head, which presents the most prominent distinguishing features, is about the size of an ordinary pin; it is spheroidal in shape, is surmounted by a blunt papilla, encircled by a double row of hooks, and, at the same time, has the four hemispherical suckers to be noticed in the beef tape-worm. The ova are somewhat smaller but otherwise identical, the scolices, long known as *Cysticercus cellulosæ*, likewise possess the double row of hooks to the head, and in this way may be distinguished from the larvæ of the other variety of tænia. The period occupied in development is about three months; the length of life probably twelve years or more.

As in the United States pork is but little used as food in comparison with beef, and when used is thoroughly cooked, the difference in the frequency of occurrence of the two species of worms can be readily explained. In regard to this point, Prof. Leidy states: "Since the writer distinctly recognized the beef tape-worm, within the last twenty years, all the specimens of tæniæ, from people of Philadelphia and its vicinity, that have been submitted to him for examination—perhaps in all about fifty—have appeared to belong solely to tænia saginata. The prevalence of this species with us is no doubt due to the common custom of eating underdone or too rare beef, while the pork tapeworm is comparatively rare, as with us pork is only used in a well-cooked condition.''

SYMPTOMS.—These may be divided into two classes: general and special.

The general symptoms are those always present, irrespective of the particular species of worm infecting the patient. They depend not so much upon the mere presence of the parasite, as upon the peculiar condition of the mucous membrane of the alimentary canal which accompanies and is, perhaps, essential to their development and existence. This condition is one of catarrh, with an excessive production of mucus, or, to give it a definite title, mucous disease; consequently, the general symptoms of worms are similar to those already studied under this head (p. 217).

The patient wastes, and the face is pale or leaden in hue, dark circles surround the orbits, the eye-balls are sunken, the pupils dilated and the upper lip swollen. The skin generally is muddy, covered with dry, epidermic scales, and devoid of natural softness and elasticity.

The lips and mucous membrane of the mouth are pale, the breath is offensive and the tongue is flabby, with the edges indented by the teeth and the dorsum covered with a thin, slimy coating, or it presents one or other of the conditions described (pp. 218, 220) as pathognomonic of mucous disease. There is often moderate hypertrophy of the tonsils and swelling of the lymphatic glands at the angles of the jaw. The appetite is capricious, sometimes almost absent, and at others insatiable. Nausea, and acid eructations and vomiting are common. Constipation is the usual state of the bowels; occasionally there is tenesmus, with constant unproductive efforts at defecation, and there is a liability to attacks of diarrhœa, attended with great straining and the passage of black, slimy, ill-smelling motions. Free mucus may be discharged from the rectum, and, in girls, from the vagina. The abdomen is always distended, feels hard on palpation, and to percussion yields a tympanitic note. Constant complaints are made of pain in the belly, especially in the neighborhood of the umbilicus. Its character varies, being in some cases tearing or cutting; in others, simply an uneasy, creeping sensation, and in others still, a sensation of coldness.

The urine is frequently voided with pain and difficulty, and may have a turbid, milky appearance.

The pulse is weak, altered in frequency and occasionally irregular; a harassing, paroxysmal cough may be present, and not uncommonly there is sighing, sobbing and hiccough.

The child's temper is altered; he becomes irritable or sullen; his sleep is broken by bad dreams or night terrors; and there are many and very diverse nervous manifestations, such as annoying itching of the nose, temporary delirium or stupor, sudden blindness, loss of voice, squinting, fixation of the eye-balls, vertigo and general convulsions.

These features, of course, are not equally marked in all cases, their degree depending upon the grade of intestinal catarrh.

Special symptoms—those due directly to the presence of the worms—differ according to the species.

The Oxyures occasion violent itching at the anus, especially at night, when they prevent sleep and lead to troublesome scratching. This irritation, transmitted to the genitalia, combined with the constant application of the hands to these parts, produces erections in the male, and may induce the habit of masturbation in both sexes. Two conditions of the bowels are observed; either forcible but ineffectual straining, often attended by prolapsus ani, or diarrhœa. Finally, the oxyures may, on inspection, be discovered moving about in the radiating folds of the anus.

Occasionally, these parasites migrate into the vagina, uterus, urethra, œsophagus and stomach. When they occupy the vagina, they give rise to leucorrhœa.

AFFECTIONS OF THE STOMACH AND INTESTINES. 319

Lumbrici occasion more or less pain in the umbilical region; also vertigo, convulsions, and even chorea. The irritation of their presence may cause chronic diarrhœa, with scanty, offensive, thin, mud-colored stools, voided with much straining, and most numerous during the night. They often migrate into the stomach, whence they are quickly expelled by vomiting. Less frequently they pass into the common bile-duct and gall-bladder; also the nose, larynx, trachea, larger bronchi, vagina, urethra and bladder; in each position giving rise to symptoms of irritation. They have been found too, in abscesses communicating with the intestine, having escaped by entering a preëxisting fistulous opening, or, perhaps, in some instances, by directly perforating the gut. These abscesses usually occupy the abdominal wall in the umbilical or inguinal regions, or are seated in the substance of the liver.

As already stated, the Tricocephalus dispar causes no special symptoms.

Tæniæ are attended by sensations of weight and gnawing in the abdomen; occasional attacks of colic, and distention, particularly of the umbilical region. With a huge appetite, there is progressive emaciation and general lassitude. A persistent headache is sometimes a feature, and there may be annoying cramps in the muscles of the legs and arms.

DIAGNOSIS.—While the occurrence of the symptoms detailed strongly indicate the presence of worms, the only positive proofs of their existence are the discovery of the parasites themselves or of their eggs in the stools; their appearance, as in the case of oxyures, at the margin of the anus; and their expulsion, as in the case of lumbrici, from the stomach in the act of vomiting. Therefore a purgative, by emptying the intestinal tract and expelling some of the parasites, is the crucial test.

In some cases the symptoms are severe enough to suggest tuberculosis or tubercular meningitis; though a mistake may be readily avoided by bearing in mind this possibility and applying the test.

PROGNOSIS.—Intestinal worms rarely cause death. When a fatal termination does occur, it results from convulsions; from the consequences of the migration of the parasite into the bile-duct and air passages, or from some secondary affection; proving dangerous because the strength of the frame attacked has been sapped by its guests.

TREATMENT.—The diagnosis having been established, either by the spontaneous appearance of the worms or by their discovery after the administration of a dose of calomel or calomel and rhubarb, remedial measures must be directed to the accomplishment of two objects: 1st, the expulsion of the worms; and 2d, the removal of the alkaline mucus—the essential nidus—and the restoration of the alimentary canal to its normal condition.

1st. For expelling the parasites, the anthelmintic to be chosen depends upon the infecting species.

Oxyures, as they inhabit the rectum, are within the reach of enemata, and are best treated by them. The object being to kill the worm, it is essential thoroughly to empty the lower bowel by an enema of warm water, immediately before injecting the parasiticide, so that the latter may come in contact with the mucous membrane, upon which the great mass of the worms lie. One or two medicated injections can be administered daily; they act best when cold, and their bulk should not be so large as to distend the gut and lead to a quick return; from one to two fluid-ounces is the proper quantity for a child of two years.* Liquor calcis; common salt and water, in the proportion of one teaspoonful to a pint; a solution of castile soap, thirty grains to a pint of water; or one of sulphuret of potassium, twelve grains to a pint; oil of turpentine in milk; half a teaspoonful to four fluid-ounces of pure olive oil; and lard beaten up with water until it becomes liquid, all constitute good injections, the last two having the property of quickly relieving itching, in addition to their parasiticide action. In my experience, however, the injection of an infusion of quassia has been most uniformly successful. It

---

* All of the succeeding directions for treatment are adapted to children of this age.

is best to order the quassia chips and have the infusion prepared in the nursery, thus:—

R. Quassiæ (rasped), ... ℨij.
S.—Place in a porcelain vessel and pour on a pint of boiling water, macerate for two hours, then strain and inject two fluidounces once or twice daily.

With children past the second year the proportion of quassia may be gradually increased to one ounce, at the age of seven.

While employing enemata it is well to aid their action and relieve itching at the anus, by anointing the parts with some mild mercurial ointment, and at the same time pushing a little into the rectum. A good preparation of this kind is—

R. Hydrargyri Chloridi Mitis, ... gr. lxxx.
Unguenti Petrolei, ... ℥j.
M.
S.—Apply morning and evening.

When there is intense rectal irritation, an injection of laudanum and starch water (gtt. iij to f℥j) and cold compresses applied to the fundament, give great comfort.

Diarrhœa and tenesmus are to be overcome by the administration of a teaspoonful of castor-oil, with five drops of paregoric once, twice or three times daily, according to circumstances. Should there be constipation, one teaspoonful of Husband's magnesia, or the appropriate dose of some other saline, must be given every morning until the symptoms disappear. Besides keeping the bowels regular, it is well to secure several free watery evacuations, at intervals of three days, for the purpose of dislodging any oxyures that may be inhabiting the upper part of the large intestine, and of clearing away accumulations of mucus; to accomplish this, saline cathartics are to be selected.

Against Lumbricoides a number of drugs bear an anthelmintic reputation. Of these, santonin, spigelia and chenopodium are the most efficient. To insure the greatest success, the patient for whom either of these medicines is ordered, must be placed on a restricted liquid diet, that the alimentary canal may be as empty as possible; and during their administration the bowels must be kept active by cathartics, that the dead worms and the ova may be swept away. Broken doses of the purgative chosen can be combined with the anthelmintic, or an occasional full dose may be given during the course of the treatment.

Santonin is almost tasteless, and when combined with sugar is readily taken by children; it may be prescribed in the following ways:—

  ℞. Santonini, ... gr. vj.
    Sacchari, ... gr. xxx.
  M. et ft. chart. No. xij.
  S.—One powder morning and evening, each second dose to be followed by two teaspoonfuls of castor-oil or a purgative dose (gr. j) of calomel.

Or—

  ℞. Santonini, ... gr. vj.
    Hydrarg. Chlorid. Mit., ... gr. vj.
    Sacchari, ... gr. xxiv.
  M. et ft. chart. No. xij.
  S.—One powder morning and evening.

Santonin sometimes produces xanthopsia, or "yellow-seeing;" this is of no importance and quickly disappears after the drug is discontinued. It is best, however, to advise the mother or nurse of the possibility of this occurrence. Occasionally, too, it increases the flow of urine and gives the fluid a reddish color.

Spigelia is a very useful remedy, though as it simply narcotizes the worm it must always be administered in association with a purge. The officinal preparation, extractum spigeliæ et sennæ fluidum, is as good a combination as can be employed; it may be given in doses of one teaspoonful three times daily. If it be desired to make success doubly sure, it is well to add santonin:—

  ℞. Santonini, ... gr. iv.
    Ext. Spigeliæ et Sennæ Fld., ... f ℥ jss.
    Syrupi, ... f ℥ ss.
    Elix. Simplicis, ... q. s. ad f ℥ iij.
  M.
  S.—Two teaspoonfuls three times daily.

Chenopodium is a very safe, non-irritant anthelmintic, being especially indicated when the evacuations are increased in number, are liquid and contain mucus or blood. The volatile oil may be administered dropped upon a lump of white sugar, in doses of five drops three times daily. A purgative is then necessary, every twenty-four or forty-eight hours, or, for convenience, both remedies may be combined in a mixture:—

  ℞. Olei Chenopodii, ... f ℨ ij.
    Olei Ricini, ... f ℥ jss.
    Olei Cinnamomi, ... ♏v.
    Syr. Acaciæ, ... q. s. ad f ℥ iij.
  M.
  S.—One teaspoonful three times daily.

Should there be reason to suspect ulceration of the bowel, five minims of oil of turpentine added to each dose of this formula will both improve the condition of the mucous membrane and increase the specific action.

The only disadvantage possessed by oil of chenopodium is that it is not so acceptable to the taste or stomach as either santonin or the liquid extract of spigelia and senna.

Whip-worms, when detected, can be removed by the same means as lumbricoides.

Tæniæ are the most difficult of the intestinal parasites to eradicate; evacuations of many feet of segments are easily brought about, but reproduction steadily continues until the head is finally expelled. This portion is obstinate in its adherence to the intestinal mucous membrane, and being minute in size is easily shielded from the action of the parasiticide by the tenacious mucus which is always secreted in excess when a tape-worm is present.

It is essential, therefore, to diminish, or, if possible, entirely remove, this secretion, before commencing the actual treatment. For one week the child* must take the following prescription:—

---

* The succeeding formulæ are adapted to children of six years.

R. Ammonii Chloridi, ... ʒij.
Ext. Sennæ Fld., ... fʒvj.
Inf. Gentianæ Comp., ... q. s. ad f℥iij.
M.
S.—One teaspoonful before each meal.

At the same time the diet must be restricted, and non-farinaceous in character; for instance:—

Breakfast at 8 A.M.—A tumblerful and a half (12 oz.) of milk, with two slices of gluten bread.*

Luncheon at 12 M.—A teacupful (4 oz.) of milk.

Dinner at 2.30 P.M.—A bowl (8–12 oz.) of beef, mutton, or chicken broth; two slices of gluten bread.

Supper at 7 P.M.—Same as breakfast.

For drink at dinner or between meals only pure filtered water in small quantities.

At the end of the week's preparation one of the anthelmintics particularly adapted to this species of worm may be ordered. Of these there are several:—

Oleoresin of male fern—oleoresina filicis—mentioned first, because most commonly used and generally efficient, should be given in one or two drachm doses, either floating upon a little (f℥ss) peppermint water or in a mixture, such as—

R. Oleoresinæ Filicis, ... fʒij-iv.
Syr. Acaciæ, ... fʒij.
Aq. Cinnamomi, ... q. s. ad f℥j.
M.
S.—Tablespoonful for a dose.

The plan of administration has much influence on the issue. For the best result, the patient, unless much debilitated, must, upon the day on which the treatment is instituted, begin a fast after his dinner; in the evening two fluidrachms of castor-oil should be given; next morning, after the bowels have been

---

* Gluten flour can be obtained in any of the larger cities, and is made into bread in the same manner as wheat flour.

thoroughly evacuated, a dose of fern ; and three hours later, a second dose of castor-oil. A few hours subsequently the worm will probably be expelled. During the interval, occasional sips of water may be allowed to relieve thirst. The nauseating taste of the oleoresin of fern may lead to its quick rejection from the stomach ; in such cases the viscus should be quieted by a few drops (3-5) of McMunn's elixir of opium, and a second dose of the anthelmintic administered after the lapse of half an hour.

Kameela (*Rottlera*) is another good remedy for tape-worm, possessing the advantage of being in itself an aperient, and hence doing its work without the aid of purgatives. The same period of absolute fasting is necessary, as when administering male fern, and on the morning of the day following the beginning of abstinence, two doses of fifteen grains of powdered kameela must be taken, at an interval of three hours. The drug may be exhibited suspended in syrup or in mucilage of acacia, a few drops of some aromatic oil being added in either case. A capital prescription, containing both kameela and male fern, is :—

℞. Kameelæ, ... gr. xxx.
    Syr. Acaciæ, ... f ℨ ij.
    Misce, et adde—
    Oleoresinæ Filicis, ... f ℨ j-ij.
    Aquæ Cinnamomi, ... f ℥ j.
    M.
    S.—To be taken in two doses at an interval of three hours.

A formula very similar to this has been in long and most successful use at the Children's Hospital, Philadelphia.

In some cases, oil of turpentine is very efficient, even when the remedies already mentioned fail. This may be given in one large dose, or in small doses frequently repeated. By the former method, from two to four fluidrachms are given in the morning after the usual fast, and followed in three hours by a dessertspoonful of castor-oil, unless the bowels have been previously relieved. To carry out the latter, the following mixture may be used :—

R. Olei Terebinthinæ,
Mellis, ... āā f℥ss.
Ol. Menth. Pip., ... gtt. vj.
Mucilag. Acaciæ, ... q. s. ad f℥iij.
M.
S.—Two teaspoonfuls every six hours.

Every second day, preferably in the morning, two grains of calomel must be administered.

Another useful drug is pumpkin seed; this may be given in the form of an electuary, six drachms of the seeds being beaten up with sugar, and taken in one or two doses; a brisk purge must be ordered after it.

Koosso and its active principle, koossin, are recommended by some authorities. One drachm of the powdered drug suspended in water, or five or ten grains of koossin in capsule are the proper doses for a child of eight years. To prevent nausea, it is better to break the dose into two or four; additional purgative is usually not required.

For a long time the bark of the pomegranate root has been known as a remedy for "tænia," or tape-worm; but the difficulty of procuring it fresh, the short time it keeps good, and the unpleasant taste of the decoction, has greatly limited its use. Besides, it has been ascertained that its action is variable, according to the season of collecting, the age and vigor of the tree, etc. It is this uncertainty that compelled Professor Laboulbene, Member of the Academy of Medicine, who has made the cure of tænia a specialty, and who considers the bark of pomegranate root the best and most efficacious remedy, to say: "I wish that some one would discover and separate from the tænicide plants a sure alkaloid always identical, and that would act with certainty, which is something we cannot obtain from pomegranate bark, or from old koosso, which is nearly inert."

Mr. Tanret has found this alkaloid, and for his discovery has been awarded the "Barbier Prize" by the Academy of Sciences. He calls it Pelletierine, in honor of the illustrious chemist, who, with Caventon, has made numerous discoveries in organic chemistry of great benefit to humanity.

AFFECTIONS OF THE STOMACH AND INTESTINES. 327

Tanret's pelletierine has given the most satisfactory results in the hospitals where it has been tried, for instance, at the Marine Hospitals of Toulon, St. Mandrier, etc., and in Paris, St. Antoine, La Charité, Necker and Beaujon, etc. Dujardin-Beaumetz, Member of the Academy of Medicine, declared to the Society of Therapeutics, that he was successful in thirty-two cases out of thirty-three treated with pelletierine, and Professor Laboulbene was successful in every case in which he used it, fourteen in all.

Pelletierine is dispensed in bottles containing the proper dose for an adult, and one dose is usually sufficient. For children from nine to twelve years, half the adult dose is sufficient. In administering the drug, certain preliminaries are indispensable to insure success.

When pieces of tape-worm are or have been ejected within a short time after some other remedy has been taken without expelling the head, pelletierine should not be taken until some pieces of the worm are again noticed.

In the evening the patient must use a large laxative injection, and place himself on milk diet. The next morning mix the contents of a bottle with a glass of sweetened water, and administer at one dose; three-quarters of an hour to an hour after, give one ounce of compound tincture of jalap, mixed with one-half a glass of sweetened water. For women, the dose should be reduced to 20 grammes, and for children a still further reduction is necessary. The purgative, compound tincture of jalap, is the best, but it can be substituted by any other cathartic.

If the bowels are not relieved in a few hours after taking the purgative, then take either another purgative or an injection made of sulphate of sodium. A few minutes after having taken pelletierine there will be a sensation of giddiness, and the entire tape-worm will be passed from two to four hours after the remedy has been taken.

After administering any anthelmintic, it is impossible to decide at once whether the tape-worm has been eradicated or not unless the head be discovered in the stools. The physician must not trust to the mother or nurse to find the head, but must look for it himself. The stools immediately following the action of the parasiticide must, therefore, be preserved until his visit; the chamber in which they are received being filled with water containing a small quantity of carbolic or salicylic acid. This is to be gently shaken in order to separate the worm from the fæces, and then allowed to stand for ten minutes; during which the parasite, from its greater specific gravity, sinks to the bottom of the vessel. Next, the supernatant liquid is poured off, the vessel refilled with water, and the process repeated until the fluid remains nearly colorless. Then the head, if present, is readily found. Should the head not be discovered, it is impossible, although all symptoms may disappear, to give a positive opinion as to complete expulsion until two or three months have passed. Any return of symptoms requires a second course of treatment.

2d. The removal of the alkaline mucus and the restoration of the normal condition of the alimentary canal are to be accomplished by the same attention to diet and the same therapeutic measures recommended when discussing chronic gastro-intestinal catarrh (p. 220).

CHAPTER III:

CASEOUS DEGENERATION AND TUBERCULOSIS OF THE MESENTERIC GLANDS—TABES MESENTERICA.

Cheesy degeneration of the mesenteric glands, existing alone, or in association with tubercular deposit, is far from being a common disease. The majority of cases occur after the third year, and under this age tabes mesenterica is rarely encountered.

MORBID ANATOMY.—The glandular lesions are identical with those so familiar in scrofulous conditions.

Usually some of the mesenteric glands remain healthy; and those involved—a variable number—do not present a uniform degree of alteration. A few are simply hyperæmic and slightly swollen from an increase in the corpuscular elements; some are increased in size and spongy in texture; others, again, are enlarged to the size of walnuts, dense, dry and anæmic looking; and still more are partially or completely converted into opaque, yellow, cheesy masses, and sometimes contain tuberculous deposits.

The diseased glands may remain isolated, but often unite into an irregular mass of variable size. This mass is situated in mid-abdomen, rests upon the vertebral column, and is movable or fixed according to the freedom or involvement of the mesentery in the swelling.

Further changes are softening of the caseous material, with suppuration and discharge through the gland capsule—an unusual event; shriveling of the gland into a fibrous mass, through thickening and contraction of the trabeculæ of the reticulum; and—quite commonly—a gradual hardening and shrinking by the absorption of fluid and the deposition of earthy salts.

Together with these changes, it is the rule to find scrofulous lesions of the superficial lymphatic or bronchial glands, cheesy or tubercular deposits in the lungs, and ulceration of the mucous membrane of the intestinal canal. Sometimes, also, tubercular peritonitis and caseous and tubercular deposits in the liver and spleen are discovered at the post-mortem examination.

ETIOLOGY.—As the disease always occurs in strumous subjects, the predisposing causes are the same as those leading to the production of that diathesis.

The exciting causes are embraced under three heads—imperfect hygiene, disease of other organs, and infection through the milk of diseased cows.

Children who live in filthy, over-crowded, dark and ill-ventilated houses, are much more likely to be affected than those born to more fortunate surroundings; but of all hygienic factors, coarse, over-stimulating or bad food is the most potent for evil. This acts by irritating the intestinal mucous membrane and producing catarrh and follicular ulceration.

Tubercular disease of the lungs, scrofulous disease of the cervical and bronchial glands or of the bones and joints, and tubercular ulceration of the bowels, are not only usual associates, but, no doubt, frequent causes of tabes mesenterica. In addition, measles and scarlet fever, from their tendency to induce inflammation of the mucous membrane of the bowels and glandular hyperæmia, must be ranked as exciting agents, and so, also, must difficult dentition and whooping-cough.

Attention has recently been directed to the possibility of the transmission of tubercle to the intestinal tract by means of the milk of diseased cows. In support of this theory, Klebs has made a number of experiments, from which he draws the conclusion that the use of the milk of cows having advanced phthisis always produces tuberculosis, which begins as an intestinal catarrh and extends to the mesenteric glands.

SYMPTOMS.—The signs elicited by physical exploration of the abdomen, and the symptoms arising from the presence of a large mass of glands are much more characteristic than the general features. The shape of the belly and the tension of its walls may be perfectly normal. Such is generally the condition during the earlier stages of the disease; later, and particularly when there is intestinal ulceration, there is considerable distention. This is due either to the accumulation of flatus in the bowel or to the large size of the glandular tumor. In the first instance, the degree of prominence varies from day to day; when marked, the wall is tense, percussion is tympanitic, and it is difficult or impossible to grasp the glands; in the second, the enlargement is constant and greatest in and about the umbilical region; here there is resistance to the palpating hand rather than tension; the tumor is easily felt, and percussion over it gives a dull sound; around it, a tympanitic one. The tumor varies in size from that of a hen's egg to that of a double fist; it is nodulated, hard, somewhat tender, slightly movable when small, and fixed when large. When of considerable size, the mass can readily be touched by placing the fingers on the umbilical region and pressing backward toward the spine. Otherwise, it is well to put one hand on either side of the abdomen and gently bring them together toward the median line, the patient being placed on his back with his shoulders and thighs elevated so as to relax the parietes; by this method, it is possible to detect a tumor as small as a walnut.

The secondary manifestations of the presence of a large mass of glands in the abdominal cavity are pains and cramps in the legs, due to pressure upon the nerves; and œdema of the legs and distention of the superficial abdominal veins, from compression of the great venous trunks. The veins are often very prominent, and ramify over the wall of the belly to join those of the thoracic wall, which are also distended.

If the glands in the notch of the liver be enlarged, direct pressure is exerted upon the portal vein and ascites results; this, however, is a very unusual symptom.

Provided the naturally prolonged course of the disease be not abridged by tubercular or other complication, the tendency is for the glands to shrink and become calcareous. This change lessens the size of the tumor, diminishes the tension of the parietes, and, by relieving pressure, leads to the disappearance of the secondary derangements.

Softening is another, but, fortunately, a rare termination. Adhesion may then take place between the gland and a loop of intestine, so that the softened matter is evacuated into the bowel without harmful result; but should the discharge be directly into the peritoneal cavity, acute peritonitis is set up, and death soon follows.

The general symptoms depend for their development upon the condition of the intestinal mucous membrane. Usually there is scrofulous ulceration, with or without general catarrh.

If ulceration and catarrh be associated, the child wastes; grows pale and feeble; presents a haggard appearance: is fretful and peevish; has a capricious appetite and much thirst; complains of wandering pains in the abdomen, and is affected with diarrhœa, attended by the expulsion of offensive, dark, watery stools, which, on standing, deposit flaky matter, mucus, and small, black blood clots. Sleep is restless, and at night the temperature rises one or more degrees above normal.

When catarrh is absent, the bowels are often constipated; the patient looks ill; is pale and languid; his muscles are flabby, and he has more or less flatulent pain in the belly; but there is no marked wasting and none of the evidences of great impairment of general nutrition.

Should there be no disease of the mucous lining of the bowels, flesh is retained; the spirits and strength are good; the appetite and bowels undisturbed; the temperature normal, and there is nothing to show ill-health save some pallor of the skin.

DIAGNOSIS.—The only positive proof of the existence of tabes mesenterica is the detection, by palpation, of a glandular tumor. Particular caution must be given against the mistake, so frequently made, of attributing every case of abdominal distention to disease of the mesenteric glands. Prominence of the belly is a frequent symptom in children, and in the vast majority of cases depends upon intestinal catarrh. In this condition there is imperfect digestion and assimilation of food, and, consequently, debility, affecting the muscles of the intestines as well as the system generally. Now, imperfectly digested food readily undergoes fermentation, with the production of carbonic-acid gas, and this distends the bowels ; the more so as they are wanting in tone from the weakness of their muscular coat. Such inflation of the gut must lead, of course, to a prominent abdomen, but one which is uniformly tympanitic on percussion, moderately soft and flaccid to the touch, and entirely free from the signs of enlargement of the mesenteric glands.

Again, distention of the superficial abdominal veins is merely an indication of obstructed circulation in the deep venous trunks, and only becomes a symptom of importance in the diagnosis of tabes mesenterica when hepatic disease can be excluded.

Even should a tumor be felt, the question arises whether it may not be an accumulation of fæces. In the latter there is no tenderness ; the mass occupies the position of the transverse or descending colon, is oblong in shape, with its long diameter corresponding to the axis of the gut in which it is placed, and is so soft that it may be somewhat moulded by the pressure of the fingers. Should there be any doubt, an enema of warm water and soapsuds must be thrown into the bowel and retained for a few moments, by firm pressure upon the anus. When expelled this will bring away a quantity of light-colored, brittle matter, if the mass be due to fæcal accumulation ; and the previously detected tumor will be found, on examination, to have disappeared or lessened in size. On the other hand, if the tumor be glandular, the expulsion of flatus and fæces, induced by the injection, only renders it still more prominent.

The diagnosis must not be considered completed by the detection of the tumor, but must extend to the discovery or elimination of the different complications—ulceration of the intestines, tubercular deposits in the lungs and tubercular peritonitis.

PROGNOSIS.—Caseation of the mesenteric glands is dangerous, but the danger does not spring from the glandular disease so much as from the affections that produce it, the conditions that accompany it, and the results that follow it. In regard to the latter, though, it may be stated that of all the glands in the body the mesenteric are least likely to be followed by ill consequences when diseased ; a fact due to their slight tendency to undergo softening.

When the sole discoverable lesion is swelling of the glands, and there is no rise in the evening temperature nor marked impairment of nutrition, the hope of subsidence of the enlargement and ultimate recovery, may be reasonably entertained. On the contrary, if there be wasting, diarrhœa and fever, indicating ulceration of the bowels, secondary, perhaps, to chronic disease of the lungs, the prognosis must be grave. Again, the occurrence of tuberculous peritonitis renders the prospect most unfavorable.

TREATMENT.—Much may be done in the direction of prophylaxis by keeping a strict watch upon the stomach and intestines in scrofulous children, so as to remove any apparently trifling disorder as quickly as possible. Supplying good food, fresh, pure air and warm clothing, and maintaining the activity of the skin are also important preventive measures.

After the disease is established, much can be accomplished by attention to the diet and general regimen. In regulating the diet, it is necessary to take into consideration the catarrhal state of the intestinal mucous membrane usually present, and the almost useless condition of, at least, a number of the mesenteric glands, and to select those articles which are absorbed in the stomach or taken directly into the blood vessels, without the intermediate action of the lacteals and mesenteric glands. The food must be sufficient to maintain the general strength, but not so abundant as to overtax the process of digestion. The following may be taken as an average daily schedule of both diet and regimen for a child of four years, in whom there is no excessive wasting or weakness:—

On waking in the morning, say at 7 A.M., a thin slice of dry, stale bread, and three fluidounces of hot veal broth.

At 8.30 A.M. a cold bath, given in this manner :—being taken from bed, the whole body is briskly shampooed with a soft towel until the surface is aglow. The child is next made to stand in a tub sufficiently filled with hot water to cover his feet and ankles, and two gallons of cold water, containing an ounce of sea-salt or concentrated sea-water, are slowly poured over his shoulders. The skin is then thoroughly dried and rubbed until reaction is established; the child is wrapped in a blanket and put back to bed for half an hour. On rising, the abdomen should be completely enveloped in a flannel binder, and the body clad in woolen underclothing from head to foot.

At 9.30 A.M., breakfast.—A soft-boiled egg and two slices of stale bread.

From 10.30 A.M. to 12 M.—A walk or romp in the open air, in good weather.

At 12 M., lunch.—Half a dozen raw oysters or a bit of sweetbread or fish, and a slice of dry, stale bread.

At 3 P.M., dinner.—Six fluidounces of beef, mutton or chicken broth; a bit of minced roast beef, beef-steak, roast mutton, chicken, or wild fowl. A moderate quantity of spinach, stewed celery, boiled cauliflower, or other non-farinaceous vegetable, and one or two slices of dry, stale bread. No dessert except junket occasionally.

At 7 P.M., supper.—Same as lunch, alternating the fish, sweetbread or oysters.

Nothing should be taken for drink but filtered water or, better still, good spring water.*

When there is much emaciation and weakness, the morning bath must be omitted or substituted by a simple warm sponging; and some stimulant, as a teaspoonful of old whiskey, should be given three times daily.

Diarrhœa demands an exclusive liquid diet, and it is advisable to artificially digest the meat broths and milk, which must form the basis of this.

The most useful drugs are cod-liver oil and the syrup of the iodide of iron, since the indications are to build up the general

---

* Directions for Philadelphia.

health and restore the glands to a healthy condition. The former can be given as an emulsion with lactophosphate of lime in two-drachm doses three times daily, after eating, at the age of four years; the latter in fifteen drop doses after meals. Often it is well to administer both preparations together.

Locally, some good may result from the daily inunction of a weak mercurial or iodine ointment, for example :—

  ℞. Ung. Iodin. Comp., ... ℨij.
    Ung. Belladonnæ, ... ℨj.
    Ung. Aquæ Rosæ, ... ℨv.
  M.

  ℞. Ung. Hydrargyri, ... ℨiij.
    Adipis, ... ℨv.
  M.

Of either a piece as large as a cherry may be rubbed into the skin over the tumor once every day.

Other remedies are, of course, required to arrest diarrhœa, or to relieve the different complications that may arise.

Should the circumstances of the patient permit, change of residence to some locality having an equable climate, a bracing atmosphere and a dry, porous soil, will greatly assist in effecting a cure.

CHAPTER IV.

AFFECTIONS OF THE LIVER.

Hepatic diseases do not occur so frequently during childhood as in adult life. Fatty and amyloid changes are the most common affections; syphilitic disease, cirrhosis, tubercular deposit and parenchymatous inflammation stand next in the order named; while echinococcus is very rare, and cancer almost unknown. Jaundice, on the contrary, is often met with, but this condition, though a complex and striking one, is simply an indication of disease of the viscus itself, or of its excretory duct. Congestion of the organ is also common.

1. JAUNDICE.

Icterus, irrespective of the age at which it occurs, is characterized by yellowness of the skin and conjunctivæ, clay-colored stools and yellow-brown urine. During the first few days of life, especially after a difficult and tedious birth, there is apt to be intense congestion of the skin, followed, as the redness fades, by a brownish-yellow discoloration. This appears on the second or third day, and disappears by the tenth.

It is not jaundice, for it is entirely independent of liver disorder, and there is no yellowness of the conjunctivæ, and no alteration in the fæces or urine. A form of true jaundice, however, does occur in the newly-born, termed icterus neonatorum, which may be studied before describing the condition as it is seen in later childhood.

ICTERUS NEONATORUM.

Both mild and dangerous types of this variety of jaundice are met with.

The *mild type* occurs in infants prematurely born, or weak from other causes; in those early exposed to the depressing action of cold, dampness, and foul air, and particularly in those who are born partly asphyxiated after tedious labor. It is difficult to understand the exact method in which these causes act. Cold undoubtedly produces catarrh of the duodenal mucous membrane, and plugging of the bile-duct by mucus; the others, the last especially, act, in all probability, by altering the hepatic circulation. At birth there is a sudden transference of the blood-supply from the umbilical to the portal vein; a change —according to Frerichs—temporarily followed by comparative emptiness of the blood vessels of the liver; a diminution of vascular tension, and the passage of bile into the blood. Weber attributes the jaundice to pressure from congestion and œdema, the result of an arrest of the circulation in the umbilical vein before the establishment of respiration; conditions present in infants born semi-asphyxiated. Birch-Hirschfeld has demonstrated that a dense areolar sheath surrounds the vessels in the notch of the liver and extends into the viscus along with the portal vein; this becomes œdematous and greatly swollen when there is venous obstruction in the liver during difficult parturition, and, by pressure, obstructs the flow of bile into the intestine.

The grade of jaundice in this type varies considerably. Sometimes the yellow discoloration is confined to the face, chest and back; the conjunctivæ are but lightly tinged; the urine and fæces are unaltered, and after three or four days the trouble is at an end. In other cases, the yellowness extends to the abdomen and arms; the conjunctivæ are distinctly yellow; the urine is dark and stains the diapers, but the stools still retain their natural color—golden yellow; the duration is about seven days. The best developed instances present universal and moderately deep discoloration of the skin; the conjunctivæ are very yellow; the urine brownish, and the stools clay-colored. With this degree of jaundice, there is malaise, loss of appetite, constipation, and enlargement of the liver; the lower edge of the right lobe often extending below the costal border as far as the umbilicus.

Occasionally, instead of constipation, there is diarrhœa, with moderate heat and tenderness of the belly, and a quick pulse, indicating severe intestinal catarrh. These cases recover after a fortnight or more, though occasionally diarrhœa arising and persisting in a feeble infant, is sufficient to determine a fatal issue.

The TREATMENT is simple. The infant must be kept in a warm, well-ventilated room; the activity of the skin must be maintained by bathing, and chilling prevented by proper clothing. Constipation is to be relieved by fifteen or twenty drops of castor-oil, a soap suppository or an enema, and, if the skin be slow in resuming its normal color, it is well to prescribe an alkali, as:—

℞. Sodii Bicarbonatis, ... gr. xxxij.
Aq. Menth. Pip.,
Syrupi, ... āā f ℥ ss.
Aquæ, ... q. s. ad f ℥ ij.
M.
S.—One teaspoonful three times daily.

The *grave type* depends upon congenital malformation of the bile-ducts and gall-bladder; compression of the bile-ducts by syphilitic inflammation and growths, and umbilical arteritis and phlebitis.

a. Congenital malformation is rare, but when it occurs is liable to affect several members of the same family in succession; boys suffer twice as often as girls. There are a number of varieties: thus, the gall-duct may be converted into a fibrous cord; the ductus communis may be contracted, obliterated or absent; the gall-bladder may be rudimentary and the ducts absent; or all the ducts may be wanting. Whatever the condition, the result is enlargement of the liver with cirrhotic change, more or less marked in proportion to the duration of life. The organ is dark green or almost black in color, feels unnaturally firm to the touch, and under the microscope shows an excess of connective tissue.

From one to two weeks after birth the retained bile begins to give rise to jaundice; this appears as a slight yellowness of the skin, and steadily grows more distinct, though it varies considerably in intensity from day to day; at the same time, the conjunctivæ are stained and the urine dark colored. After a day or two, the liver begins to encroach upon the abdominal cavity and rapidly enlarges; the spleen, too, increases in size, and these two lesions, together with flatulent distention of the bowels and occasional ascites, produce decided prominence of the belly. In spite of a uniformly good appetite, there is constant wasting. The bowels act sluggishly, the fæces are offensive, clay colored or dark green, from the presence of altered blood, and dilated hæmorrhoidal veins can often be seen by inspecting the anus. Another frequent symptom is oozing of blood, either arterial or venous, from the umbilicus. This hemorrhage is capillary in nature, and usually begins at night, and soon after the fall of the navel string; an event that occurs between the fifth and ninth day. It may be combined with bleeding from the nose, mouth, stomach or bowels, and is exhausting and always difficult to control.

This form of jaundice ends in death. When umbilical hemorrhage occurs, the course is short, varying from a few hours to six or seven days; in other cases, life may be prolonged as many months, and death result from some intercurrent disease. In the latter class, the secreting elements of the liver are so far crippled by the constantly progressing cirrhosis that little bile is found, and the yellowness of the surface fades, or almost entirely disappears, before life ends.

b. Syphilitic inflammation of the liver with its lesions and symptoms will be referred to in another place (see page 351).

c. Inflammation of the umbilical blood vessels is due to blood poisoning. The infecting material in the infant is apparently identical with that producing puerperal fever in the mother, and is possibly caused by bacteria, as two forms of these—spherical and rod-shaped—have been discovered in the blood of infants so affected. In consequence, the liver undergoes marked degenerative changes; the connective tissue about the portal vein and its branches becomes swollen and presses upon the bile-ducts, and from this, as well as from alterations in the crasis of the blood, jaundice results.

Discoloration of the skin makes its appearance a few days after birth and rapidly increases; the urine is very dark, and the stools are scanty and passed at long intervals. The face is livid and pinched; the hands and feet are purple; petechiæ appear under the skin; the abdomen is distended by flatus and by enlargement of the liver and spleen; there is tenderness with fluctuation on palpation, and blood or bloody pus exudes from the umbilicus. The tongue is dry, there is little appetite, and the stomach rejects what food is taken, together with quantities of greenish mucus. Pyrexia is noticeable from the beginning, and becomes more marked as the disease progresses; the pulse is quick and the breathing hurried.

The course is always short, and the invariably fatal termination may be preceded by convulsions and coma.

Treatment in either variety is most unsatisfactory; little can be done beyond the employment of measures to maintain the vital forces as long as possible. Umbilical hemorrhage may be arrested by the application of Monsel's solution, or, if this fail, by inserting two hare-lip pins through the skin at the root of the navel and twisting a ligature tightly around them in the form of a figure of eight.

Syphilitic inflammation demands appropriate constitutional remedies, and in pyæmic cases the abdominal tenderness must be relieved by warm fomentations and sedative applications.

ICTERUS IN OLDER CHILDREN.

Jaundice in late infancy and childhood usually depends upon catarrh, extending from the mucous membrane of the duodenum into the ductus communis; sometimes it is due to plugging of the duct by inspissated bile; and, again, to occlusion by the entrance of a lumbricoid worm. Certain structural lesions of the liver, poisoning by phosphorus and miasmatic influences, also produce it.

Catarrhal jaundice—the only form necessary to consider in this connection—presents the features so common to, and so characteristic of, the same condition in adults. Briefly stated, there is more or less yellow or brownish-yellow discoloration of the skin, with troublesome itching, yellowness of the conjunctivæ, porter-like urine, and clay-colored stools, devoid of the natural fæcal odor. Other symptoms are anorexia, craving for acid drinks, a yellow-furred tongue, disordered digestion, listlessness, slowness of the pulse, slight reduction of the surface temperature, and disturbed sleep. The liver may be somewhat enlarged, projecting two inches or more below the costal border, and tender, or even painful, on pressure. The result is always favorable, and the duration rarely longer than two or three weeks.

TREATMENT.—Warm clothing, daily bathing followed by gentle friction to promote the activity of the skin, and a diet based on the same plan as for intestinal catarrh, are the first requisites.

The medicinal treatment can be begun by a moderate dose of calomel, followed by a saline; but if a laxative be required later, the drugs that stimulate the secretion of the liver and act upon the upper bowel must be excluded, and those selected which affect the lower segment, as aloës and castor-oil.

Duodenal catarrh—the causal factor—is most speedily removed by alkalies. Four fluidounces of some saline water, as Kissingen or Vichy, should be drunk at each meal, and the following mixture taken:—

℞. Ammonii Chloridi, ... ʒij.
Aq. Menth. Pip., ... f℥iij.
M.
S.—One teaspoonful, diluted, three times daily after meals,
for a child six years old.

Nux vomica is also useful, and two or three drops of the tincture may be administered thrice daily before eating.

2. CONGESTION OF THE LIVER.

Congestion of the liver is quite common, especially in children of four years of age and upward.

MORBID ANATOMY.—There is an increase in the size, weight and density of the organ, and its peritoneum is tense and shining. On incision, blood flows freely, and the section presents a mottled or "nutmeg" appearance, partly from dilatation of the intra-lobular veins and partly from staining of the cells by retained bile. In long-standing cases, those due to cardiac disease, for example, the cells in immediate proximity to the dilated intra-lobular veins atrophy; those near them are stained with bile, and those most distant undergo fatty degeneration. In time the atrophied cells disappear; their place is taken by connective tissue, which shrinks and produces a cirrhotic condition, the surface of the liver becoming granular and the capsule thickened.

ETIOLOGY.—Even in health the amount of blood in the hepatic vessels varies from time to time, and there is always a temporary increase during the process of digestion. This normal hyperæmia readily becomes abnormal and continuous when there is habitual over-feeding; when the food is highly spiced and too stimulating; and when insufficient exercise is taken. Congestion is often produced by chills, whether resulting from exposure to cold or from the poison of malarial fever, since, in either case, the blood is driven from the surface to the interior of the body. Again, cardiac disease, by obstructing the return of blood from the lung and overfilling the vena cava and portal vein, is an active cause.

SYMPTOMS.—The skin is sallow, or together with the conjunctivæ distinctly jaundiced. There is malaise; headache; a yellow, furred tongue; anorexia; nausea; relaxed bowels with clay-colored, offensive stools, and dark-colored urine loaded with lithates. Pain in the right hypochondrium is usually present, and, as this is increased by turning upon either side, the patient maintains a dorsal position; there is also tenderness in this region, and the suffering is increased by coughing or deep breathing. On palpation, the right lobe of the liver can be detected, extending two or three inches beyond the costal border, while at its edge is felt the gall-bladder distended into a pyriform tumor of variable size. At the same time the upper limit of percussion dulness begins in the third interspace, or at the level of the third rib, instead of the fourth interspace, as in health.

Should the congestion depend upon heart disease, albuminuria and œdema of the feet and legs are associated.

DIAGNOSIS.—Many instances of disordered digestion, with the expulsion of putty-like, undigested material from the bowels, are attributed to congestion of the liver, when in reality the gastro-intestinal tract alone is at fault. Such a mistake can be avoided, if it be remembered that to establish the existence of the hepatic disease it is necessary to have enlargement of the organ, with pain and tenderness; jaundice and clay-colored, offensive stools; combined with disturbance of the functions of the stomach and intestines.

Extension of the liver a finger's-breadth or more, below the costal border, does not absolutely indicate enlargement, since this often occurs without disease, in short-chested children, and in those whose chests are contracted and deeply grooved by rickets. Downward displacement and apparent enlargement may also be caused by pleuritic and pericardial effusions, and by emphysema of the lungs. On the other hand, an enlarged liver may be completely under cover of the ribs, for, in addition to being normally high in the thorax, it may be pushed upon by a collection of fluid or a growth in the abdominal cavity, or drawn up through the shrinking of a collapsed or indurated lung. It is essential, therefore, to fix the position of the upper limit by percussion, as well as the lower edge by palpation, before forming a conclusion.

PROGNOSIS.—The course of the affection is short, and there is no danger unless the child be greatly reduced by previous ill-health, or there be cardiac disease. In the latter case, the duration and result correspond to, and depend upon, the gravity of the heart lesion.

TREATMENT.—The child may be put to bed, or, if not ill enough to be so confined, should be kept within doors. The abdomen must be protected by a flannel binder or a layer of cotton batting covered with oiled silk, and the skin kept active by a daily warm bath, administered, in walking cases, just before retiring to bed. Too much food of any kind is bad; meat and highly-seasoned dishes are to be excluded from the diet; and it is best not to extend the list beyond milk, mutton or veal broth, fish, bread and plain light puddings, as rice and milk.

In the beginning, a child of six or eight years should get the following powder:—

   ℞. Hydrargyri Chlorid. Mit., ... gr. ij.
   Pulv. Ipecacuanhæ, ... gr. ss.
   Sacchari, ... gr. v.
  M. et ft. chart. No. j.
  S.—To be taken in the evening and followed, next morning,
   by a teaspoonful of magnesia.

Subsequently, five grains of chloride of ammonium should be given after food, and a small tumbler (five fluidounces) of Vichy taken with each meal.

Aloës and the salines are the best remedies to relieve constipation during the course of the attack.

In cardiac cases, treatment must be directed chiefly to the heart; and in those due to malarial poisoning, antiperiodics are of little avail until the hepatic congestion is relieved.

When convalescence is established, regular exercise in the open air must be insisted upon and a plain diet maintained. Change of air is often most useful to break up the "bilious habit."

3. FATTY LIVER.

This condition presents itself in two distinct forms, namely, fatty infiltration and fatty degeneration.

FATTY INFILTRATION.

In fatty infiltration, the quantity of fat in the hepatic cells is greatly increased without any alteration in the walls of the cells.

MORBID ANATOMY.—The liver is increased in all its dimensions, its surface is yellowish and oily, its margins rounded and its texture doughy. On section, the cut surface is distinctly yellow, mottled with brownish-red spots, and if a bit be put under the microscope, abundant granules and globules of fat are seen.

ETIOLOGY.—One cause of fatty infiltration is an excess of farinaceous food. Then the deposition is physiological and transitory, the excess of hydrocarbons supplied from without being deposited in the liver in the form of fat. The second cause is chronic, exhausting disease, such as tubercle, scrofula, rickets, caries of bone, intestinal catarrh and syphilis. Here the fat is absorbed from the subcutaneous and other fat-containing tissues of the body.

SYMPTOMS.—It is only in well-marked cases that these are developed. An increase in the bulk of the liver with a rounded, inferior margin may be detected by percussion and palpation; but this is frequently impossible on account of the tendency the organ has, from its softness, to fall away from the abdominal wall. There is a sense of weight in the right hypochondriac region and disturbed gastro-intestinal function, due to portal obstruction. Jaundice and ascites are absent, and there is neither pain nor tenderness over the viscus.

The DIAGNOSIS is not difficult when enlargement, softness and blunting of the edge of the viscus can be detected by examination.

The PROGNOSIS depends upon the cause rather than the degree of change; occurring in the course of a protracted, wasting disease, fatty infiltration shows dangerous impairment of nutrition.

TREATMENT.—Beyond a rigid exclusion of farinaceous and fatty foods from the dietary, all remedies must be directed to the relief of the originating disease.

FATTY DEGENERATION.

Fatty degeneration is a much rarer lesion in children than fatty infiltration.

MORBID ANATOMY.—The liver appears normal to the unassisted eye, but with the aid of the microscope the cells are found to be filled with minute protein or fatty granules, and tend to fragmentation and destruction. The whole, or only isolated portions of the viscus, may be changed in this way.

ETIOLOGY.—The lesion is produced by acute affections, as measles, variola, scarlatina and typhoid fever; by chronic, exhausting diseases, as tubercle, scrofula and rickets; and by accidental poisoning with arsenic or phosphorus.

There are no characteristic symptoms, the result is invariably unfavorable, and no special indications for treatment are presented.

4. AMYLOID LIVER.

This lesion is moderately common in childhood, and usually occurs as a factor of general amyloid degeneration.

MORBID ANATOMY.—The disease consists in a more or less complete infiltration of the cells by a peculiar translucent, refracting substance, possessing the property of fixing iodine and assuming a mahogany-brown color, which, on the application of sulphuric acid, changes to green, blue, violet or red. The infiltration begins in the hepatic arterioles and capillaries, and at first is limited to the middle zone of the lobules; thence it extends to the periphery and centre, destroying the normal elements of the cells and converting them into irregularly shaped, glassy-looking blocks. Fatty infiltration is often associated. Uniform enlargement; increased density; yellowish-gray color;

smooth, shining peritoneum; thin edges, and the exposure, on section, of dry, homogeneous, glistening surfaces, are the gross characteristics.

The spleen, kidneys and lymphatic glands are often similarly altered, and sometimes the mucous membrane of the stomach and intestines.

ETIOLOGY.—Amyloid degeneration of the liver is always produced by some chronic disease, attended by suppuration and purulent discharge. Empyema with a fistulous opening in the chest-wall; dilated bronchi with copious muco-purulent expectoration; scrofulous abscess; chronic pulmonary tuberculosis; suppurative diseases of the bones and joints, and constitutional syphilis, are the most frequent causes. It occurs at any age, but is more frequent after the fifth year, and in boys than girls.

SYMPTOMS.—There are few rational symptoms other than those belonging to the originating disease. Tenderness and pain in the hepatic region are absent, and so, too, are jaundice, distention of the superficial abdominal veins and ascites; except the glands in the fissure of the liver be coincidently enlarged by waxy deposit, when, from pressure upon the portal vein and bile-ducts, the last three phenomena may be developed. The patient complains of weight, discomfort in the right hypochondrium, and is weak, wasted and anæmic, with pale, sallow skin, clubbed fingers and œdematous feet and ankles. When the kidneys are involved, the urine is increased in quantity, has a low specific gravity (about 1014), is pale, lemon colored, and contains albumen, and, at times, hyaline tube casts. Dropsy of the extremities is due in great part to this complication. If the stomach and intestines be implicated, there is a tendency to vomiting and diarrhœa.

Physical examination yields very characteristic signs. The abdomen is prominent, especially over the upper third, and both percussion and palpation show that the liver is greatly and uniformly enlarged. The upper margin of dulness is higher, by an inch or more, than normal; while the lower edge of the right lobe, somewhat blunted, but perfectly well defined, can often be

AFFECTIONS OF THE LIVER. 349

felt as low down as the level of the umbilicus. The portion uncovered by the ribs feels very dense and firm, and perfectly smooth, except where broken by the natural fissures.

The spleen can often be detected projecting as a hard mass from beneath the left costal border. The absence of enlargement, however, is no proof against the existence of amyloid change in the organ; in about half the cases there is no alteration in size.

In course, the disease is always slow.

DIAGNOSIS.—This is readily made from the physical signs furnished by the liver and spleen; the absence of jaundice and ascites; the previous history of cachexia and suppuration; the character of the urine; the anæmia, and the gastro-intestinal symptoms.

Congestion of the liver with consequent enlargement has a different clinical history, rarely occurring in cachectic or anæmic cases. A fatty liver, while large, is soft and yielding to the touch, and is unattended by increase in the size of the spleen or albuminuria.

PROGNOSIS.—The prospect of ultimate recovery is better in children than in adults, for, provided the cause of the degeneration can be removed, it is quite possible for the liver to return to its natural dimensions and to an apparently healthy condition, through the active reparative power always present in early life. Nevertheless, amyloid change in the liver adds greatly to the danger of the originating disease, and is fatal in most cases.

TREATMENT.—It is almost needless to state that attention must first be given to the removal or amelioration of the cause. It is much more difficult to cure the disease when once developed, than to prevent it by checking chronic suppuration, removing carious bone, healing diseased joints, energetically treating constitutional syphilis, and building up the health in cachectic subjects.

To combat the disease itself, the diet must be as nutritious as the activity of digestion will permit; a moderate quantity of alcoholic stimulants must be taken daily; the child must be properly clothed, to prevent chilling, and must live as much as possible in the sunlight and open air, or, if confined to the house, in a light, airy room. Alkalies, iron and iodine are the most useful drugs.

Of alkalies, chloride of ammonium is the best, and it may be given in combination with a bitter, as:—

 ℞. Ammonii Chloridi, ... ʒij.
  Inf. Gentianæ Comp., ... f ℥ iij.
 M.
 S.—One teaspoonful four times daily at the age of six years.

It is often well to combine iron with the ammonia salt, for example:—

 ℞. Tr. Ferri Chloridi, ... f ʒ j.
  Ammonii Chloridi, ... ʒij.
  Inf. Calumbæ, ... q. s. ad f ℥ iij.
 M.
 S.—One teaspoonful three times daily.

Another good way of administering iron is in the form of a modified Basham's mixture:—

 ℞. Tr. Ferri Chloridi, ... f ʒ j.
  Acid. Acetici dil., ... f ʒ jss.
  Liq. Ammonii Acetatis, ... f ʒ x.
  Elix. Aurantii, ... f ʒ v.
  Syrupi, ... f ℥ j.
  Aquæ, ... q. s. ad f ℥ vj.
 M.
 S.—One tablespoonful four times a day.

This formula is particularly useful when there is kidney complication with œdema.

Iodine is most efficient if there be a syphilitic taint; it may be given in the form of iodide of potassium, five grains or more three times a day, with a bitter infusion; or liquor iodinii comp. can be employed in doses of two drops, well diluted, thrice daily.

Complications must be met as they arise. Vomiting, by ice,

cold Apollinaris water, bismuth and counter-irritation to the epigastrium; diarrhœa, by vegetable astringents, with small doses of opium; and dropsy, by diaphoretics and diuretics.

5. SYPHILITIC INFLAMMATION OF THE LIVER.

Syphilitic hepatitis is frequently encountered in the newly-born, though rare in more advanced childhood.

MORBID ANATOMY.—The liver may be the seat of acute swelling, which, without showing marked gross alteration, is associated with a diffused growth of connective tissue elements; again, there may be a localized gummatous change; and, finally, the inflammatory process may be confined to the septa—peripylephlebitis syphilitica. The proliferation of connective tissue takes place both between the hepatic islands and in their interior, thus differing from cirrhosis, where the increase is only between the lobules. When jaundice occurs, the small bile-ducts are thickened and occluded by epithelial cells, and the organ is enlarged, and brownish-yellow in color.

SYMPTOMS.—In mild cases these are few and uncharacteristic; in those that are grave there are jaundice, ascites, hemorrhage from the umbilicus and intestines, ecchymosis of the skin, subnormal temperature, rapid wasting, and often syphilitic lesions of the skin and mucous membranes. On abdominal exploration, the liver is found to be enlarged and hard, and the spleen increased in size.

DIAGNOSIS.—The early age, the history of an inherited taint, the association of enlargement of the liver with jaundice and ascites, make this a matter of little difficulty in cases that are at all marked.

PROGNOSIS is unfavorable, though the opinion must rest upon the degree of cachexia. Goodhart states that all of his cases proved remarkably amenable to mercurial treatment, but this does not correspond with the experience of other observers.

Should deep jaundice, ascites and hemorrhage occur, death is the almost invariable end.

TREATMENT.—As in other syphilitic affections, mercurials must be followed by tonics. One-eighth of a grain of calomel, or one grain of mercury with chalk; may be administered morning and evening, or ten grains of mercurial ointment may be rubbed into the skin once a day, either directly by the fingers of the nurse, or by being smeared upon the flannel binder.

After the liver has been reduced in bulk and other manifestations of the poison are under control, syrup of the iodide of iron, in two-drop doses three times daily, is the most efficient tonic.

Iodide of potassium is also useful; it acts best when combined with chloride of ammonium, as:—

   ℞. Potassii Iodidi, ... gr. xxiv.
     Ammonii Chloridi, ... gr. xxxvj.
     Syrup. Sarsaparillæ Comp., ... f ℥ ss.
     Aquæ, ... q. s. ad f ℥ iij.
   M.
   S.—Teaspoonful three times daily for an infant of one month.

In those fortunate instances that yield to treatment, splenic enlargement disappears less rapidly than that of the liver, and requires the daily application of compound iodine ointment diluted in the proportion of one part to seven of lard.

6. CIRRHOSIS OF THE LIVER.

In childhood, cirrhosis must be classed among the uncommon diseases of the liver; the fact of its occasional occurrence, however, has been abundantly proved by post-mortem examinations.

MORBID ANATOMY.—There are two forms, namely, the atrophic and the hypertrophic.

In atrophic cirrhosis or hob-nailed liver, the organ is contracted and dense in texture, with nodulated surfaces, thin edges and thickened capsule; on incision the cut surface is grayish-yellow in color, and traversed by a distinct fibrous network. The lesion begins as a chronic inflammatory condition of the branches of the portal vein, and consists of a rapid development of embryonic cells, with subsequent conversion into fibrous tissue. The new formed tissue follows the branches of the portal vein within the substance of the gland; extends into the inter-lobular spaces and forms meshes of variable size, but always embraces several lobules. Some enlargement may attend the primary formation of embryonal tissue, but the shrinking of cicatricial contraction invariably follows; the cells become flattened and atrophied; there is a marked reduction in size, and the circulation in the hepatic portal vessels is greatly obstructed. The smaller bile-ducts are little affected, and blood for the nourishment of the organ and for the formation of bile is carried by vessels developed in the neoplasm.

In hypertrophic or biliary cirrhosis, the liver is usually enlarged, perhaps to twice its normal dimensions. It has a smooth surface, a thin edge, and its section is orange-yellow or green. The fibroid growth begins around the intra-lobular branches of the bile-duct, and envelops and isolates separate lobules; it follows the ramifications of the bile-ducts; is more diffused than in the atrophic form, and denser and thicker in some portions than in others. The portal circulation is not necessarily embarrassed, but the biliary ducts are obstructed and dilated, and have their epithelial lining increased in thickness.

In both forms there is enlargement of the spleen, and in some cases there is an association of the characteristic lesions.

ETIOLOGY.—The causes are, as yet, ill-determined. Alcoholic excess, the prime factor in adults, is, of course, inoperative in children, except in very rare cases; some authorities, however, are inclined to look upon the intemperance of parents as, at least, a predisposing element, and regard the vice of drunkenness as one of the sins of the fathers visited upon their offspring. Congenital deficiency of the bile-duct is always attended by cirrhosis. Constitutional syphilis frequently, and general tuberculosis occasionally; precede it. It is not limited to any sex or age, though more frequent in boys than girls, and oftener met with between the sixth and twelfth years than at an earlier period of life.

SYMPTOMS.—Both forms are preceded, for a variable time, by the evidences of defective nutrition, but, as might be expected from the different pathological conditions, the after symptoms are dissimilar. With atrophic cirrhosis the child is peevish and restless, sleeps badly at night; has indigestion, flatulence and costive bowels; a pale and pasty complexion, and dark circles about the eyes. His muscles grow flabby, there is general wasting, and the urine is thick with lithates, or is very acid and deposits a brick-dust sediment of uric acid. After these symptoms have been present for a period—usually a long one—pain in the region of the liver and ascites are developed. With the ascites there is prominence of the abdomen, dilatation of the superficial abdominal veins, and, at first, enlargement of both the liver and spleen. Soon the liver begins to decrease in size, but the spleen continues to enlarge.* Weakness and loss of flesh are progressive; the ascites becomes more marked; there is œdema of the feet and legs; the skin is sallow, and harsh to the touch; the tongue is coated; the appetite impaired; the stomach irritable; the bowels alternately confined and relaxed; there is abdominal pain; hemorrhoidal swellings are noticeable; hemorrhages occur from the stomach, bowels, nose and gums, and petechial spots appear beneath the skin.

The course is prolonged and interrupted by periods of apparent improvement, during which the ascites diminishes and the patient is free from discomfort, and in some degree recovers health and spirits.

General dropsy, severe diarrhœa or hemorrhages indicate that the end is near. Sometimes intercurrent inflammation of the pleura or lungs is the direct cause of death.

---

* If ascites be extreme, it is often difficult to detect the spleen by palpation when the patient is in the ordinary dorsal position, or on the right side. In such cases, placing the patient upon the hands and knees entirely removes the difficulty.

In hypertrophic cirrhosis, the skin, conjunctivæ and urine are deeply stained by bile, and the stools, which vary greatly in number and consistency, are clay-colored. The liver and spleen are enlarged, but there is no distention of the superficial abdominal veins, and no ascites. At times the jaundice and enlargement of the liver increase rapidly; then there is moderate fever, with much pain in the right hypochondrium. As the end approaches the pulse becomes markedly irregular; the tongue grows dry and brown; the teeth are covered with sordes; there is complete anorexia; rapid wasting; bleeding from the gums, from the stomach or beneath the skin; apyrexia, drowsiness, stupor, and, finally, convulsions. The course is more rapid than in the former variety, but still protracted. Should both forms exist together, there is a combination of jaundice, ascites and distention of the veins in the abdominal wall.

DIAGNOSIS.—The characteristic features of atrophic cirrhosis are diminution in the area of liver dulness, following a temporary increase in the bulk of the organ; enlargement of the spleen; dilatation of the superficial veins; ascites; hemorrhoids; a dry, earthy skin, and gastro-intestinal hemorrhages, occurring, without fever, in a child who has a history of prolonged ill-health, feebleness and wasting.

The second and more uncommon variety, while having very much the same preliminary history, presents as its distinguishing marks enlargement of the liver and spleen without ascites; jaundice, with fever; pain in the hepatic region; and, subsequently, malignant jaundice, with typhoid symptoms, rapid wasting, coma and convulsions.

Acute yellow atrophy, which has many of the symptoms of the final stage of the biliary cirrhosis, is distinguished by its abrupt onset and rapid course, and is among the rarest of diseases in children.

PROGNOSIS.—The result is almost invariably unfavorable, and it is only under the most fortunate conditions that even a temporary improvement can be obtained.

TREATMENT.—Before a diagnosis is established, and while the

patient is merely suffering from ill-defined symptoms of bad health, with imperfect digestion, hygienic and therapeutic measures are to be directed to the restoration and preservation of the general strength, and to correcting any disorder of the organic functions.

When the hepatic affection declares itself, an alkaline or a purely tonic treatment may be adopted. Alkalies are indicated when the hepatic and gastro-intestinal symptoms are in excess of the wasting and general debility; tonics under opposite circumstances. In the former case, the following prescription is useful:—

  ℞. Sodii Bicarb., ... ʒij.
    Tr. Nucis Vom., ... ♏xviij.
    Inf. Calumbæ, ... q. s. ad f℥iij.
  M.
  S.—Two teaspoonfuls three times daily, for a child of ten years.

In the latter, Basham's mixture may be employed, or a combination of iron and quinine, as:—

  ℞. Quiniæ Sulph., ... gr. xij.
    Tr. Ferri Chloridi, ... fʒj.
    Syr. Zingib., ... f℥j.
    Aquæ, ... q. s. ad f℥iij.
  M.
  S.—Two teaspoonfuls three times daily.

Both plans must be followed out steadily and continuously, to obtain any beneficial results.

To relieve constipation, from two to four fluidounces of Hunyadi water should be taken every morning on an empty stomach. Diarrhœa can be controlled by sub-carbonate of bismuth, and hemorrhage by gallic acid or aromatic sulphuric acid.

It is important to order a liberal diet—milk, eggs, meat and farinaceous foods in full proportion to the capacity of digestion. As in other diseases of the liver, the skin must be kept active by daily warm baths, and chilling prevented by flannel underclothing.

If ascites be so great as to impede the action of the diaphragm, paracentesis must be resorted to at once. A fine trocar or one of Southey's tubes may be used.* The operation should be repeated so soon, and as often, as reaccumulation renders it necessary. When performed early enough, it sometimes has, as in adults, more than a merely palliative effect.

7. SUPPURATIVE HEPATITIS.

Abscess of the liver is an extremely uncommon disease in children. The only case that has ever come under my notice, presented the following clinical history:—

George ———, æt. 5 years, was first brought to the Dispensary of the Children's Hospital on April 27th, 1875, during the service of Dr. George S. Gerhard. Though residing in a malarious locality, and in a poor and filthily-kept house, he had always had good health up to one week previous to the above date, when he began to complain of pain in the region of the umbilicus. Under appropriate treatment he passed several lumbricoid worms, and the pain disappeared. A week later, however, it returned, and as his bowels were constipated, his father administered a tablespoonful of castor-oil; this produced a free evacuation, containing from twenty to thirty lumbrici, many being of large size. After this he seemed to be perfectly well until May 9th, when the pain in the abdomen reappeared; he now began to lose his appetite, and a swelling was noticed in the right hypochondriac region.

When I saw him first, on May 15th, his general appearance was good; his cheeks having a healthy color, and his body being sufficiently stout. His tongue was lightly coated, and his father stated that his appetite was poor, and that, though his bowels were moved daily, the passages were small. There was no heat of skin or jaundice, the pulse and respiratory movements were

---

* See section on Ascites.

normal in frequency; he had no cough, and, on physical examination, no pulmonary or cardiac affection could be detected. His abdomen was tympanitic, the whole of its upper third was tender to the touch, and in the upper part of the right hypochondrium there was an oval tumor, about as large as a turkey's egg, having its long diameter directed transversely, and projecting at its most prominent part nearly an inch from the surface of the abdomen. The skin covering this tumor was somewhat œdematous, but was freely movable, and natural in color and temperature, while the tumor itself was hard, tender, completely immovable, and the seat of neither fluctuation nor pulsation. It was surrounded by an area of induration, the boundary of which could not be accurately ascertained on account of the pain produced by palpation, though it appeared to extend from the costal border to the lower third of the right hypochondriac region, and from the median line of the abdomen to the right side.

The right hypochondrium was dull, except just below the margin of the ribs, where there was slight, probably transmitted, tympanitic resonance, detected only by deep percussion. The liver dulness began in the ordinary position above, and, on light percussion, was continuous below with that in the hypochondriac region. The patient did not complain of pain except when the swelling was touched, or when the whole body was jarred, as in walking down stairs. On a level surface he was able to walk easily. No history of an injury could be obtained, but on careful questioning it was discovered that throughout the winter he had "coasted" a great deal on his sled, and always rode "belly-bumpers." He was ordered to be kept quiet, and to have a liquid diet, with poultices over the abdomen, and a dose of castor-oil.

May 16th. Had a large passage from the bowels, the evacuation being dark-colored and lumpy; during the night was restless and feverish.

May 17th. Tongue somewhat cleaner and abdomen less tympanitic. The tumor was more prominent; there was deep-seated fluctuation, and the skin covering the mass was less œdematous, not so freely movable as before, and of a dusky-red hue. Patient walked with his body bent forward, as if a more upright posture was painful. Prescribed f℥ss of tinct. cinchon. comp., three times daily, and an increased diet, at the same time directing his parents not to allow him to get out of bed, and to apply warm poultices continuously to the belly.

May 19th. Visited him at his home, and found him entirely free from fever; his tongue was clean, his appetite had returned, and his bowels had been opened; the stool, which had been kept for inspection, was copious, well formed, and in every way natural. The induration around the tumor, or more properly the *abscess*, for such it now appeared to be, had extended so as to fill nearly all of the right hypochondrium, being almost five inches in transverse diameter. There was well-marked fluctuation, and the skin investing the abscess was tightly adherent over a space about four inches in circumference. The abdominal respiratory movements were restricted, and any effort at full inspiration caused pain. There was no sensation of throbbing in the abscess, and the patient seemed to be perfectly comfortable as long as he remained quiet. The abdomen was moderately distended.

May 22d. No change, except that the fluctuation was more superficial, and the integument adherent over a large surface. The former treatment was continued, and as the pulse was more frequent than before, and as he was pale and languid, a teaspoonful of brandy thrice daily and full diet were ordered.

May 26th. Found him up and playing about as if nothing was the matter. Having him stripped and placed in bed, the following observations were made: Abscess more prominent than at last note, but more localized; in its centre there is very superficial fluctuation, extending over an area an inch and a half in diameter and bounded by a firm margin. The skin covering this space is dark red in color, feels very thin, almost as if it could be broken by the pressure of the finger, and is somewhat hotter than the surrounding integument. About the abscess there is a mass of induration which does not project beyond the level of the rest of the abdomen, but which extends from the lower border of the ribs to the middle of the right lumbar region, and from the mid-line of the abdomen to the right side; its outline is semicircular, the edge being smooth and well defined, so much so that the fingers can be inserted beneath it. The skin is adherent over the whole mass, but most so immediately around the position of fluctuation. Both palpation and percussion indicate that it is connected with the right lobe of the liver. There is some pain excited by palpation, though this is much less than formerly. There is no jaundice. No change was made in the treatment.

May 29th. The abscess was more circumscribed, being about the size of an English walnut, and the pus was still nearer to the surface. As it was impossible to keep the patient quiet, and fearing lest he might rupture the abscess in his play, aspiration was determined upon. Accordingly, a large aspirator needle was introduced to the depth of half an inch. About two drachms of thick, grumous pus, mixed with blood, escaped into the receiver, when the canula became plugged, and no more could be withdrawn.* A poultice was applied, and the patient ordered to remain in bed during the rest of the day. The operation was followed by no bad symptoms, and the next day he was up, amusing himself as before. There was, however, considerable discharge of thick pus from the opening left by the aspirator needle.

June 1st. The wound made by the needle closed. Scarcely any fluctuation could be detected, and there was but little redness of the skin. On passing the finger over the position of the abscess a cup-shaped depression was felt, bordered by a well-defined edge of dense tissue. The induration was reduced; its lower margin was still semicircular, and could be easily isolated,

---

* A microscopical examination of a portion of this material revealed pus cells, compound granule cells, blood corpuscles, and numerous polygonal cells having well-defined nuclei and resembling liver cells.

while the upper margin, on the other hand, could not be discovered, as the mass extended under the ribs. There was hardly any pain on manipulation, and the boy's general condition was very good.

June 5th. There were no signs of the abscess, except a small spot of dusky redness, and slight retraction and puckering of the skin at the point of puncture; in this situation, also, the integuments were adherent to the parts below. The induration was diminishing, and its edge, which could still be distinctly felt, was approaching the right costal border. All treatment was suspended.

On October 30th, the child was in excellent health. The skin, for a short distance about the seat of puncture, was somewhat discolored and puckered, and was less freely movable than that of the remainder of the abdomen. Percussion and palpation showed that the right lobe of the liver was slightly contracted.

In reviewing the preceding history, the question that naturally suggests itself is, whether the disease was hepatic abscess, or merely an abscess of the abdominal wall. In the early stage of the former affection, the general symptoms are similar to those observed in acute hepatitis; jaundice being present only in exceptional instances, while the formation of pus gives rise to rigors, frequency of the pulse, night sweats and fever, the latter often resembling the pyrexia of quotidian or tertian intermittents. The almost entire absence of constitutional disturbance in this case, however, is no argument against the existence of hepatic abscess; as it is generally admitted that the symptoms are often very latent, and that in many instances no suspicion of an abscess has been entertained until its discovery by manual exploration, or by the discharge of pus in various directions, and sometimes even not until revealed by post-mortem examination.

The local symptoms, on the contrary, were well marked; thus there was localized, though extensive, enlargement of the right lobe of the liver, and toward the upper part of this enlargement there was an ill-defined, oblong tumor extending beyond the level of the abdomen. The skin covering this tumor was at first slightly œdematous, but perfectly movable and normal in color and temperature. From day to day, as the tumor became more circumscribed and approached nearer the surface, the hepatic enlargement increased, and conjointly with the appearance of fluctuation the œdema disappeared, the skin became dusky-red in hue, hotter than the surrounding integument, and adherent. There was also tenderness on pressure; pain excited by deep inspiration or any jarring movement, and a peculiar bending forward of the body in walking. Again, after the opening of the abscess, all these symptoms subsided, and there was puckering of the skin and rapid reduction in the size of the liver; its projecting margin remaining semicircular, smooth, and well defined. Finally, there was slight contraction of the right lobe.

There are two other points of importance, viz., the detection, by palpation, of a smooth edge of dense tissue bordering the area of fluctuation, which gave the impression that the fluid was contained in a cup-shaped cavity in a solid organ, and the microscopical characters of the pus which was removed.

Now, all these symptoms are characteristic of an hepatic abscess, so situated on the convexity of the liver as to point toward the surface of the abdomen; the adhesion of the integuments being, of course, due to local peritonitis.

Abscesses of the abdominal wall, on the other hand, besides being superficial from the outset, have a different position, being usually seated in the rectus muscle or adjoining connective tissue, and in the neighborhood of the umbilicus; at the same time there is generally violent throbbing pain; the redness and tumefaction of the skin are earlier and better developed, and the constitutional symptoms accompanying the formation of pus are more constantly observed than in abscess of the liver.

The general management of circumscribed hepatitis, prior to the formation of pus—if the symptoms be such as to lead to a diagnosis at this time—simply requires careful regulation of the diet, rest, and attention to the various functions of the body, particularly that of the bowels; for even if the existence of inflammation be ascertained, it is hardly probable that anything can be done to prevent suppuration.

In relation to the propriety of evacuating hepatic abscesses, the bulk of authority is in favor of so doing, when they point externally so as to be detected by palpation, when firm adhesions have formed, and when the pus is near the surface. As to the method of evacuation, a free incision is perhaps preferable to puncture with an aspirator needle ; first, because the pus is often mingled with shreds of connective tissue and broken-down liver substance, liable to obstruct the needle and render it useless ; or, even if all the fluid be withdrawn, to remain and prolong the process of suppuration ; second, as the inelasticity of the walls of the cavity cannot prevent the entrance of air, it is much better to provide a free way of exit, than to have the air confined, as it would likely be in the event of the small opening made by the needle becoming closed. For the purposes of exploration, however, the aspirator may be used with advantage. After being opened, the abscess is to be dressed in the ordinary manner, while strict rest should be enjoined, and tonic and supporting measures employed. Subsequently, nutritious diet and exercise in the open air, the latter adapted to the strength of the patient, are much more important than mere medication.

TUBERCULOSIS of the liver is sometimes associated with tubercular peritonitis, and is commonly encountered at the autopsies of children who have succumbed to acute general tuberculosis. In such cases the liver is anæmic, yellowish and small. Semi-transparent granules (miliary tubercles) are seen upon the capsule and detected by the microscope in the connective tissue that surrounds the branches of the portal vein ; there is, too, an interstitial hepatitis, with the formation of embryonal and fibrous tissue. There are no definite symptoms, and a diagnosis is hardly possible without post-mortem section.

HYDATID DISEASE and CANCER are so infrequent in childhood, and when they do occur present so nearly the symptoms of the same conditions in adults, that it is unnecessary to devote space to their consideration.

CHAPTER V.

AFFECTIONS OF THE PERITONEUM.

1. PERITONITIS.

Children, like adults, are subject to attacks of inflammation of the peritoneum. These may be primary or secondary in origin, acute or chronic in course, and general or local in distribution.

The affection occurs at any age from birth to puberty, and there are indisputable evidences on record of its developing during the later months of intra-uterine life. The primary or essential form is almost uniformly acute and general. Secondary peritonitis, on the contrary, may be either general or local, the inflammation often beginning in a limited area and gradually extending over the whole surface. It is also more common than the primary variety, and, while often acute, more frequently runs a chronic course.

MORBID ANATOMY.—In acute general peritonitis, the blood vessels of the sub-serous tissue of the peritoneum are engorged with blood, and the membrane is reddened, either generally or in patches; mottled by isolated spots of ecchymosis, and opaque and thickened. Serum, sometimes clear, sometimes mixed with pus and flakes of fibrin, fills the abdominal cavity; or, again, the effusion may be purulent; in either case, it is most abundant in the pelvis and between the mesenteric folds.

Acute local peritonitis occasions connective tissue hyperplasia, omental and intestinal adhesions, and, at times, localized suppuration.

Chronic general peritonitis gives rise to a sero-fibrinous exudate; this may be sufficiently abundant to appear as a thick membrane, and in time may undergo fatty, caseous or calcareous degeneration.

Chronic local peritonitis results in the formation of circumscribed adhesions, membranous exudations of limited extent, and sacculated collections of pus.

ETIOLOGY.—Fœtal peritonitis is caused by syphilis or some specific infection of the mother. During the first few days of life it may be due to inflammation, suppuration or gangrene of the umbilicus; to congenital occlusion of the anus; or to infection from a mother ill with puerperal fever. Later in childhood, primary peritonitis arises from blows upon, or other injuries to the abdomen, and from sudden chilling of the body after violent exercise; a number of cases having been noted in which the attack could be traced to the act of lying to rest, after an exciting or fatiguing game, prone upon damp ground. The secondary form may result from the escape of fæcal matter into the abdominal cavity through a perforation of the intestine— called perforative peritonitis. It may also occur during the course of one of the exanthemata, scarlatina especially—septic peritonitis. Finally, it may be occasioned by extension of inflammation from some one of the abdominal viscera, or from the pleura; in the last instance there may be an element of sepsis.

Chronic peritonitis sometimes follows an acute attack, but is most often an attendant of tuberculosis and presents the characters of chronicity from the outset.

SYMPTOMS.—In primary peritonitis, and in other cases of the acute general disease not due to perforation or sepsis, the attack begins with more or less rigor, abdominal pain and vomiting. The pain is stinging or lancinating in character, and is limited, at first, to one flank, to the supra-pubic region, or the neighborhood of the umbilicus, but soon becomes general; it is increased by pressure or by any act calling the abdominal muscles into play, as deep breathing, sneezing, coughing and vomiting. The vomiting is frequent and very violent, producing so much distress and fatigue that after each effort the patient falls back on the pillow with pale, haggard, and sweat-bedewed face; the material rejected consists, in the beginning, of partially digested food; later, of bile-stained mucus.

Fever quickly follows the shivering, and as soon as inflammation is fully established, the axillary temperature may reach 104° F., although the usual range is from 101° to 102°. With the pyrexia there is a frequent, small, wiry pulse, and the breathing assumes the superior costal type ; in some cases (where there is a large effusion) growing hurried and difficult. The child ceases to move his legs, or takes to bed early and lies immovably upon his back, with the knees drawn up. The face is pale and anxious, the nose sharp and the nostrils thin and expanded. The abdomen is distended and passive, so far as respiratory movements are concerned ; palpation yields a certain sense of resistance, sometimes develops fluctuation,* and always excites intense pain ; percussion elicits tympany over the upper anterior portion of the belly and dulness over the dependent parts, and on auscultation, friction sounds may be heard when there is a fibrinous exudation.

The tongue is pointed, red at the tip and edges, and covered in the centre with a dry, moderately heavy, brown-white fur. There is anorexia and increased thirst. Constipation is the rule if the intestinal peritoneum be involved; then, too, there are frequent attacks of severe griping pain ; on the other hand, there may be diarrhœa, with watery evacuations, if the inflammation be attended by œdema of the sub-mucosa with transudation of serum into the bowel.

The urine is high colored and somewhat reduced in quantity, and, while ordinarily passed with freedom, is retained when the serous coat of the bladder is involved in the inflammation.

Sleep is disturbed and restless; in infants there may be convulsions ; in older children, delirium.

---

* When fluctuation is indistinct, Duparcque recommends that the child be placed on one side for a few moments, so that the whole quantity of fluid may gravitate to the depending flank; then quickly turned upon the back, when dulness and temporary fluctuation will be found at the site of accumulation.

During the course of the attack, which usually extends over a period of seven days, the strength steadily fails; there is considerable loss of flesh, and the symptoms present at the onset continue unabated and unchanged. As death approaches vomiting usually stops, but the other symptoms become more and more grave. The patient lies in an apathetic condition, with sunken eyes and half-closed lids; his face is drawn and either pale or cyanosed; the tongue is dry, brown and pointed; there is marked tympanites and the pulse is extremely small and frequent.

Occasionally this variety of acute peritonitis ends in recovery; the exuded fluid being either reabsorbed or spontaneously evacuated through the umbilicus or abdominal wall.* In the first instance, the symptoms subside gradually; in the second, rapidly; though in both, the course is protracted; the fistulous openings left after the discharge of pus rarely closing under four or five weeks.

Perforative peritonitis requires separate description, since it has a set of symptoms entirely its own. It is the most common form of the disease in children, and in the majority of cases results from rupture of the vermiform appendix or cæcum; perforation of typhus or tubercular ulcers being exceptional in this class of patients.

The attack begins suddenly with intense pain in the abdomen, quickly followed by profound collapse. The face soon becomes pale and haggard; the eyes are deeply sunken, and the hands and feet cold, though the body heat is increased; the rectal temperature ranging to 103° or 104°. Other features are great thirst, occasional vomiting, a dry, red and pointed tongue; locked bowels; a rapid, small, thready pulse; thoracic respiration, often hurried and difficult, and suppression of urine. From the beginning the belly is greatly distended by gas escaping from the intestine; the abdominal respiratory movements cease; palpation is very painful, and percussion yields a uniformly drum-

---

* M. Gauderon mentions ten such cases, eight of which recovered.

like tympany that extends high up under the ribs, and completely masks the liver dulness. Death almost invariably takes place either on the third or fourth day of illness, and is usually preceded by a few hours' freedom from suffering.

While this is the ordinary course of perforative peritonitis, it happens sometimes that the shock is so great that the patient neither feels pain nor complains of tenderness when the abdomen is touched, and there is a general latency in the symptoms. Again, extravasation being limited by preformed adhesions, the inflammatory action is circumscribed, and the resulting abscess, by pointing and discharging through the abdominal wall or into the intestine, may either end in recovery, or in the production of a permanent fæcal fistula.

In septic peritonitis the symptoms are either inherently latent, or are masked by the collapse that follows the onset of a new inflammation in a patient already debilitated by disease. There is usually rapid prostration, restlessness, and delirium, with a tendency to stupor; a pale, anxious face; swollen belly; persistent watery diarrhœa; a frequent, wiry pulse, and quick, costal breathing. Pain, tenderness, tension of the abdominal walls, dulness on percussion and fluctuation may be entirely absent. Without care, such attacks are readily overlooked.

Should peritoneal inflammation become chronic the pain lessens and is more paroxysmal in character; the fever is remittent, with evening exacerbations; constipation alternates with diarrhœa; there is great emaciation, and death occurs from exhaustion. However, on account of the usual tubercular origin, the symptoms of this form will be more appropriately studied under the head of "tubercular peritonitis."

Local peritonitis is almost uniformly secondary; that attending inflammation of the cæcum and vermiform appendix being the most common in children.

DIAGNOSIS.—An immovable dorsal decubitus; a pale, haggard face; a frequent wiry pulse; distention, pain and tenderness of the belly, and inactivity of the abdominal muscles in respiration, suffice to render the diagnosis of acute general peritonitis easy. Intense pain, sudden collapse and rapid and extreme meteorism characterize the perforative variety.

In colic there is constipation and vomiting, with severe pain; but between the paroyxsms there is no abdominal tenderness, and the pulse is never so rapid, small, and wiry, nor is there the fear of movement so noticeable in peritonitis.

Rheumatism of the abdominal muscles is attended by tenderness on pressure; distressed facial expression; dorsal decubitus with knees drawn up, and constipation, and thus simulates peritonitis; but the face is never haggard, there is no vomiting nor hiccough, nor distention of the belly, neither is tenderness extreme. The pulse is soft, compressible, and only moderately frequent; the temperatnre nearly normal, and the urine scanty, high-colored, acid and scalding.

It is important to remember that constipation is the rule in peritonitis when the inflammation involves and paralyzes the muscular coat of the bowel; diarrhœa, when it spreads through the muscular coat to the mucous membrane.

The great difficulty in diagnosis is experienced with latent peritonitis, whether septic or due to other causes. Suspicion of its existence may be entertained when, in the course of any predisposing disease, the patient suddenly grows pale and haggard, and has a full belly, with a tendency on the part of the abdominal muscles to become rigid on palpation. Restlessness, delirium and stupor, a change in the type of respiration and in the character of the pulse, all strengthen the suspicion. Under these circumstances it is well to practice Duparcque's method for detecting the presence of fluid, and this, if successful, leaves no further doubt.

In the words of Eustace Smith: "In cases of chronic empyema we should always be on the watch for the occurrence of peritonitis. If the child, after a period of improvement, cease all at once to gain ground, and begins to look pale and distressed, with an elevated temperature, a more or less distended belly, and a rapid, wiry pulse, we are justified in suspecting peritonitis, although there be no tension, tenderness or other sign connected with the abdomen to give support to this opinion."

PROGNOSIS.—This must always be most grave. Perforative peritonitis is invariably fatal. The primary variety, when due to cold, exceptionally ends in recovery, and so, too, does the partial form.

TREATMENT.—Absolute rest in bed and quiet surroundings are essential. Hot applications, in the form of light flax-seed poultices and of turpentine stupes, should be made to the surface of the belly; or, if these fail to give relief, cloths wrung out of ice-water may be applied; they must be frequently changed, to secure the constant action of cold. Leeching is sometimes of great service in subduing pain, but it is only to be employed with robust subjects and in an early stage of the attack.

Of drugs, opium alone can be relied upon. It may be exhibited by the mouth, the rectum, or subcutaneously, and can safely be pushed to the point of producing drowsiness, with decided contraction of the pupils, provided ease from suffering be not attained before. For a child of six years, three drops of laudanum every two hours, by the mouth or rectum; and, by hypodermic injection, one-eighteenth of a grain of sulphate of morphia, repeated as required, are the average commencing doses.

Under no circumstances is a purge to be given. Should constipation be obstinate, and the indications urgent to unload the bowels, a simple enema may be employed. It is a good rule, however, to interfere as little as possible in this way.

The patient's strength must be sustained by concentrated liquid food in small quantities and at short intervals. Three fluidounces of milk and two fluidounces of beef-tea, alternating, every two hours, with the occasional substitution of the yolk of a soft-boiled egg for one or the other, would be a proper diet for a child of six years; stimulants are also necessary, and so soon as there is evidence of failing strength a teaspoonful of good whiskey must be added to each portion of milk. Bits of ice may be allowed from time to time to allay thirst and quiet the stomach.

Should the inflammation subside, the opium is to be gradually withdrawn and its place supplied by sorbefacients and tonics ; at first mercury in alterative doses, or iodide of potassium, with quinine ; and, later, syrup of the iodide of iron. At the same time, the hot or cold application being removed, a weak mercurial ointment should be rubbed into the skin of the belly once or twice daily ; for example :—

    ℞. Ung. Hydrargryi,
        Ung. Belladonnæ, ... āā ʒij.
        Adipis, ... ʒiv.
    M.
    S.—Use locally as directed.

A most important point is to make no change in the diet, except, perhaps, to increase gradually the quantity of liquid food, until convalescence be fully established.

2. TUBERCULAR PERITONITIS.

As a rule, peritonitis due to the presence of tubercle in the abdominal cavity runs a chronic course, and is associated with tuberculosis of some other organ of the body—of the brain or lungs, for instance ; less frequently it occurs as an isolated affection. Acute tubercular peritonitis is not unknown ; it is detected with difficulty during life, and is invariably an element of general tuberculosis. The disease is quite common after the age of seven years, but is rare in earlier childhood and almost never met with in infancy.

MORBID ANATOMY.—At the autopsy of a child dead from tubercular peritonitis, the intestines will be found covered by a layer of yellow, greenish or gray lymph, varying in thickness, and either loose and soft in texture or tough. Lymph having the same characters also covers the parietal peritoneum, and extends between the intestinal coils, binding them, more or less firmly, together. The exudate contains caseous masses of variable size ; its meshes are filled with greenish-yellow, sometimes bloody, sero-purulent effusion, and a quantity of the same material is usually found in the dependent portions of the abdominal cavity. Removal of the layer of lymph discloses gray and yellow tubercles, studding the surface of the peritoneum, together with masses and broad plates of caseous consistence and tuberculous nature. The thickness and extent of the exudation, the number of granulations, and the size of the caseous masses increase with the chronicity of the attack. In acute cases the cheesy collections are absent; the exudate is comparatively thin, soft and translucent, and the granulations, which vary in size from that of a pin's head to that of a pea, are scattered at intervals through its substance.

The omentum is shriveled, hard and often firmly bound to the abdominal wall; the mesentery is firm and contracted; the mesenteric glands are enlarged and show evidences of cheesy or tubercular alteration; tubercular ulceration of the bowels is common, and the liver may be increased in size from amyloid or fatty change, and in some instances is cirrhotic. Inspection of other organs of the body usually leads to the discovery of tubercle, though this is not uniformly the case.

Tubercular peritonitis is not always general in distribution; when localized the inflammation and inflammatory products are usually to be observed in the neighborhood of the diaphragm.

ETIOLOGY.—The factors leading to peritoneal tuberculosis are identical with those producing other tubercular affections. The age at which the disease is most prone to occur has already been mentioned. Male children seem to suffer more frequently than those of the opposite sex.

SYMPTOMS.—The onset is slow and insidious, and the physician is apt to have his attention diverted from the abdomen by more striking manifestations of tuberculosis of the lungs or other organs. Unless such features be present and precedent, there is but little evidence of failing health in the beginning, and the first symptom to attract notice is an abnormal prominence of the belly. The patient gradually grows dull and listless, looks ill, and, on account of abdominal tenderness and the pain produced by jarring, becomes slow and guarded in his movements.

Often after the disease is fully developed the child "keeps about," but the face is drawn and wears an expression of anxiety and suffering; the frame slowly wastes and the skin becomes dry and harsh and loses its healthy hue. Complaints are made of tenderness and griping pains in the abdomen, and the little sufferer takes very characteristic precautions to lessen his ills by steadying his belly with his hands in walking, and by moving down stairs backward so as to pass from step to step on his toes, to avoid jolting. The symptoms denoting disturbance in the functions of the gastro-intestinal tract are inconstant at this stage of the disease; the tongue either shows little alteration or is lightly frosted or more pointed and red than natural; nausea and vomiting may be entirely absent, and are never so persistent and severe as in simple peritonitis; the appetite often remains unimpaired, and the bowels are alternately relaxed and confined. On the other hand the signs to be detected by abdominal exploration are very constant and characteristic. The belly is oval in shape and somewhat irregularly distended, the greatest enlargement occupying the epigastric and umbilical regions; the natural folds and furrows are obliterated; the superficial veins are prominent; and the integument has a smooth, shining appearance, as if smeared with oil. When the hand is applied to the surface, the recti muscles become tense, in an involuntary effort to protect the tender parts beneath; some portions of the abdomen feel soft and flaccid; in others, firm masses are perceptible to the touch; tenderness on pressure is universal, though most marked over the firm masses. Palpation also reveals fluctuation; this is usually indistinct, though occasionally, when enlarged glands or cheesy masses exert pressure on the portal vein, there is a large collection of fluid in the peritoneal cavity, and the fluctuation wave is readily elicited and very distinct. The edge of the right lobe of the liver can often be felt extending half an inch or more beyond the right costal border. On percussion, tympany will be elicited over the flaccid portions of the abdomen; dulness over the firm masses and flatness over the flanks—in the recumbent position—while, if the patient be rolled to one side, the note on the flank turned uppermost becomes tympanitic.

The respiratory movements are somewhat increased in frequency and thoracic in type; the pulse is quickened and feeble in proportion to the general weakness; the axillary temperature ranges from 98° F. in the morning to 101° in the evening; and there is dysuria with high-colored, but otherwise unaltered, urine. Sometimes, with a large collection of fluid in the peritoneum, there is œdema of the feet and legs; then, too, the urine may be slightly albuminous.

In time the patient is forced to go to bed, where he lies on his back, or partially turned on one side, with his legs drawn up; this position is rigidly maintained, for every movement is painful. Now, the wasting is rapid; the face wears a haggard expression; the cheeks and temples are hollow, and the skin becomes inelastic and dotted with purpuric spots. The tongue is dry, heavily coated, or red and smooth; the appetite fails and there is urgent thirst. The bowels are in one of two conditions: relaxed, with watery, offensive stools, containing flaky matter and small black clots of blood when there is tubercular ulceration; obstinately confined, when the intestines are pressed upon, or obstructed by adhesions. In the latter case the belly becomes greatly distended, and there are frequent attacks of severe colicky pain. Under other circumstances, however, the size of the belly may diminish, and then hard, tender lumps are felt in contact with the abdominal wall. The pulse is more frequent and feeble; the evening temperature ranges as high as 103° and 104°, and night sweats are common.

Death occurs after a lapse of time varying from several months to a year or more.

The course of the disease is not uniformly progressive, being interrupted by remissions and exacerbations. During the former the tenderness and distention of the abdomen diminish, the appetite returns, nutrition improves, and false hopes arise of rapid recovery.

Sometimes before death an abscess forms, and pus is discharged through the abdominal wall in the neighborhood of the umbilicus; in other cases the intestines may be perforated from without, but this complication scarcely hastens the fatal termination, for extravasation is limited by adhesions between the knuckles of the intestines. Such complications as tuberculosis of the lungs and cerebral meninges, however, certainly hasten death.

Acute tubercular peritonitis always occurs as an element of disseminated tuberculosis, and presents the general features of that condition; usually there are no local manifestations other than abdominal fulness and slight pain—symptoms sufficiently common in children to be altogether indefinite. The course of these acute attacks is measured in little more than a week.

DIAGNOSIS.—Ordinarily the formation of a correct opinion is not difficult. The distinctive features are the irregular distention of the abdomen; the smooth, shiny appearance of the investing skin; tenderness; unequal resistance to the touch in different positions, and indistinct fluctuation, combined with alterations in the temperature; impairment of nutrition; an insidious onset; a family record of tuberculosis or scrofula; the presence of the tubercular diathesis, and the existence of symptoms of tubercular deposit in some other organ of the body. In doubtful cases, where there is little distention or tenderness, and fluctuation is absent, it is well to try the effect of a sudden jar; this may be done by directing the child to jump from a low chair to the floor. Free fluctuation is to be regarded as a point in the negative.

Many children have prominent bellies and suffer severely from abdominal pain, both due to the accumulation of flatus in the intestines, the consequence of a chronic catarrh of the mucous lining. These patients, though pale and flabby, are but little wasted, and express in their faces no trace of severe illness; they are lively in action; their temperature is normal; there is no tenderness or involuntary contraction of the recti muscles on palpation; the abdominal distention disappears spontaneously at times, and subsides entirely when a non-farinaceous diet is ordered. There can be no greater mistake than that of attributing every instance of abdominal distention to tuberculosis.

As already stated, the diagnosis of the rare acute form is very difficult, and is often only made at the post-mortem table.

Typhoid fever is the disease most likely to be confounded with it, but the absence of rash and splenic enlargement, and the difference in the degree and course of the fever, should prevent error.

PROGNOSIS.—This must always be unfavorable. Recovery, though possible, is extremely rare.

TREATMENT.—While little is to be expected from therapeutic measures, the physician's ambition will be to obtain a favorable result if he can. To accomplish this end it is necessary, first, to keep the child at perfect rest in bed; and second, to select a diet that will meet the capacity for digestion, excluding as nearly as may be the farinaceous foods so prone to cause acidity and flatulence, with their attendant suffering. The following is a sample diet list for a patient of seven years :—

For breakfast, at 7.30 A.M.—The yolk of a soft-boiled egg, a slice of well-toasted bread lightly buttered, and a tumblerful (f℥viij) of warm milk.

For luncheon, at 12 M.—The soft parts of a dozen oysters or a bit of fish, or a bowl (f℥vj) of good beef-tea, with a biscuit.

For dinner, at 3 P.M.—Two to four tablespoonfuls of minced mutton or chicken, one or two thin slices of stale buttered bread, eight tablespoonfuls of rice and milk or junket.

For supper, at 7 P.M.—Two slices of milk-toast and a tumblerful of warm milk.

Such a list can only be used in the earlier stages of the disease; later, when the appetite fails, it is necessary to resort to liquid food, milk and meat broths, administered in small quantities at short intervals.

Stimulants—and whiskey is the best—are required from the beginning, and must be given in increasing quantities as the strength fails.

Of drugs, opium, quinine, and syrup of the iodide of iron with cod-liver oil, when the stomach will bear them, are the most useful. Opium must be given sufficiently freely to relieve pain, and quinine in doses large enough to maintain the flagging forces. Constipation is to be remedied only by simple enemata, while excessive diarrhœa may be checked by full doses of bismuth combined with ipecacuanha and opium, as :—

    ℞. Pulv. Ipecacuanhæ Comp., ... gr. xxiv.
        Bismuth. Sub-carb., ... ʒj.
        Pulv. Aromat., ... gr. xij.
    M. et ft. chart. No. xij.
    S.—One powder every two or three hours for a child of seven years.

A good formula for the same purpose is :—

    ℞. Ext. Hæmatoxyli, ... gr. xxx.
        Tr. Opii Deod., ... ♏xxiv.
        Vin. Ipecacuanhæ, ... ♏xxxvj.
        Mist. Cretæ, ... q. s. ad f℥iij.
    M.
    S.—Two teaspoonfuls every three hours.

Externally, light flax-seed poultices are useful in relieving pain. Sometimes even the lightest poultice is uncomfortable, then the abdomen may be anointed once daily with—

    ℞. Ext. Belladonnæ, ... ʒij.
        Glycerinæ, ... f℥vj.
    M.

and covered with a thick layer of cotton batting.

Should the quantity of fluid in the peritoneum be large, diuretics and diaphoretics are indicated; if excessive, paracentesis is required.

3. ASCITES.

The collection of a quantity of transparent serum in the sack of the peritoneum is not of very common occurrence during childhood. The condition is, probably, always secondary, and must be regarded rather as a symptom than a disease proper;

it is of sufficient import to warrant a brief, separate consideration.

ETIOLOGY.—Ascites is sometimes produced by simple or tubercular inflammation of the peritoneum; more frequently it depends upon obstruction to the return of venous blood, due to diseases of the liver or heart; to enlargement of the mesenteric glands, and, occasionally, to disease of the lungs; again, it may be the result of a general hydræmic state of the blood, attending affections of the kidneys and anæmia. It is occasionally impossible to decide upon the preëxisting lesion.

SYMPTOMS.—In a well-developed case the abdomen is distended and globular, the exact shape depending upon the position of the patient, being broader in the recumbent than in the erect posture, as, then, the fluid tends to spread and collect in the flanks. The integument is smooth and shining; the superficial veins are very distinct, and the normal depression at the umbilicus is either effaced or there is a projection at this point. There is a sense of fullness, with moderate resistance, but no tenderness on palpation; and if a hand be placed on either side of the belly, and a sharp tap given with one of the fingers, a distinct impulse—fluctuation wave—is felt by the other hand; this is not interrupted by pressure, made by an assistant, on the median line. While the child lies upon its back, percussion is tympanitic over the upper anterior parts of the belly, where the intestines float free, and dull elsewhere; a change in position alters the relation of the areas of tympany and dullness, and the extent of the latter depends entirely upon the amount of fluid present.

Pain is not a prominent symptom; if present, it is paroxysmal, and has the griping character of the colic of intestinal indigestion. Such attacks are often attended or followed by moderate diarrhœa; in the intervals the bowels may be confined. Should the effusion be large the mere weight of the fluid causes discomfort; then, too, respiration is embarrassed, even to the extent of orthopnœa; micturition is painful; the urine is scanty, high-colored and albuminous, and there may be œdema of the genitalia and legs, resulting solely from pressure.

Cases having an obscure etiology furnish few additional features; there are no constant or characteristic alterations of the tongue, appetite, appearance of the skin or temperature; for these, with other rational symptoms, depend upon the determining disease.

When due to inflammation of the peritoneum, the amount of effusion is small; the abdomen is tense and tender; the temperature is usually elevated, and the general symptoms of acute or chronic peritonitis are more or less marked.

In hepatic disease, especially cirrhosis, the effusion is great; the superficial abdominal veins are very prominent; the hemorrhoidal veins are distended; the spleen is often enlarged; the digestive functions are impaired, and the general integument has a sallow hue or is decidedly jaundiced.

Cardiac disease causes anasarca and hydrothorax as well as ascites, and these conditions are apt to be associated; the face is livid; the lips and finger tips blue; the jugular veins are distended and pulsating; there is dyspnœa, and a scanty, albuminous urine, with the physical signs of heart lesion.

DIAGNOSIS.—There is little difficulty in detecting ascites, unless the effusion be so small that it sinks away into the pelvis or between the folds of the intestine beyond the reach of the examiner's hand. Under these circumstances it is well to try Duparcque's method (see page 366) of increasing the distinctness of fluctuation, or to put in practice another plan for the same purpose, namely, placing the patient on the hands and knees so that the fluid may gravitate to the most dependent portion of the abdomen—now the neighborhood of the umbilicus—and come within the range of palpation.

A large belly, produced by flatulent distention of the intestines, may yield indistinct fluctuation, the palpation stroke being transmitted through the bowels; but the imperfect wave is readily interrupted by pressure in the median line, and the results of percussion are quite different from those obtained in ascites.

The collection of a large quantity of fluid in the pelvis of one or other kidney—hydronephrosis—is attended by abdominal distention, fluctuation, and percussion dulness. The enlargement, however, is more noticeable on the side of the affected kidney; here, also, there is more resistance and greater dullness, the opposite flank being often tympanitic; changes of position have little effect in altering the percussion sounds, the umbilicus rarely protrudes, a kidney-shaped outline can often be detected, and tapping liberates a liquid charged with urea.

The PROGNOSIS depends chiefly upon the nature of the originating disease. When this cannot be discovered, the forecast must be based upon the general strength and nutrition, the condition of the skin, the temperature, and the character of the urine. If the strength be moderately preserved, the appetite and digestion fairly good, the skin natural in texture and color, the temperature normal, and the urine free and non-albuminous, the prognosis for an ultimate recovery is good, irrespective of the amount of effusion.

TREATMENT.—This must, in the main, be regulated by a consideration of the primary disease. Cases of obscure origin, as well as those depending upon anæmia or disease of the liver, are much benefited by full doses of iron. The tincture of the chloride of iron or the dried sulphate are, perhaps, the best preparations to use, and their effect is increased by the addition of quinine. The following is a serviceable formula :—

> ℞. Ferri Sulph. Exsiccat., . . . gr. xxiv (to be increased to ʒj).
> Quiniæ Sulph., ... gr. xij.
> Acid. Sulphurici dil., ... ♏xij.
> Syrupi, ... f℥j.
> Aquæ Menth. Pip., . . q. s. ad f℥iij.
> M.
> S.—Two teaspoonfuls three times daily, taken diluted and after eating, for a child of six years.

Diuretics can be employed at the same time, if there be no kidney complication, for example :—

℞. Potassii Acetatis, ... ℨij.
Spt. Juniperis Comp., ... f ℨss.
Spt. Ætheris Nitrosi, ... f ℨvj.
Syrupi, ... f ℨij.
Aquæ, ... q. s. ad f ℨvj.
M.
S.—Two teaspoonfuls every three hours.

A combination of acetate of potassium, squill and digitalis is often useful.

Should this class of remedies fail, much may be accomplished by a properly regulated course of purgatives. For this purpose, thirty grains of compound licorice powder, or ten grains of compound jalap powder, may be given from two to three times daily. Sometimes it is advisable to begin this treatment by two grains of calomel, administered at bedtime, and followed next morning by a teaspoonful of magnesia.

It is always important to keep the skin active by a daily warm bath, and to maintain an equal surface temperature by woolen underclothing.

The diet should contain as little fluid as possible; thus the child may eat—

For Breakfast.—A saucer of oatmeal porridge or cracked wheat with cream; a soft-boiled egg; two slices of stale bread or toast with butter; a teacupful (four fluidounces) of milk.

For Dinner.—A bit of roast chicken, or tenderloin of beefsteak, or roast beef or mutton; mashed potatoes with gravy, or spinach or cauliflower; two or three slices of stale bread; rice pudding or junket and a glass of filtered water.

For Supper.—A poached egg on toast, or a bowl of cream toast and a cup of milk.

Between meals, some water must be taken to relieve thirst, but the less the better.

When the fluid does not diminish after a thorough trial of ordinary remedies, the peritoneal cavity must be tapped. It is best to make the puncture with a very fine canula; the instrument having been inserted is left in position; a rubber tube is attached, and the fluid allowed to drain away slowly for some eight or ten hours, constant and equal pressure being maintained in the meanwhile by a broad bandage. After the canula is removed, the abdomen must either be strapped or carefully bandaged. The effusion is never entirely removed in this way, but enough is taken to relieve pressure and allow absorption to go on. This method of operation causes so little pain, that, if necessary, but slight objection is offered to its repetition; in very timid subjects, though, it is well to lessen the sensibility of the skin by the momentary application of ice and salt to the point selected for puncture. Paracentesis is often a remedial agent of much value; though in some cases it is merely palliative.

INDEX.

A.

Abdomen, barrel-shaped, 49
   distention of, 48
   examination of, 47
   scaphoid, 50
   tenderness of, 50
Abdominal respiration, 39
Abnormal dentition, 149
Abscess, mammary, 67
   retropharyngeal, 197
Absorption of fat, 47
Accelerated breathing, 40
Acute gastric catarrh, 199
   anatomical lesion of, 206
   symptoms of, 200
   diagnosis of, 201
   prognosis of, 201
   treatment of, 201
   intestinal catarrh, 227
   etiology of, 228
   symptoms of, 229
   diagnosis of, 230
   prognosis of, 230
   treatment of, 231
Acute pleuritis, 27
Affections of the liver, 337
   of the mouth and throat, 124
   of the peritoneum, 364
   of the stomach and intestines, 199
Air insufflation of, 309
Alæ nasi, dilatation of, 21
Alantoin, 33
Albumin, 35
Ammoniacal breath, 31
Amyloid liver, 347
   morbid anatomy of, 347
   etiology of, 348
   symptoms of, 348

Amyloid liver, diagnosis of, 349
   prognosis of, 349
   treatment of, 349
Anæmia, 44, 121–123
Analysis of breast-milk, 73
Anatomical lesion of acute gastric catarrh, 200
   of aphthous stomatitis, 126
   of catarrhal stomatitis, 124
   of simple pharyngitis, 183
   of ulcerative stomatitis, 131
Anterior fontanelle, 46
Antiseptics in entero-colitis, 257
Apex beat in pleuritis, 52
Apparatus for gavage, 113
   for hand feeding, 96
Aphthous stomatitis, 126
   anatomical lesions of, 126
   etiology of, 126
   symptoms of, 127
   diagnosis of, 128
   treatment of, 128
Arrowroot water, 232
Artificial feeding, 71
Ascaris lumbricoides, 312
   egg of, 313
   symptoms of, 319
   treatment of, 321
Ascites, 377
   etiology of, 378
   symptoms of, 378
   diagnosis of, 379
   prognosis of, 380
   treatment of, 380
Aspirating hepatic abscess, 363
Asses' milk, 73

A.

Asthma, 41
Astringent bath, 104
Atrophic cirrhosis, 352
Atrophy, simple, 279
Attendant, questioning the, 18
Auscultation of the chest, 52
Auvard's hatching cradle, 111

B.

Bandage, abdominal, 105
Barley water, 80
Barrel-shaped abdomen, 49
    chest, 51
Basham's mixture, modified, 350
Bath, astringent, 104
    bran, 104
    cold, 103
    cooled, 103
    hot, 103
    mercurial, 104
    mustard, 104
    nitro-muriatic acid, 104
    salt water, 104
    soda, 104
Bathing, 102
    mode of, 102
Bed clothes, 108
Beef juice, raw, 92
    tea, 84
Bethlehem oatmeal, in constipation, 276
Bicarbonate of sodium, 80, 87, 88
Blennorrhœa, 178
Boiled milk, 85
Boracic acid, 68
Bothriocephalus latus, 311
Bottle, graduated nursing, 95
Bottle tip, 96
Bran bath, 104
Brandy and egg mixture in intussusception, 308
Breast-feeding, 60
    proper, number per day, 62
    milk, analysis of, 73
    spec. grav. of, 72
Breath, the, 29
    fetor of, 31
    ill smelling, 29
Breathing, different forms of, 41
    puerile, 52
Bridge of nose, broadness of, 23

Bright's disease, 34
Brinton's theory of fæcal vomiting, 303
Bronchitis during dentition, 179
Broth, veal, 92
    veal, with barley water, 208
Brows, contraction of, 22

C.

Calculi, intestinal, 288
Cancer of the liver, 363
Cardiac disease, 56
Caseous degeneration and tuberculosis of the mesenteric glands, 329
Casts in the urine, 36
Catarrh, acute intestinal, 227
    chronic gastro-intestinal, 213
    of the bladder, 36
Catarrhal stomatitis, 124
    anatomical lesions of, 124
    etiology of, 124
    symptoms of, 125
    treatment of, 125
Causes of ill smelling breath, 29
Cereal foods, 77
Cerebral disease, 24
Cestodes, 311
Cheese poison in milk, 99
Chest, examination of, 50
    inspection of, 51
    barrel-shaped, 51
    auscultation of, 52
    palpation of, 54
    percussion of, 55
Cheyne-Stoke's respiration, 41
Child, inspecting the, 20
    position of, during feeding, 97
Children, general management of, 60
Childhood, 100
Chlorate of potassium, 134, 135
Cholera infantum, 258
    morbid anatomy of, 258
    etiology of, 259
    symptoms of, 259
    diagnosis of, 261
    prognosis of, 262
    treatment of, 262
Chorea, 122, 182
Chronic diarrhœa, 235
    enlargement of tonsils, 24, 194
    entero-colitis, 235

Chronic entero colitis, morbid anatomy
  of, 235
  etiology of, 235
  symptoms of, 236
  diagnosis of, 240
  prognosis of, 240
  treatment of, 241
  gastric catarrh, 202
    morbid anatomy of, 203
    etiology of, 203
    symptoms of, 204
    diagnosis of, 206
    prognosis of, 206
    treatment of, 207
  gastro-intestinal catarrh, 213
  hydrocephalus, 46
  intussusception, 304
  lung disease, 26
  peritonitis, 364
Cirrhosis of the liver, 352
    morbid anatomy of, 352
    etiology of, 353
    symptoms of, 354
    diagnosis of, 355
    prognosis of, 355
    treatment of, 355
Clinical investigation of disease, 17
  thermometer, 43
Clothing, 105
  change of, 105
Clotting, to prevent, 79
Clubbing of the finger tips, 26
Cold bath, 103
Colic, 270
  etiology of, 270
  symptoms of, 270
  treatment of, 271
Collapse, 260, 263, 302
Colon, 299
Condensed milk, 75-85
    reared children, 76
Congestion of the liver, 343
    morbid anatomy of, 343
    etiology of, 343
    symptoms of, 343
    diagnosis of, 344
    prognosis of, 344
    treatment of, 345
Conjunctival blennorrhœa, 151
    during primary dentition, 178

Constipation, (see Habitual Constipation, 273)
Convulsions caused by teething, 161
  in chronic entero-colitis, 238
Cooled bath, 103
Cough, varieties of, 28
  stomach, 227
Cows' milk, analysis of, 74
  spec. grav. of, 74
  sound, 98
Cream, whey and barley water mixture, 207
Crib, the, 108
Croup, 59
Cyanosis, 112
Crying, different characters of, 27
Crusta lactea, 156
Cysticercus cellulosæ, 316
Cystitis, tubercular, 361

D.

Day nursery, 107
Decubitus, 23
Defecation, frequency of, 32
Deformity of sternum caused by hypertrophy of tonsils, 195
Dental paralysis, 162
Dentition, 148
  delayed, 150
  difficult, 150
  irregular, 150
Dermatitis, 155
Development, 45
Diabetes, 34
Diachylon ointment, 159
Diagnosis of acute gastric catarrh. 201
  of acute intestinal catarrh, 230
  of amyloid liver, 349
  of aphthous stomatitis, 128
  of ascites, 379
  of cholera infantum, 261
  of chronic entero-colitis, 240
  of chronic gastric catarrh, 206
  of cirrhosis of the liver, 335
  of congestion of the liver, 344
  of dysentery, 265
  of entero-colitis, 252
  of fatty infiltration of the liver, 346
  of follicular tonsillitis, 189

Diagnosis of habitual constipation, 274
  of habitual indigestion, 217
  of intussusception, 305
  of jaundice, 339
  of mucous disease, 220
  of noma, 139
  of peritonitis, 368
  of simple atrophy, 284
  of simple pharyngitis, 184
  of suppurative hepatitis, 362
  of suppurative tonsillitis, 192
  of syphilitic hepatitis, 351
  of tabes mesenterica, 332
  of thrush, 145
  of tubercular peritonitis, 375
  of typhlitis, 291
  of worms, 319
Diagram showing eruption of milk teeth, 148
  showing method of lancing gums, 163
  showing relation between the permanent and temporary teeth, 170
Diarrhœa, chronic, 235
  during 2d dentition, 177
Diathesis, tuberculous, 22
Diet during 2d dentition, 175
  during the first week, 81
  during the sixth month, 82
  for 8th and 9th months, 83
  for 7th month, 82
  for six weeks, 84
  for tenth month, 65
  for 10th and 11th months, 84
  from 18 months to 2½ years, 93
  from 2d to the 6th week, 81
  from 6th week to the end of 2d month, 82
  from 3d to the 6th month, 82
  from 3½ years up, 101
  from 12th to the 18th month, 93
  in amyloid liver, 349
  in aphthous stomatitis, 129
  in ascites, 381
  in acute intestinal catarrh, 232
  in chronic entero-colitis, 242, 243
  in cirrhosis of the liver, 356
  in colic, 271
  in congestion of the liver, 345
  in constipation, 276

Diet in dysentery, 266
  in entero-colitis, 255
  in intussusception, 308
  in mucous disease, 221
  in peritonitis, 370
  in simple pharyngitis, 185
  in suppurative hepatitis, 362
  in tabes mesenterica, 335
  in tapeworm, 324
  in tubercular peritonitis, 376
  in typhlitis, 293, 294
  up to 3½ years, 100
Difficult dentition, 150
  complications during, 161
  local affections of, 150
  sympathetic effects of, 151
Diphtheria, urine in, 37
Disease, features of, 20
  investigation of, 17
  of the digestive organs, 124
Disorders of the digestive system during 2d dentition, 176
Distention of abdomen, 48
  of bladder, 50
Drinking, mode of, 26
Dysentery, 264
  morbid anatomy of, 264
  etiology of, 264
  symptoms of, 264
  diagnosis of, 265
  prognosis of, 265
  treatment of, 265
Dyspnœa, expiratory, 41
  inspiratory, 41

E.

Ear-ache, 24, 27
Ears, nerve supply of, 168
Eating between meals, 101
Eczema during primary dentition, 154
  during second dentition, 178
  of the scalp, 156
  treatment of, 157
Effleurage, 116
Egg of ascaris lumbricoides, 313
  of oxyuris vermicularis, 312
  of tricocephalus dispar, 314
Electricity in paralysis, 121
Electro-cautery, 196
Emphysema, 31, 51

En chien de fusil, 24
Enemata in entero-colitis, 256
  purgative, 274
Entero-colitis (summer diarrhœa), 248
  morbid anatomy of, 248
  etiology of, 249
  symptoms of, 250
  diagnosis of, 252
  prognosis of, 253
  treatment of, 253
  chronic, 235
Epidemic cholera, 262
Eruption of milk teeth, 57
  of permanent teeth, order of, 169
  of permanent teeth, 164
  of temporary teeth, 148
Etiology of acute intestinal catarrh, 228
  of amyloid liver, 348
  of aphthous stomatitis, 126
  of ascites, 378
  of catarrhal stomatitis, 124
  of cholera infantum, 259
  of chronic entero-colitis, 235
  of chronic gastric catarrh, 203
  of cirrhosis of the liver, 353
  of colic, 270
  of congestion of the liver, 343
  of dysentery, 264
  of entero-colitis, 249
  of fatty degeneration of the liver, 347
  of fatty infiltration of the liver, 346
  of follicular tonsillitis, 187
  of gangrenous stomatitis, 136
  of habitual constipation, 273
  of habitual indigestion, 213
  of hypertrophy of the tonsils, 194
  of intussusception, 300
  of jaundice, 338
  of mucous disease, 217
  of peritonitis, 365
  of simple atrophy, 279
  of simple pharyngitis, 183
  of tabes mesenterica, 330
  of thrush, 142
  of tubercular peritonitis, 372
  of tubercular ulceration of the intestines, 268
  of typhlitis, 289
  of ulcerative stomatitis, 131

Evacuations, fecal. 31
Examination, physical, 39
Exercise, 108
Exhaustion, 299
Expiratory respiration, 41
Explanation of Plate I, 165
Eyelids, incomplete closure of, 21
  lividity of, 25
  puffiness of, 22
  twitching of, 21
Eyes, nerve supply of, 167
Eye teeth, 151

F.

Face, the, 21
  the change of features in disease, 21
Fæcal abscess, 287
  accumulation, 49
  evacuations, 31
  tumor, 31
Faradism, 122
Farinaceous food, 76
Fatty degeneration of the liver, 347
  liver, 346
  infiltration of the liver, morbid anatomy of, 346
  of the liver, symptoms of, 346
  of the liver, etiology of, 346
  of the liver, prognosis of, 346
  of the liver, diagnosis of, 346
Fauces, the, 58
Febrile diarrhœa, 248
Features of disease, 20
Feeding, 60
  apparatus, care of, 97
  artificial, 71
  breast, 60
  by a wet nurse, 69
  general rules for, 79
  intervals of, 64
  mistake of constant, 63
Fever, temperature in, 44
Feverish breath, 30
Filtered water, 101
Finger-nails, blueness of, 26
  deformity of, 26

Fissure of nipple, 67
    treatment for, 68
Flour ball, 83
Follicular tonsillitis, 187
    etiology of, 187
    symptoms of, 187
    diagnosis of, 187
    prognosis of, 188
    treatment of, 189
Fontanelle, 46
    bulging of, 46
Food, farinaceous, 76
    Horlick's, 81
    Mellin's, 81
    preparation of, 97
    quantity per diem, 77
Forced enema in intussusception, 309
Frænum linguæ, ulceration of, 150
Friction, 117
Formula for acute gastric catarrh, 201
    for an alkali in jaundice, 339
    for catarrhal stomatitis, 126
    for chronic gastric catarrh, 210
    for congestion of the liver, 345
    for convulsions, 162
    for enlarged glands during 2d dentition, 178
    for entero-colitis, 257
    for jaundice, 342
    for painting about loose teeth, 172
    for peritonitis, 371
    for second dentition, 175
    for softening the gums, 172
    for tubercular ulceration of the intestines, 269
    for urticaria, 154
    for vomiting, 153
Formulæ for acute intestinal catarrh, 233, 234
    for amyloid liver, 350
    for aphthous stomatitis, 130
    for ascaris lumbricoides, 322, 323
    for ascites, 380, 381
    for cholera infantum, 263
    for chronic entero-colitis, 244, 245
    for cirrhosis of the liver, 356
    for colic, 272
    for constipation, 277
    for dysentery, 266, 267
    for eczema, 158, 159, 160
    for follicular tonsillitis, 189, 190

Formulæ for hypertrophy of the tonsils, 195, 196
    for intussusception, 307
    for laxative, 157
    for mucous disease, 224, 225, 226, 227
    for oxyuris vermicularis, 321
    for simple pharyngitis, 185, 186
    for suppurative tonsillitis, 193
    for syphilitic hepatitis, 352
    for tabes mesenterica, 336
    for tapeworm, 324, 325, 326
    for thrush, 147
    for tubercular peritonitis, 377
    for typhlitis, 293, 294
    for ulcerative stomatitis, 134
Furrows, facial, 22

G.

Gangrenous stomatitis, 136
    etiology of, 136
    symptoms of, 136
    treatment of, 137
Gastric catarrh, acute, 199 (see Acute Gas. Cat.)
    chronic (see Chron. Gas. Cat.)
Gastro-intestinal catarrh, 260
Gastro-malacia, 212
Gavage, 113
    de renfort, 114
Gelatine, 80
Genal furrows, 22
General development, 45
Gluten flour, 324
Glycerine suppositories in constipation, 276
Goats' milk, 73
Gradual weaning, 64
Graduated nursing bottle, 95
Growing pains, 123
Growth, 45
Gums, condition of during dentition, 149
"Gun-hammer" decubitus, 284

H.

Habitual constipation, 273
    etiology of, 273
    symptoms of, 274
    diagnosis of, 274

Habitual constipation, prognosis of, 274
  treatment of, 274
  indigestion, 213
    etiology of, 213
    symptoms of, 215
    diagnosis of, 217
    prognosis of, 217
    treatment of, 217
Halitosis, 29
Hand-feeding, success in, 96
  to insure success in, 72
Hands, movement of, 24
Hard palate, 58
Hatching cradle, 109
Headache during second dentition, 179
Head, shape of, 23
Heavy breath, 30
Hebra's diachylon ointment, 159
Hemorrhage, renal, 35
Hepatitis suppurative, 357
  syphilitic, 351
Herpes of the lips during second dentition, 178
Hip-joint disease, 46
Hob-nailed liver, 352
Horlick's food, 81
Hot bath, 103
Hydrocephalus, spurious, 206
Human milk, substitute for, 73.
Humanized milk, analysis of, 88
Hunger, 27
Hydatid disease of the liver, 363.
Hydrencephalic cry, 27
Hydrocephalus, 46
Hydronephrosis, 36
Hypertrophic cirrhosis, 353,
Hypertrophy of the tonsils, 194
  etiology of, 194
  symptoms of, 194
  treatment of, 195
Hypostatic pneumonia in chronic entero-colitis, 238

I.

Icterus, 337
  neonatorum, 338
  in older children, 341
  treatment of, 342

Idiopathic form of acute gastric catarrh, 199
Ileo-cæcal intussusception, 305
Immature infants, management of, 109
Incontinence, 36
Incubator, 109.
Incubators, description of, 111
Indican, 34, 37
Indigestion, 200
Infants' food, type of, 72
  foods, 77
Inflammation of the colon and rectum (see Dysentery), 264
Injections, medicated, 320
Inspection of chest, 51
  of child, 20
Inspiratory respiration, 41
Insufflation of air in intussusception, 309
Intertrigo in simple atrophy, 283
Intestinal concretions, 288
  worms, 311
Intestines, nerve supply of, 166
Intussusception, 296
  varieties of, 296
  morbid anatomy of, 297
  without symptoms, 297
  with symptoms, 297
  results of, 298.
  strangulation in, 298
  etiology of, 300
  symptoms of, 301
  diagnosis of, 305
  prognosis of, 306
  treatment of, 306
  reduction of, 309
Invagination, 306
Investigation of disease, 17
Inward spasms, 283

J.

Jadelot's lines, 22
Jaundice, 337
  etiology of, 338
  grade of severity, 338
  diagnosis of, 338
  due to congenital malformation of the bile ducts, 339
  treatment of, 339
Junket, 65

K.

Kidney, sarcoma of, 36
Kidneys, amyloid degeneration of, 38
   lesions of, 29

L.

Labial furrows, 22
Lactation, 66
Lactometer, 74
Lancing the gums, 163
Laparotomy in intussusception, 309
Laryngeal stenosis, 41
Larynx, nerve supply of, 168
Lavage, 114
Laxative confection, 279
Leeds' analysis of breast milk, 73
Leucorrhœa, 318
Lids, incomplete closure of, 21
Lime, saccharated solution of, 80
   water, 80
Lines of Jadelot, 22
Lips, herpes of, 178
   puffing of, 51
Lithæmia, 36
Lithuria, 35
Liver, abscess of, 357
   affections of, 337
   amyloid, 347
   cancer of, 363
   cirrhosis of, 352
   congestion of, 343
   fatty degeneration of, 347
   hydatid disease of, 363
   fatty infiltration of, 346
   syphilitic inflammation of, 351
   tuberculosis of, 363
Lividity of eyelids, 25
Local treatment for simple pharyngitis, 186
Loss of taste during second dentition, 172
Lungs, nerve supply of, 167

M.

Malarial fever, 37
Mammary abscess, 67
Management of weak and immature infant, 109
Marasmus, 47, 305
Massage, 116
   à frictions, 117
   effects of, 117
   in chorea, 122
   in chronic gastro-intestinal catarrh, 119
   in colic, 120
   in constipation, 120
   in general debility and anæmia, 121
   in infantile paralysis, 121
   in pleurisy, 123
   in pseudo-hypertrophic paralysis, 123
Masturbation, 35
Maxillary bones, necrosis of, 31
Meckel's ganglion, 166
Meigs' food, 85
Mellin's food, 81
Membranous croup, urine in, 37
Menstruation in nursing woman, 68
Mercurial bath, 104
Method of gavage, 113
   of giving suck, 62
Microscopic examination in thrush, 143
Micturition, painful, 33
Milk, asses', 73
   boiled, 85
   care of, 98
   condensed, 75
   cows', analysis of, 74
   goats', 73
   mixture for chronic gastric catarrh, 208
   mode of drinking, 26
   peptonized, 86
   poisoning, 100
   scanty secretion of, 67
   secretion of, 61
   sterilized, 88
   teeth, 57
   teeth, the eruption of, 148
   transportation of, 98
Morbid anatomy of amyloid liver, 347
   of cholera infantum, 258
   of chronic entero colitis, 235
   of chronic gastric catarrh, 203
   of cirrhosis of the liver, 352

INDEX. 391

Morbid anatomy of congestion of the liver, 343
of dysentery, 264
of entero-colitis, 248
of fatty degeneration of the liver, 347
of fatty infiltration of the liver, 346
of intussusception, 297
of peritonitis, 364
of simple atrophy, 279
of suppurative tonsillitis, 190
of syphilitic hepatitis, 351
of tabes mesenterica, 329
of thrush, 142
of tubercular peritonitis, 371
of tubercular ulceration of the intestines, 268
of typhlitis, 287
Morbus cœruleus, 25
Mortality from laparotomy in intussusception, 310
Motor paralysis, 25
Mouth and fauces, examination of, 57
  inspection of, during second dentition, 177
  soreness of, 26
Mucous disease, 217
  etiology of, 217
  symptoms of, 218
  diagnosis of, 220
  prognosis of, 220
  diet for, 221
  treatment of, 220
Mustard bath, 104

N.

Nails, deformity of, 26
Naphthalin in entero-colitis, 257
Nasal catarrh during second dentition, 178
  treatment of, 179
Nausea, 39
Necrosis, 135
Nematodes, 311
Nephritis, 43
Nervous disorders in dentition, 179
Night-dress, 105
Nipple, fissures of, 67
Nitro-muriatic acid bath, 104

Noma (see Gangrenous stomatitis), 139
  pathology and morbid anatomy of, 138
  diagnosis of, 139
  prognosis of, 140
  treatment of, 140
Normal capacity of infant's stomach, 78
Nostrils, sharpness of, 22
Nursing-bottle, 95
  mother's diet, 67
  regularity in, 62

O.

Oculo-zygomatic furrows, 22
Œdema, 112
Oïdium albicans, 142
Oil inunction for constipation, 275
Oral pain in second dentition, 170
Otitis, 151, 178
Oxaluria, 35
Oxyuris vermicularis, 312
  egg of, 312
  symptoms of, 318
  treatment of, 320
Ozæna, 179

P.

Painful micturition, 33
Palpation of the chest, 54
Pancreatin, 86–87, 88, 92
Papillæ, 58
Paracentesis in ascites, 382
Paralysis, dental, 162
  during dentition, 181
Parasitic stomatitis (see Thrush), 141
Parenchymatous nephritis, 43
Pathology and morbid anatomy of noma, 138
Peptogenic milk, powder, 88
Peptonization, partial, 87
Peptonized milk, 86
Percussion of the chest, 55
Perforation of the cæcum, 291
Peritoneum, affections of, 364
Peritonitis, 364
  morbid anatomy of, 364
  etiology of, 365
  symptoms of, 365

Peritonitis, diagnosis of, 368
  prognosis of, 370
  treatment of, 370
  tubercular, 371
Parasites, intestinal, 33
Permanent teeth, 58
  eruption of, 164
  order of eruption, 169
Perityphlitis (see Typhlitis, p. ri.), 287
Pertussis, 28
Petrissage, 116
Pharyngitis, simple, 183
Photophobia, 24
Phthisis, 54
Physical examination, 39
Platt's chloride, 141
Pneumonia, hypostatic, 238
Premature weaning, 66
Process for peptonizing milk, 86
Prognosis of acute gastric catarrh, 201
  of acute intestinal catarrh, 230
  of amyloid liver, 349
  of ascites, 380
  of cholera infantum, 262
  of chronic entero-colitis, 240
  of chronic gastric catarrh, 206
  of cirrhosis of the liver, 355
  of congestion of the liver, 344
  of dysentery, 265
  of entero-colitis, 253
  of fatty infiltration of the liver, 346
  of follicular tonsillitis, 188
  of habitual constipation, 274
  of habitual indigestion, 217
  of intussusception, 306
  of mucous disease, 220
  of noma, 140
  of peritonitis, 370
  of simple atrophy, 285
  of syphilitic hepatitis, 351
  of tabes mesenterica, 333
  of thrush, 145
  of tubercular peritonitis, 376
  of typhlitis, 292
  of ulcerative stomatitis, 134
  of worms, 319
Pseudo hypertrophic paralysis, 123
Puerile breathing, 53
Puffiness of eyelids, 22
Pulse, variations in, 42
Purpura hemorrhagica, 36

Q.

Questioning the attendants, 18
Quinsy, 190

R.

Raw beef juice, 92
Reaction of the urine, 33
Red gum, 154
Reflex spasm during dentition, 180
Regimen in acute intestinal catarrh, 232
  in amyloid liver, 350
  in cholera infantum, 263
  in chronic entero-colitis, 241
  in chronic gastric catarrh, 209
  in dysentery, 266
  in entero-colitis, 254
  in mucous disease, 222
  in simple atrophy, 286
Renal calculus, 36
  hemorrhage, 35
Resorcin in entero-colitis, 257
Respiration, 39
  character of, 40
  expiratory, 41
  inspiratory, 41
Retention of urine, 37
Retro-pharyngeal abscess, 197
  symptoms of, 197
  treatment of, 198
Rheumatism, 47
Rice pudding, 100
Rickets, 59, 69, 119
Rubber shoes, 104
Rules for feeding, 79

S.

Saccharated solution of lime, 80, 294
Salicylate of sodium in entero-colitis, 257
Salt-water bath, 104
Sarcoma of kidney, 36
Scarlet fever, 59, 128
Sclerema, 112
Secondary thrush, 144
Second dentition, disorders of, 170
  as a cause of mucous disease, 177
Secretion of milk, 61

Serous effusion in chronic entero-
  colitis, 238
Shoes, 106
Simple atrophy, 279
  morbid anatomy of, 279
  etiology of, 279
  manner of preparing food in, 281
  symptoms of, 282
  diagnosis of, 284
  prognosis of, 285
  treatment of, 285
  regimen in, 286
  diarrhœa of dentition, 152
  pharyngitis, 183
  anatomical lesion of, 183
  etiology of, 183
  symptoms of, 183
  diagnosis of, 184
  treatment of, 185
Skin, discoloration of, in jaundice, 338
  the, 25
  conditions of the, 46
Sleep, 106
Sleeping, different characters of, 24
Sleeping room, 107
Soda bath, 104
Softening of the stomach, 212
Sound cows' milk, 98
Sour breath, 30
Spasms during dentition, 161
  inward, 283
Spec. grav. of breast milk, 72
  cows' milk, 74
Spinal irritability, 123
Stationary washstand, 107
Statistics from the Maternité, in Paris, 112
Stercoraceous breath, 31
  vomiting, 303
Sterilized milk, 88
  rules to be observed in its use, 91
  uses of, 91
Sterilizer, the author's, 89
Stomach, measurements of infants, 78
  nerve supply of, 165
  softening of, 212
  ulcer of, 211
Stomatitis, aphthous, 126
  catarrhal, 124
  gangrenous, 136

Stomatitis, parasitic (see Thrush), 141
  ulcerative, 131
Stools, characters of, 32
Strippings, 85, 208
Strophulus during dentition, 154
Strumous diathesis, 23
Submaxillary gland, enlargement of,
  during second dentition, 177
Sudden weaning, 66
Summer diarrhœa, 248
Sunstroke, 261
Superficial catarrh of the tonsils, 186
Suppurative hepatitis, 357
  report of case, 357
  symptoms of, 357
  diagnosis of, 362
  treatment of, 362
  tonsillitis, 190
  morbid anatomy of, 190
  symptoms of, 191
  diagnosis of, 192
  treatment of, 192
Symptoms of acute gastric catarrh, 200
  of acute intestinal catarrh, 229
  of amyloid liver, 348
  of aphthous stomatitis, 127
  of ascaris lumbricoides, 319
  of ascites, 378
  of cholera infantum, 259
  of chronic entero-colitis, 237
  of chronic gastric catarrh, 204
  of cirrhosis of the liver, 354
  of colic, 270
  of congestion of the liver, 343
  of dysentery, 264
  of entero-colitis, 250
  of fatty infiltration of the liver, 346
  of follicular tonsillitis, 187
  of mucous disease, 218
  of gangrenous stomatitis, 136
  of habitual constipation, 274
  of habitual indigestion, 215
  of hypertrophy of the tonsils, 194
  of intussusception, 301
  of oxyuris vermicularis, 318
  of peritonitis, 365
  of retro-pharyngeal abscess, 197
  of septic peritonitis, 368
  of simple atrophy, 282

Symptoms of simple pharyngitis, 183
 of suppurative hepatitis, 357
 of suppurative tonsillitis, 191
 of tabes mesenterica, 330
 of tænia, 319
 of thrush, 143
 of tubercular peritonitis, 372
 of tubercular ulceration of the intestines, 269
 of typhlitis, 289
 of ulcerative stomatitis, 132
 of ulcer of the stomach, 211
 of syphilitic hepatitis, 351
 of worms, 317
Syphilitic hepatitis, 351
 morbid anatomy of, 351
 symptoms of, 351
 diagnosis of, 351
 prognosis of, 351
 treatment of, 352

T.

Tabes mesenterica, 329
 morbid anatomy of, 329
 etiology of, 330
 symptoms of, 330
 diagnosis of, 332
 prognosis of, 333
 treatment of, 334
Tæniæ, 314
 saginata, 314
 egg of, 315
 solium, 316
 symptoms of, 319
 treatment of, 323
Tanret's Pelletierine for tapeworm, 326
Tapôtement, 117
Tarnier's hatching cradle, 110
Taste, loss of, 172
Taxis in intussusception, 309
Tears, formation of, 28
 suppression of, 28
Teéth, children born with, 149
 — eruption of the temporary, 148
 milk, 57
 permanent, 58
 premature appearance of, 149
Teething cough, 178
Temperature, 43
 of room, 108
 variations in, 44

Thermometer, clinical, 43
Throat affections during second dentition, 172
Thrombosis of the sinuses of the brain during chronic entero-colitis, 238
Thrush, 141
 morbid appearances of, 142
 etiology of, 142
 symptoms of, 143
 secondary, 144
 diagnosis of, 145
 prognosis of, 145
 treatment of, 146
Tongue, 58
 in disease, 59
Tonsillitis, follicular, 187
 suppurative, 190
Tonsils, excision of, 196
 hypertrophy of, 194
Tooth rash, 154
Treatment of acute gastric catarrh, 201
 of acute intestinal catarrh, 231
 of amyloid liver, 349
 of aphthous stomatitis, 128
 of ascaris lumbricoides, 321
 of ascites, 380
 of catarrhal stomatitis, 125
 of cholera infantum, 262
 of chronic entero-colitis, 241
 of chronic gastric catarrh, 207
 of cirrhosis of the liver, 355
 of colic, 271
 of congestion of the liver, 345
 of convulsions during teething, 162
 during second dentition, 175
 of dysentery, 265
 of eczema, 157
 of entero colitis, 253
 of fatty infiltration of the liver, 347
 of fissure of nipple, 68
 of follicular tonsillitis, 189
 of habitual constipation, 274
 of headache during dentition, 180
 of hypertrophy of the tonsils, 195
 of intussusception, 306
 of icterus in older children, 342
 of jaundice, 339
 of mucous disease, 220
 of nasal catarrh, 179
 of noma, 140

Treatment of oxyuris vermicularis, 320
of peritonitis, 370
of retro-pharyngeal abscess, 198
of simple atrophy, 285
of simple pharyngitis, 185
of superficial ulcers of the tongue, 173
of suppurative hepatitis, 362
tonsillitis, 192
of syphilitic hepatitis, 352
of tabes mesenterica, 334
of tænia, 323
of thrush, 146
of tubercular peritonitis, 376
of tubercular ulceration of the intestines, 269
of typhlitis, 292
of ulcerative stomatitis, 134
of ulcer of the stomach, 212
of worms, 320
Trichocephalus dispar, 314
egg of, 314
True intussusception, 297
Tubercular meningitis, 284
peritonitis, 371
morbid anatomy of, 371
etiology of, 372
symptoms of, 372
diagnosis of, 375
prognosis of, 376
treatment of, 376
ulceration of the intestines, 268
ulceration of the intestines, morbid anatomy of, 268
ulceration of the intestines, etiology of, 268
ulceration of the intestines, symptoms of, 269
ulceration of the intestines, treatment of, 269
Tuberculosis of the liver, 363
Tuberculous tendency, signs of, 22
Tumor, fæcal, 31
Typhlitis, 287
morbid anatomy of, 287
etiology of, 289
stercoralis, 289
symptoms of, 289
diagnosis of, 291
prognosis of, 292
treatment of, 292
Tyrotoxicon, 99

U.

Ulceration of the appendix, 291
of the lungs, 31
Ulcerative stomatitis, 131
anatomical lesions of, 131
etiology of, 131
symptoms of, 132
diagnosis of, 134
prognosis of, 134
treatment of, 134
Ulcer of the stomach, 211
symptoms of, 211
treatment of, 212
Ulcers of the tongue during second dentition, 171
Uræmic poisoning, 31
Uric acid, 33, 37
Urine the, 33
spec. grav. of, 33
characters of, 33
daily amount voided, 34
abnormal ingredients of, 35
of different diseases, 36
Urinometer, 74
Urticaria during dentition, 153
Uvula the, 58

V.

Varicella, 59
Variola, 128
Veal broth, 92
with barley water, 208
Ventilation, 107
Vertigo, 319
Vesicles, herpetic, 26
Vocal fremitus, 54
Vomiting, 38
chronic, 203
during dentition, 153
stercoraceous, 303

W.

Walking, delay in, 46
Weak and immature infants, 109
Weaning, 64
sudden, 66
premature, indications for, 66
Wet-nurse, feeding by, 69
proper woman for a, 70

Wet-nurse, examination of, 70
   diet of, 71
Whey, 81, 207
Whip worms, 314
White gum 154
Whooping cough, 128, 177, 217
Worms, 311
   mode of entering the body, 312
   symptoms of, 317

Worms, diagnosis of, 319
   prognosis of, 319
   treatment of, 320

Y.

Yawning, 41
Yellow discoloration of the skin in jaundice, 338

CATALOGUE No. 7.   DECEMBER, 1890.

A CATALOGUE
OF
BOOKS FOR STUDENTS.
INCLUDING THE
? QUIZ-COMPENDS ?

CONTENTS.

	PAGE		PAGE
New Series of Manuals,	2,3,4,5	Obstetrics,	10
Anatomy,	6	Pathology, Histology,	11
Biology,	11	Pharmacy,	12
Chemistry,	6	Physiology,	11
Children's Diseases,	7	Practice of Medicine,	12
Dentistry,	8	Prescription Books,	12
Dictionaries,	8	? Quiz-Compends ?	14, 15
Eye Diseases,	8	Skin Diseases,	12
Electricity,	9	Surgery,	13
Gynæcology,	10	Therapeutics,	9
Hygiene,	9	Urine and Urinary Organs,	13
Materia Medica,	9	Venereal Diseases,	13
Medical Jurisprudence,	10		

PUBLISHED BY

P. BLAKISTON, SON & CO.,
Medical Booksellers, Importers and Publishers.

LARGE STOCK OF ALL STUDENTS' BOOKS, AT
THE LOWEST PRICES.

1012 Walnut Street, Philadelphia.

∗ For Sale by all Booksellers, or any book will be sent by mail, postpaid, upon receipt of price. Catalogues of books on all branches of Medicine, Dentistry, Pharmacy, etc., supplied upon application.

☞ Gould's New Medical Dictionary Just Ready. *See page 16.*

"*An excellent Series of Manuals.*"—*Archives of Gynæcology.*

A NEW SERIES OF
STUDENTS' MANUALS

On the various Branches of Medicine and Surgery.

Can be used by Students of any College.

Price of each, Handsome Cloth, $3.00. Full Leather, $3.50.

The object of this series is to furnish good manuals for the medical student, that will strike the medium between the compend on one hand and the prolix text-book on the other—to contain all that is necessary for the student, without embarrassing him with a flood of theory and involved statements. They have been prepared by well-known men, who have had large experience as teachers and writers, and who are, therefore, well informed as to the needs of the student.

Their mechanical execution is of the best—good type and paper, handsomely illustrated whenever illustrations are of use, and strongly bound in uniform style.

Each book is sold separately at a remarkably low price, and the immediate success of several of the volumes shows that the series has met with popular favor.

No. 1. SURGERY. 236 Illustrations.

A Manual of the Practice of Surgery. By WM. J. WALSHAM, M.D., Asst. Surg. to, and Demonstrator of Surg. in, St. Bartholomew's Hospital, London, etc. 228 Illustrations.

Presents the introductory facts in Surgery in clear, precise language, and contains all the latest advances in Pathology, Antiseptics, etc.

"It aims to occupy a position midway between the pretentious manual and the cumbersome System of Surgery, and its general character may be summed up in one word—practical."—*The Medical Bulletin.*

"Walsham, besides being an excellent surgeon, is a teacher in its best sense, and having had very great experience in the preparation of candidates for examination, and their subsequent professional career, may be relied upon to have carried out his work successfully. Without following out in detail his arrangement, which is excellent, we can at once say that his book is an embodiment of modern ideas neatly strung together, with an amount of careful organization well suited to the candidate, and, indeed, to the practitioner."—*British Medical Journal.*

Price of each Book, Cloth, $3.00; Leather, $3.50.

No. 2. DISEASES OF WOMEN. 150 Illus.
NEW EDITION.

The Diseases of Women. Including Diseases of the Bladder and Urethra. By DR. F. WINCKEL, Professor of Gynæcology and Director of the Royal University Clinic for Women, in Munich. Second Edition. Revised and Edited by **Theophilus Parvin**, M.D., Professor of Obstetrics and Diseases of Women and Children in Jefferson Medical College. 150 Engravings, most of which are original.

"The book will be a valuable one to physicians, and a safe and satisfactory one to put into the hands of students. It is issued in a neat and attractive form, and at a very reasonable price."—*Boston Medical and Surgical Journal.*

No. 3. OBSTETRICS. 227 Illustrations.

A Manual of Midwifery. By ALFRED LEWIS GALABIN, M.A., M.D., Obstetric Physician and Lecturer on Midwifery and the Diseases of Women at Guy's Hospital, London; Examiner in Midwifery to the Conjoint Examining Board of England, etc. With 227 Illus.

"This manual is one we can strongly recommend to all who desire to study the science as well as the practice of midwifery. Students at the present time not only are expected to know the principles of diagnosis, and the treatment of the various emergencies and complications that occur in the practice of midwifery, but find that the tendency is for examiners to ask more questions relating to the science of the subject than was the custom a few years ago. * * * The general standard of the manual is high; and wherever the science and practice of midwifery are well taught it will be regarded as one of the most important text-books on the subject."—*London Practitioner.*

No. 4. PHYSIOLOGY. Fourth Edition.
321 ILLUSTRATIONS AND A GLOSSARY.

A Manual of Physiology. By GERALD F. YEO, M.D., F.R.C.S., Professor of Physiology in King's College, London. 321 Illustrations and a Glossary of Terms. Fourth American from second English Edition, revised and improved. 758 pages.

This volume was specially prepared to furnish students with a new text-book of Physiology, elementary so far as to avoid theories which have not borne the test of time and such details of methods as are unnecessary for students in our medical colleges.

"The brief examination I have given it was so favorable that I placed it in the list of text-books recommended in the circular of the University Medical College."—*Prof. Lewis A. Stimson*, M.D., *37 East 33d Street, New York.*

Price of each Book, Cloth, $3.00; Leather, $3.50.

No. 5. ORGANIC CHEMISTRY.

Or the Chemistry of the Carbon Compounds. By Prof. VICTOR VON RICHTER, University of Breslau. Authorized translation, from the Fourth German Edition. By EDGAR F. SMITH, M.A., PH.D.; Prof. of Chemistry in University of Pennsylvania; Member of the Chem. Socs. of Berlin and Paris.

"I must say that this standard treatise is here presented in a remarkably compendious shape."—*J. W. Holland*, M.D., *Professor of Chemistry, Jefferson Medical College, Philadelphia.*

" This work brings the whole matter, in simple, plain language, to the student in a clear, comprehensive manner. The whole method of the work is one that is more readily grasped than that of older and more famed text-books, and we look forward to the time when, to a great extent, this work will supersede others, on the score of its better adaptation to the wants of both teacher and student."—*Pharmaceutical Record.*

"Prof. von Richter's work has the merit of being singularly clear, well arranged, and for its bulk, comprehensive. Hence, it will, as we find it intimated in the preface, prove useful not merely as a text-book, but as a manual of reference."—*The Chemical News, London.*

No. 6. DISEASES OF CHILDREN.

SECOND EDITION.

A Manual. By J. F. GOODHART, M.D., Phys. to the Evelina Hospital for Children; Asst. Phys. to Guy's Hospital, London. Second American Edition. Edited and Rearranged by LOUIS STARR, M.D., Clinical Prof. of Dis. of Children in the Hospital of the Univ. of Pennsylvania, and Physician to the Children's Hospital, Phila. Containing many new Prescriptions, a list of over 50 Formulæ, conforming to the U. S. Pharmacopœia, and Directions for making Artificial Human Milk, for the Artificial Digestion of Milk, etc. Illus.

" The author has avoided the not uncommon error of writing a book on general medicine and labeling it ' Diseases of Children,' but has steadily kept in view the diseases which seemed to be incidental to childhood, or such points in disease as appear to be so peculiar to or pronounced in children as to justify insistence upon them. * * * A safe and reliable guide, and in many ways admirably adapted to the wants of the student and practitioner."—*American Journal of Medical Science.*

Price of each Book, Cloth, $3.00 ; Leather, $3.50.

THE NEW SERIES OF MANUALS.

No. 6. Goodhart and Starr :—Continued.

"Thoroughly individual, original and earnest, the work evidently of a close observer and an independent thinker, this book, though small, as a handbook or compendium is by no means made up of bare outlines or standard facts."—*The Therapeutic Gazette.*

"As it is said of some men, so it might be said of some books, that they are 'born to greatness.' This new volume has, we believe, a mission, particularly in the hands of the younger members of the profession. In these days of prolixity in medical literature, it is refreshing to meet with an author who knows both what to say and when he has said it. The work of Dr. Goodhart (admirably conformed, by Dr. Starr, to meet American requirements) is the nearest approach to clinical teaching without the actual presence of clinical material that we have yet seen."—*New York Medical Record.*

No. 7. PRACTICAL THERAPEUTICS.
FOURTH EDITION, WITH AN INDEX OF DISEASES.

Practical Therapeutics, considered with reference to Articles of the Materia Medica. Containing, also, an Index of Diseases, with a list of the Medicines applicable as Remedies. By EDWARD JOHN WARING, M.D., F.R.C.P. Fourth Edition. Rewritten and Revised by DUDLEY W. BUXTON, M.D., Asst. to the Prof. of Medicine at University College Hospital.

"We wish a copy could be put in the hands of every Student or Practitioner in the country. In our estimation, it is the best book of the kind ever written."—*N. Y. Medical Journal.*

No. 8. MEDICAL JURISPRUDENCE AND TOXICOLOGY.
NEW, REVISED AND ENLARGED EDITION.

By John J. Reese, M.D., Professor of Medical Jurisprudence and Toxicology in the University of Pennsylvania; President of the Medical Jurisprudence Society of Phila.; 2d Edition, Revised and Enlarged.

"This admirable text-book."—*Amer. Jour. of Med. Sciences.*
"We lay this volume aside, after a careful perusal of its pages, with the profound impression that it should be in the hands of every doctor and lawyer. It fully meets the wants of all students. . . . . He has succeeded in admirably condensing into a handy volume all the essential points."—*Cincinnati Lancet and Clinic.*

Price of each Book, Cloth, $3,00; Leather, $3.50.

ANATOMY.

Macalister's Human Anatomy. 816 Illustrations. A new Text-book for Students and Practitioners, Systematic and Topographical, including the Embryology, Histology and Morphology of Man. With special reference to the requirements of Practical Surgery and Medicine. With 816 Illustrations, 400 of which are original. Octavo. Cloth, 7.50; Leather, 8.50

Ballou's Veterinary Anatomy and Physiology. Illustrated. By Wm. R. Ballou, M.D., Professor of Equine Anatomy at New York College of Veterinary Surgeons. 29 graphic Illustrations. 12mo. Cloth, 1.00; Interleaved for notes, 1.25

Holden's Anatomy. A manual of Dissection of the Human Body. Fifth Edition. Enlarged, with Marginal References and over 200 Illustrations. Octavo. Cloth, 5.00; Leather, 6.00
Bound in Oilcloth, for the Dissecting Room, $4.50.
"No student of Anatomy can take up this book without being pleased and instructed. Its Diagrams are original, striking and suggestive, giving more at a glance than pages of text description. * * * The text matches the illustrations in directness of practical application and clearness of detail."—*New York Medical Record.*

Holden's Human Osteology. Comprising a Description of the Bones, with Colored Delineations of the Attachments of the Muscles. The General and Microscopical Structure of Bone and its Development. With Lithographic Plates and Numerous Illustrations. Seventh Edition. 8vo. Cloth, 6.00

Holden's Landmarks, Medical and Surgical. 4th ed. Clo., 1.25

Heath's Practical Anatomy. Sixth London Edition. 24 Colored Plates, and nearly 300 other Illustrations. Cloth, 5.00

Potter's Compend of Anatomy. Fifth Edition. Enlarged. 16 Lithographic Plates. 117 Illustrations.
Cloth, 1.00; Interleaved for Notes, 1.25

CHEMISTRY.

Bartley's Medical Chemistry. Second Edition. A text-book prepared specially for Medical, Pharmaceutical and Dental Students. With 50 Illustrations, Plate of Absorption Spectra and Glossary of Chemical Terms. Revised and Enlarged. Cloth, 2.50

Trimble. Practical and Analytical Chemistry. A Course in Chemical Analysis, by Henry Trimble, Prof. of Analytical Chemistry in the Phila. College of Pharmacy. Illustrated. Third Edition. 8vo. Cloth, 1.50

☞ *See pages 2 to 5 for list of Students' Manuals.*

STUDENTS' TEXT-BOOKS AND MANUALS. 7

Chemistry:—Continued.

Bloxam's Chemistry, Inorganic and Organic, with Experiments. Seventh Edition. Enlarged and Rewritten. 330 Illustrations.
Cloth, 4.50; Leather, 5.50

Richter's Inorganic Chemistry. A text-book for Students. Third American, from Fifth German Edition. Translated by Prof. Edgar F. Smith, PH.D. 89 Wood Engravings and Colored Plate of Spectra. Cloth, 2.00

Richter's Organic Chemistry, or Chemistry of the Carbon Compounds. Illustrated. Cloth, 3.00; Leather, 3.50

Symonds. Manual of Chemistry, for the special use of Medical Students. By BRANDRETH SYMONDS, A.M., M.D., Asst. Physician Roosevelt Hospital, Out-Patient Department; Attending Physician Northwestern Dispensary, New York. 12mo.
Cloth, 2.00; Interleaved for Notes, 2.40

Tidy. Modern Chemistry. 2d Ed. Cloth, 5.50

Leffmann's Compend of Chemistry. Inorganic and Organic. Including Urinary Analysis. Third Edition. Revised.
Cloth, 1.00; Interleaved for Notes, 1.25

Leffmann and Beam. Progressive Exercises in Practical Chemistry. 12mo. Illustrated. Cloth, 1.00

Muter. Practical and Analytical Chemistry. Second Edition. Revised and Illustrated. Cloth, 2.00

Holland. The Urine, Common Poisons, and Milk Analysis, Chemical and Microscopical. For Laboratory Use. 3d Edition, Enlarged. Illustrated. Cloth, 1.00

Van Nüys. Urine Analysis. Illus. Cloth, 2.00

Wolff's Applied Medical Chemistry. By Lawrence Wolff, M.D., Dem. of Chemistry in Jefferson Medical College. Clo., 1.00

CHILDREN.

Goodhart and Starr. The Diseases of Children. Second Edition. By J. F. Goodhart, M.D., Physician to the Evelina Hospital for Children; Assistant Physician to Guy's Hospital, London. Revised and Edited by Louis Starr, M.D., Clinical Professor of Diseases of Children in the Hospital of the University of Pennsylvania; Physician to the Children's Hospital, Philadelphia. Containing many Prescriptions and Formulæ, conforming to the U. S. Pharmacopœia, Directions for making Artificial Human Milk, for the Artificial Digestion of Milk, etc. Illustrated. Cloth, 3.00; Leather, 3.50

Hatfield. Diseases of Children. By M. P. Hatfield, M.D., Professor of Diseases of Children, Chicago Medical College. Colored Plate. 12mo. Cloth, 1.00; Interleaved, 1.25

Day. On Children. A Practical and Systematic Treatise. Second Edition. 8vo. 752 pages. Cloth, 3.00; Leather, 4.00

☞ *See pages 14 and 15 for list of ? Quiz-Compends ?*

8 STUDENTS' TEXT-BOOKS AND MANUALS.

Children:—Continued.

Meigs and Pepper. The Diseases of Children. Seventh Edition. 8vo. Cloth, 5.00; Leather, 6.00

Starr. Diseases of the Digestive Organs in Infancy and Childhood. With chapters on the Investigation of Disease, and on the General Management of Children. By Louis Starr, M.D., Clinical Professor of Diseases of Children in the University of Pennsylvania. Illus. Second Edition. *In Press.*

DENTISTRY.

Fillebrown. Operative Dentistry. 330 Illus. Cloth, 2.50

Flagg's Plastics and Plastic Filling. 3d Ed. *Preparing.*

Gorgas. Dental Medicine. A Manual of Materia Medica and Therapeutics. Third Edition. Cloth, 3.50

Harris. Principles and Practice of Dentistry. Including Anatomy, Physiology, Pathology, Therapeutics, Dental Surgery and Mechanism. Twelfth Edition. Revised and enlarged by Professor Gorgas. 1028 Illustrations. Cloth, 7.00; Leather, 8.00

Richardson's Mechanical Dentistry. Fifth Edition. 569 Illustrations. 8vo. Cloth, 4.50; Leather, 5.50

Sewill. Dental Surgery. 200 Illustrations. 3d Ed. Clo., 3.00

Stocken's Dental Materia Medica. Third Edition. Cloth, 2.50

Taft's Operative Dentistry. Dental Students and Practitioners. Fourth Edition. 100 Illustrations. Cloth, 4.25; Leather, 5.00

Talbot. Irregularities of the Teeth, and their Treatment. Illustrated. 8vo. Second Edition. Cloth, 3.00

Tomes' Dental Anatomy. Third Ed. 191 Illus. Cloth, 4.00

Tomes' Dental Surgery. 3d Edition. Revised. 292 Illus. 772 Pages. Cloth, 5.00

Warren. Compend of Dental Pathology and Dental Medicine. Illustrated. Cloth, 1.00; Interleaved, 1.25

DICTIONARIES.

Gould's New Medical Dictionary. Containing the Definition and Pronunciation of all words in Medicine, with many useful Tables etc. ½ Dark Leather, 3.25; ½ Mor., Thumb Index 4.25

Cleaveland's Pronouncing Pocket Medical Lexicon. 31st Edition. Giving correct Pronunciation and Definition. Very small pocket size. Cloth, red edges .75; pocket-book style, 1.00

Longley's Pocket Dictionary. The Student's Medical Lexicon, giving Definition and Pronunciation of all Terms used in Medicine, with an Appendix giving Poisons and Their Antidotes, Abbreviations used in Prescriptions, Metric Scale of Doses, etc. 24mo. Cloth, 1.00; pocket-book style, 1.25

☞ *See pages 2 to 5 for list of Students' Manuals.*

EYE.

Arlt. Diseases of the Eye. Including those of the Conjunctiva, Cornea, Sclerotic, Iris and Ciliary Body. By Prof. Von Arlt. Translated by Dr. Lyman Ware. Illus. 8vo. Cloth, 2.50
Hartridge on Refraction. 4th Ed. Cloth, 2.00
Meyer. Diseases of the Eye. A complete Manual for Students and Physicians. 270 Illustrations and two Colored Plates. 8vo. Cloth, 4.50; Leather, 5.50
Fox and Gould. Compend of Diseases of the Eye and Refraction. 2d Ed. Enlarged. 71 Illus. 39 Formulæ.
Cloth, 1.00; Interleaved for Notes, 1.25

ELECTRICITY.

Mason's Compend of Medical and Surgical Electricity. With numerous Illustrations. 12mo. Cloth, 1.00

HYGIENE.

Parkes' (Ed. A.) Practical Hygiene. Seventh Edition, enlarged. Illustrated. 8vo. Cloth, 4.50
Parkes' (L. C.) Manual of Hygiene and Public Health. 12mo. Cloth, 2.50
Wilson's Handbook of Hygiene and Sanitary Science. Sixth Edition. Revised and Illustrated. Cloth, 2.75

MATERIA MEDICA AND THERAPEUTICS.

Potter's Compend of Materia Medica, Therapeutics and Prescription Writing. Fifth Edition, revised and improved.
Cloth, 1.00; Interleaved for Notes, 1.25
Biddle's Materia Medica. Eleventh Edition. By the late John B. Biddle, M.D., Professor of Materia Medica in Jefferson Medical College, Philadelphia. Revised, and rewritten, by Clement Biddle, M.D., Assist. Surgeon, U. S. N., assisted by Henry Morris, M.D. 8vo., illustrated. Cloth, 4.25; Leather, 5.00
Headland's Action of Medicines. 9th Ed. 8vo. Cloth, 3.00
Potter. Materia Medica, Pharmacy and Therapeutics. Including Action of Medicines, Special Therapeutics, Pharmacology, etc. Second Edition. Cloth, 4.00; Leather, 5.00
Starr, Walker and Powell. Synopsis of Physiological Action of Medicines, based upon Prof. H. C. Wood's "Materia Medica and Therapeutics." 3d Ed. Enlarged. Cloth, .75
Waring. Therapeutics. With an Index of Diseases and Remedies. 4th Edition. Revised. Cloth, 3.00; Leather, 3.50

☞ *See pages 14 and 15 for list of ? Quiz-Compends ?*

MEDICAL JURISPRUDENCE.

Reese. A Text-book of Medical Jurisprudence and Toxicology. By John J. Reese, M.D., Professor of Medical Jurisprudence and Toxicology in the Medical Department of the University of Pennsylvania; President of the Medical Jurisprudence Society of Philadelphia; Physician to St. Joseph's Hospital; Corresponding Member of The New York Medico-legal Society. 2d Edition. Cloth, 3.00; Leather, 3.50

Woodman and Tidy's Medical Jurisprudence and Toxicology. Chromo-Lithographic Plates and 116 Wood engravings.
Cloth, 7.50; Leather, 8.50

OBSTETRICS AND GYNÆCOLOGY.

Byford. Diseases of Women. The Practice of Medicine and Surgery, as applied to the Diseases and Accidents Incident to Women. By W. H. Byford, A.M., M.D., Professor of Gynæcology in Rush Medical College and of Obstetrics in the Woman's Medical College, etc., and Henry T. Byford, M.D., Surgeon to the Woman's Hospital of Chicago; Gynæcologist to St. Luke's Hospital, etc. Fourth Edition. Revised, Rewritten and Enlarged. With 306 Illustrations, over 100 of which are original. Octavo. 832 pages. Cloth, 5.00; Leather, 6.00

Cazeaux and Tarnier's Midwifery. With Appendix, by Mundé. The Theory and Practice of Obstetrics; including the Diseases of Pregnancy and Parturition, Obstetrical Operations, etc. By P. Cazeaux. Remodeled and rearranged, with revisions and additions, by S. Tarnier, M.D., Professor of Obstetrics and Diseases of Women and Children in the Faculty of Medicine of Paris. Eighth American, from the Eighth French and First Italian Edition. Edited by Robert J. Hess, M.D., Physician to the Northern Dispensary, Philadelphia, with an appendix by Paul F. Mundé, M.D., Professor of Gynæcology at the N. Y. Polyclinic. Illustrated by Chromo-Lithographs, Lithographs, and other Full-page Plates, seven of which are beautifully colored, and numerous Wood Engravings. *Students' Edition.* One Vol., 8vo. Cloth, 5.00; Leather, 6.00

Lewers' Diseases of Women. A Practical Text-Book. 139 Illustrations. Second Edition. Cloth, 2.50

Parvin's Winckel's Diseases of Women. Second Edition. Including a Section on Diseases of the Bladder and Urethra. 150 Illus. Revised. *See page 3.* Cloth, 3.00; Leather, 3.50

Morris. Compend of Gynæcology. Illustrated. Cloth, 1.00

Winckel's Obstetrics. A Text-book on Midwifery, including the Diseases of Childbed. By Dr. F. Winckel, Professor of Gynæcology, and Director of the Royal University Clinic for Women, in Munich. Authorized Translation, by J. Clifton Edgar, M.D., Lecturer on Obstetrics, University Medical College, New York, with nearly 200 handsome illustrations, the majority of which are original with this work. Octavo.
Cloth, 6.00; Leather, 7.00

Landis' Compend of Obstetrics. Illustrated. 4th edition, enlarged. Cloth, 1.00; Interleaved for Notes, 1.25

☞ *See pages 2 to 5 for list of New Manuals.*

STUDENTS' TEXT-BOOKS AND MANUALS. 11

Obstetrics and Gynæcology:—Continued.
Galabin's Midwifery. By A. Lewis Galabin, M.D., F.R.C.P. 227 Illustrations. *See page 3.* Cloth, 3.00; Leather, 3.50
Glisan's Modern Midwifery. 2d Edition. Cloth, 3.00
Rigby's Obstetric Memoranda. 4th Edition. Cloth, .50
Meadows' Manual of Midwifery. Including the Signs and Symptoms of Pregnancy, Obstetric Operations, Diseases of the Puerperal State, etc. 145 Illustrations. 494 pages. Cloth, 2.00
Swayne's Obstetric Aphorisms. For the use of Students commencing Midwifery Practice. 8th Ed. 12mo. Cloth, 1.25

PATHOLOGY. HISTOLOGY. BIOLOGY.

Bowlby. Surgical Pathology and Morbid Anatomy, for Students. 135 Illustrations. 12mo. Cloth, 2.00
Davis' Elementary Biology. Illustrated. Cloth, 4.00
Gilliam's Essentials of Pathology. A Handbook for Students. 47 Illustrations. 12mo. Cloth, 2.00
** The object of this book is to unfold to the beginner the fundamentals of pathology in a plain, practical way, and by bringing them within easy comprehension to increase his interest in the study of the subject.
Gibbes' Practical Histology and Pathology. Third Edition. Enlarged. 12mo. Cloth, 1.75
Virchow's Post-Mortem Examinations. 2d Ed. Cloth, 1.00

PHYSIOLOGY.

Yeo's Physiology. Fourth Edition. The most Popular Students' Book. By Gerald F. Yeo, M.D., F.R.C.S., Professor of Physiology in King's College, London. Small Octavo. 758 pages. 321 carefully printed Illustrations. With a Full Glossary and Index. *See Page 3.* Cloth, 3.00; Leather, 3.50
Brubaker's Compend of Physiology. Illustrated. Fifth Edition. Cloth, 1.00; Interleaved for Notes, 1.25
Stirling. Practical Physiology, including Chemical and Experimental Physiology. 142 Illustrations. Cloth, 2.25
Kirke's Physiology. New 12th Ed. Thoroughly Revised and Enlarged. 502 Illustrations. Cloth, 4.00; Leather, 5.00
Landois' Human Physiology. Including Histology and Microscopical Anatomy, and with special reference to Practical Medicine. Third Edition. Translated and Edited by Prof. Stirling. 692 Illustrations. Cloth, 6.50; Leather, 7.50
" With this Text-book at his command, no student could fail in his examination."—*Lancet.*
Sanderson's Physiological Laboratory. Being Practical Exercises for the Student. 350 Illustrations. 8vo. Cloth, 5.00
Tyson's Cell Doctrine. Its History and Present State. Illustrated. Second Edition. Cloth, 2.00

☞ *See pages 14 and 15 for list of ? Quiz-Compends !*

12   STUDENTS' TEXT-BOOKS AND MANUALS.

PRACTICE.

Taylor. Practice of Medicine. A Manual. By Frederick Taylor, M.D., Physician to, and Lecturer on Medicine at, Guy's Hospital, London; Physician to Evelina Hospital for Sick Children, and Examiner in Materia Medica and Pharmaceutical Chemistry, University of London.   Cloth, 4.00

Roberts' Practice. New Revised Edition. A Handbook of the Theory and Practice of Medicine. By Frederick T. Roberts, M.D.; M.R.C.P., Professor of Clinical Medicine and Therapeutics in University College Hospital, London. Seventh Edition. Octavo.   Cloth, 5.50; Sheep, 6.50

Hughes. Compend of the Practice of Medicine. 4th Edition. Two parts, each, Cloth, 1.00; Interleaved for Notes, 1.25

PART I.—Continued, Eruptive and Periodical Fevers, Diseases of the Stomach, Intestines, Peritoneum, Biliary Passages, Liver, Kidneys, etc., and General Diseases, etc.

PART II.—Diseases of the Respiratory System, Circulatory System and Nervous System; Diseases of the Blood, etc.

Physician's Edition. Fourth Edition. Including a Section on Skin Diseases. With Index. 1 vol. Full Morocco, Gilt, 2.50

From John A. Robinson, M.D., Assistant to Chair of Clinical Medicine, now Lecturer on Materia Medica, Rush Medical College, Chicago.
"Meets with my hearty approbation as a substitute for the ordinary note books almost universally used by medical students. It is concise, accurate, well arranged and lucid, . . . just the thing for students to use while studying physical diagnosis and the more practical departments of medicine."

PRESCRIPTION BOOKS.

Wythe's Dose and Symptom Book. Containing the Doses and Uses of all the principal Articles of the Materia Medica, etc. Seventeenth Edition. Completely Revised and Rewritten. *Just Ready.* 32mo.   Cloth, 1.00; Pocket-book style, 1.25

Pereira's Physician's Prescription Book. Containing Lists of Terms, Phrases, Contractions and Abbreviations used in Prescriptions Explanatory Notes, Grammatical Construction of Prescriptions, etc., etc. By Professor Jonathan Pereira, M.D. Sixteenth Edition. 32mo.   Cloth, 1.00; Pocket-book style, 1.25

PHARMACY.

Stewart's Compend of Pharmacy. Based upon Remington's Text-Book of Pharmacy. Third Edition, Revised. With new Tables, Index, Etc.   Cloth, 1.00; Interleaved for Notes, 1.25

Robinson. Latin Grammar of Pharmacy and Medicine. By H. D. Robinson, PH.D., Professor of Latin Language and Literature, University of Kansas, Lawrence. With an Introduction by L. E. Sayre, PH.G., Professor of Pharmacy in, and Dean of, the Dept. of Pharmacy, University of Kansas. 12mo.   Cloth, 2.00

SKIN DISEASES.

Anderson, (McCall) Skin Diseases. A complete Text-Book, with Colored Plates and numerous Wood Engravings. 8vo.   Cloth, 4.50; Leather, 5.50

☞ *See pages 2 to 5 for list of New Manuals.*

Skin Diseases :—Continued.

Van Harlingen on Skin Diseases. A Handbook of the Diseases of the Skin, their Diagnosis and Treatment (arranged alphabetically). By Arthur Van Harlingen, M.D., Clinical Lecturer on Dermatology, Jefferson Medical College; Prof. of Diseases of the Skin in the Philadelphia Polyclinic. 2d Edition. Enlarged. With colored and other plates and illustrations. 12mo. Cloth, 2.50

Bulkley. The Skin in Health and Disease. By L. Duncan Bulkley, Physician to the N. Y. Hospital. Illus. Cloth, .50

SURGERY AND BANDAGING.

Jacobson. Operations in Surgery. A Systematic Handbook for Physicians, Students and Hospital Surgeons. By W. H. A. Jacobson, B A., Oxon. F.R.C.S. Eng.; Ass't Surgeon Guy's Hospital; Surgeon at Royal Hospital for Children and Women, etc. 199 Illustrations. 1006 pages. 8vo. Cloth. 5.00; Leather, 6.00

Heath's Minor Surgery, and Bandaging. Ninth Edition. 142 Illustrations. 60 Formulæ and Diet Lists. Cloth, 2.00

Horwitz's Compend of Surgery, Minor Surgery and Bandaging, Amputations, Fractures, Dislocations, Surgical Diseases, and the Latest Antiseptic Rules, etc., with Differential Diagnosis and Treatment. By ORVILLE HORWITZ, B.S., M.D., Demonstrator of Surgery, Jefferson Medical College. 4th edition. Enlarged and Rearranged. 136 Illustrations and 84 Formulæ. 12mo. Cloth, 1.00; Interleaved for the addition of Notes, 1.25

** The new Section on Bandaging and Surgical Dressings, consists of 32 Pages and 41 Illustrations. Every Bandage of any importance is figured. This, with the Section on Ligation of Arteries, forms an ample Text-book for the Surgical Laboratory.

Walsham. Manual of Practical Surgery. For Students and Physicians. By WM. J. WALSHAM, M.D., F.R.C.S., Asst. Surg. to, and Dem. of Practical Surg. in, St. Bartholomew's Hospital, Surgeon to Metropolitan Free Hospital, London. With 236 Engravings. *See Page 2.* Cloth, 3.00; Leather, 3.50

URINE, URINARY ORGANS, ETC.

Holland. The Urine, and Common Poisons and The Milk. Chemical and Microscopical, for Laboratory Use. Illustrated. Third Edition. 12mo. Interleaved. Cloth, 1.00

Ralfe. Kidney Diseases and Urinary Derangements. 42 Illustrations. 12mo. 572 pages. Cloth, 2.75

Marshall and Smith. On the Urine. The Chemical Analysis of the Urine. By John Marshall, M.D., Chemical Laboratory, Univ. of Penna; and Prof. E. F. Smith, PH.D. Col. Plates. Cloth, 1.00

Thompson. Diseases of the Urinary Organs. Eighth London Edition. Illustrated. Cloth, 3.50

Tyson. On the Urine. A Practical Guide to the Examination of Urine. With Colored Plates and Wood Engravings. 6th Ed. Enlarged. 12mo. Cloth, 1.50

Van Nüys, Urine Analysis. Illus. Cloth, 2.00

VENEREAL DISEASES.

Hill and Cooper. Student's Manual of Venereal Diseases, with Formulæ. Fourth Edition. 12mo. Cloth, 1.00

☞ *See pages 14 and 15 for list of ? Quiz-Compends!*

NEW AND REVISED EDITIONS.

? QUIZ-COMPENDS ?

The Best Compends for Students' Use in the Quiz Class, and when Preparing for Examinations.

Compiled in accordance with the latest teachings of prominent lecturers and the most popular Text-books.

They form a most complete, practical and exhaustive set of manuals, containing information nowhere else collected in such a condensed, practical shape. Thoroughly up to the times in every respect, containing many new prescriptions and formulæ, and over two hundred and fifty illustrations, many of which have been drawn and engraved specially for this series. The authors have had large experience as quiz-masters and attachés of colleges, with exceptional opportunities for noting the most recent advances and methods.

Cloth, each $1.00. Interleaved for Notes, $1.25.

No. 1. HUMAN ANATOMY, "Based upon Gray." Fifth Enlarged Edition, including Visceral Anatomy, formerly published separately. 16 Lithograph Plates, New Tables and 117 other Illustrations. By SAMUEL O. L. POTTER, M.A., M.D., late A. A. Surgeon U. S. Army. Professor of Practice, Cooper Medical College, San Francisco.

Nos. 2 and 3. PRACTICE OF MEDICINE. Fourth Edition. By DANIEL E. HUGHES, M.D., Demonstrator of Clinical Medicine in Jefferson Medical College, Philadelphia. In two parts.

PART I.—Continued, Eruptive and Periodical Fevers, Diseases of the Stomach, Intestines, Peritoneum, Biliary Passages, Liver, Kidneys, etc. (including Tests for Urine), General Diseases, etc.

PART II.—Diseases of the Respiratory System (including Physical Diagnosis), Circulatory System and Nervous System; Diseases of the Blood, etc.

** These little books can be regarded as a full set of notes upon the Practice of Medicine, containing the Synonyms, Definitions, Causes, Symptoms, Prognosis, Diagnosis, Treatment, etc., of each disease, and including a number of prescriptions hitherto unpublished.

No. 4. PHYSIOLOGY, including Embryology. Fifth Edition. By ALBERT P. BRUBAKER, M.D., Prof. of Physiology, Penn'a College of Dental Surgery; Demonstrator of Physiology in Jefferson Medical College, Philadelphia. Revised, Enlarged and Illustrated.

No. 5. OBSTETRICS. Illustrated. Fourth Edition. By HENRY G. LANDIS, M.D., Prof. of Obstetrics and Diseases of Women, in Starling Medical College, Columbus, O. Revised Edition. New Illustrations.

BLAKISTON'S ? QUIZ-COMPENDS ?

No. 6. MATERIA MEDICA, THERAPEUTICS AND PRESCRIPTION WRITING. Fifth Revised Edition. With especial Reference to the Physiological Action of Drugs, and a complete article on Prescription Writing. Based on the Last Revision of the U. S. Pharmacopœia, and including many unofficinal remedies. By SAMUEL O. L. POTTER, M.A., M.D., late A. A. Surg. U. S. Army; Prof. of Practice, Cooper Medical College, San Francisco. Improved and Enlarged, with Index.

No. 7. GYNÆCOLOGY. A Compend of Diseases of Women. By HENRY MORRIS, M.D., Demonstrator of Obstetrics, Jefferson Medical College, Philadelphia. 45 Illustrations.

No. 8. DISEASES OF THE EYE AND REFRACTION, including Treatment and Surgery. By L. WEBSTER FOX, M.D., Chief Clinical Assistant Ophthalmological Dept., Jefferson Medical College, etc., and GEO. M. GOULD, M.D. 71 Illustrations, 39 Formulæ. Second Enlarged and Improved Edition. Index.

No. 9. SURGERY, Minor Surgery and Bandaging. Illustrated. Fourth Edition. Including Fractures, Wounds, Dislocations, Sprains, Amputations and other operations; Inflammation, Suppuration, Ulcers, Syphilis, Tumors, Shock, etc. Diseases of the Spine, Ear, Bladder, Testicles, Anus, and other Surgical Diseases. By ORVILLE HORWITZ, A.M., M.D., Demonstrator of Surgery, Jefferson Medical College. Revised and Enlarged. 84 Formulæ and 136 Illustrations.

No. 10. CHEMISTRY, Inorganic and Organic. For Medical and Dental Students. Including Urinary Analysis and Medical Chemistry. By HENRY LEFFMANN, M.D., Prof. of Chemistry in Penn'a College of Dental Surgery, Phila. Third Edition, Revised and Rewritten, with Index.

No. 11. PHARMACY. Based upon " Remington's Text-book of Pharmacy." By F. E. STEWART, M.D., PH.G., Quiz-Master at Philadelphia College of Pharmacy. Third Edition, Revised.

No. 12. VETERINARY ANATOMY AND PHYSIOLOGY. 29 Illustrations. By WM. R. BALLOU, M.D., Prof. of Equine Anatomy at N. Y. College of Veterinary Surgeons.

No. 13. DENTAL PATHOLOGY AND DENTAL MEDICINE. Containing all the most noteworthy points of interest to the Dental student. By GEO. W. WARREN, D.D.S., Clinical Chief, Penn'a College of Dental Surgery, Philadelphia. Illus.

No. 14. DISEASES OF CHILDREN. By DR. MARCUS P. HATFIELD, Prof. of Diseases of Children, Chicago Medical College. Colored Plate.

Bound in Cloth, $1. Interleaved, for the Addition of Notes, $1.25.

☞ *These books are constantly revised to keep up with the latest teachings and discoveries, so that they contain all the new methods and principles. No series of books are so complete in detail, concise in language, or so well printed and bound. Each one forms a complete set of notes upon the subject under consideration.*

Illustrated Descriptive Circular Free.

JUST PUBLISHED.

GOULD'S NEW
MEDICAL DICTIONARY

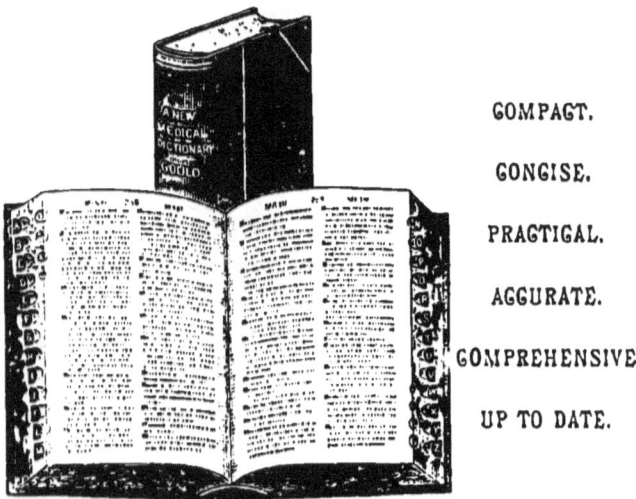

COMPACT.

CONCISE.

PRACTICAL.

ACCURATE.

COMPREHENSIVE

UP TO DATE.

It contains Tables of the Arteries, Bacilli, Ganglia, Leucomaïnes, Micrococci, Muscles, Nerves, Plexuses, Ptomaïnes, etc., etc., that will be found of great use to the student.

Small octavo, 520 pages, Half-Dark Leather, . $3.25
With Thumb Index, Half Morocco, marbled edges, 4.25

From J. M. DaCOSTA, M. D., Professor of Practice and Clinical Medicine, Jefferson Medical College, Philadelphia.

"*I find it an excellent work, doing credit to the learning and discrimination of the author.*"

www.ingramcontent.com/pod-product-compliance
Lightning Source LLC
Chambersburg PA
CBHW022117290426
44112CB00008B/713